£11.90

551.5

60

CHEMISTRY OF THE
LOWER ATMOSPHERE

CONTRIBUTORS

Richard D. Cadle
National Center for Atmospheric Research
Boulder, Colorado

James P. Friend
New York University
Department of Meteorology and Oceanography
New York, New York

G. M. Hidy
North American Rockwell Science Center
Thousand Oaks, California
and
California Institute of Technology
Pasadena, California

Charles D. Keeling
Scripps Institution of Oceanography
University of California at San Diego
La Jolla, California

William W. Kellogg
National Center for Atmospheric Research
Boulder, Colorado

Hans R. Pruppacher
Department of Meteorology
University of California
Los Angeles, California

Stephen H. Schneider
National Center for Atmospheric Research
Boulder, Colorado

CHEMISTRY OF THE LOWER ATMOSPHERE

EDITED BY

S. I. RASOOL

Deputy Director, Planetary Programs
National Aeronautics and Space Administration
Washington, D. C.

PLENUM PRESS · NEW YORK–LONDON · 1973

Library of Congress Catalog Card Number 72-90336
ISBN 0-306-30591-7

© 1973 Plenum Press, New York
A Division of Plenum Publishing Corporation
227 West 17th Street, New York, N.Y. 10011

United Kingdom edition published by Plenum Press, London
A Division of Plenum Publishing Company, Ltd.
Davis House (4th Floor), 8 Scrubs Lane, Harlesden, London,
NW10 6SE, England

Printed in the United States of America

PREFACE

About three years ago Catherine de Berg and I published a short article in *Nature* in which we attempted to explain why the chemistry of the atmosphere of the Earth is today so completely different from that of our two neighboring planets, Mars and Venus. Our atmosphere is composed mainly of N_2 and O_2 with traces of A, H_2O, CO_2, O_3, etc., while the atmospheres of both Mars and Venus are almost entirely made up of CO_2. Also, the Earth appears to be the only one of the three planets which has oceans of liquid water on the surface. Since the presence of liquid water on Earth is probably an essential requirement for life to have originated and evolved to its present state, the question of the apparent absence of liquid water on Mars and Venus suddenly acquires significant proportions.

In our paper in *Nature*, and later in a more detailed discussion of the subject (Planetary Atmospheres, in *Exobiology*, edited by C. Ponnamperuma, North Holland Publishing Co.), we tried to describe why we believe that in the early history of the solar system all the terrestrial planets lost the atmospheres of H_2 and He which they had acquired from the solar nebula at the time of their formation. These planets, completely devoid of atmospheres, like the Moon today, started accumulating new gases which were exhumed from the interior by the commencement of volcanic activity. On the Earth we know that these gases were largely water vapor (steam), carbon dioxide, and nitrogen, with a number of other constituents like chlorine, hydrogen, sulfur, etc. The Earth soon acquired a thin atmosphere of water vapor and CO_2 with traces of N_2, H_2, and other constituents. The temperature at the surface of the Earth at this time was probably about 260°K, mainly governed by the Earth's distance from the Sun and the albedo of the solid surface. However, both H_2O and CO_2 are extremely efficient in trapping infrared radiation so that the surface temperature of the Earth began to rise. Because of continued replenishment of the atmosphere from the interior, the temperature soon rose above 273°K and the pressure above 6.1 mb so that atmospheric water vapor started to condense. This marked the beginning of the accumulation of the oceans on the surface of the Earth. With the atmosphere now containing nitrogen, hydrogen, oxides of carbon, and traces of water vapor (or NH_3, CH_4, and H_2O, depending on the amount of hydrogen relative to nitrogen, carbon, and oxygen), the solar ultraviolet radiation could penetrate close to the surface of the Earth and deposit enough energy to allow synthesis of organic molecules like HCN and HCHO. These, when dissolved in the oceans, continued to combine and form more complex molecules, ultimately producing the complex array of molecules which gave rise to the first "living" systems, and started the train of biological evolution. The presence of liquid water on Earth was therefore a key in determining the subsequent evolution of the atmosphere and of life itself. Gradually, biological activity changed the chemistry of the atmosphere by adding to it substantial amounts of oxygen, paving the way to the highly oxidized atmosphere of today. As the atmosphere became more and more oxidized, the methane and carbon monoxide were converted to carbon dioxide, which, being highly soluble in water at these temperatures, dissolved in the oceans, reacted with the silicates in the rocks, and was deposited on the ocean floor as limestone. Only N_2 and O_2 therefore accumulated in the atmosphere and gave rise to the present-day conditions.

Venus, being about 30% closer to the Sun than Earth is, was initially at a higher temperature, and therefore the first emanations of steam from the early volcanoes were not condensed at the surface to form oceans. Water accumulated in the atmosphere as vapor and, being an extremely efficient absorber of infrared radiation, accentuated the greenhouse effect and raised the ground temperature to such high values that oceans could never accumulate. The absence of oceans and the relatively high temperature of the surface prevented CO_2 from entering the crust as limestone, and therefore it continued to accumulate in the atmosphere. Water, a light molecule, eventually thermally evaporated from a rather hot exosphere, leaving behind an atmosphere predominantly made up of CO_2, as observed today.

Mars, being further away from the Sun than the Earth, was initially too cold for the volcanic steam to liquefy; rather, it froze. This again inhibited the transfer of CO_2 from the atmosphere to the crust. At the same time Mars is only half the size of Venus and Earth, and that is why, although Mars is volcanically active, the rate of emanation is slow and has so far accumulated only a very thin atmosphere of CO_2.

In summary, therefore, it seems to us that the size of the planet and its distance from the Sun are the two crucial parameters which determine the nature of the atmosphere and the oceans and therefore the nature of the living organisms that originate and evolve on the planet. To illustrate this point we showed in our paper that if Earth were only 6% closer to the Sun, the increased solar radiation would not have allowed the volcanic steam to condense at the surface as oceans, and today we would probably have conditions on the surface of the Earth just as hostile as they are on Venus: a 700°K temperature at the surface, heavy CO_2 atmosphere, and probably no life at all.

This is my view of the evolution of our atmosphere during its long history of 3 or 4 billion years. However, when I expounded this thesis at various seminars and colloquia around the country, the questions that were invariably asked were not so much concerned with the history of our atmosphere but with its future. What is the effect of man's input to the atmosphere with regard to the total picture of atmospheric evolution? Is man undoing what nature did by creating Earth where it is located in the solar system? Will the increasing pollution of the atmosphere change the course of atmospheric evolution and will Earth become like Venus? What is its impact on the climate? Although at times I have tried to answer these questions, both the questions and the answers have left me very uneasy. The reason is very simple. We do not know enough about the subject because it involves complex interaction among a number of disciplines: atmospheric dynamics, chemical processes, radiative transfer, and—most significantly—the exchange of gas between the crust, oceans, and the atmosphere. Talking with some of my colleagues in these fields, I realized that many of the questions I had attempted to answer cannot even be formulated properly because we do not know enough about, for example, the physics, chemistry, and dynamics of atmospheric aerosols, the CO_2 and sulfur cycles, and the total interaction of these constituents with the radiation environment of the planet. On one thing, however, all of us did agree, that the literature in this field desperately needs a thorough review of the state of art in each of the above areas, and this is what this book is all about.

S. I. RASOOL

CONTENTS

Chapter 3
Removal Processes of Gaseous and Particulate Pollutants
    G. M. HIDY

Chapter 4
The Global Sulfur Cycle
    JAMES P. FRIEND

Chapter 5
The Chemical Basis for Climate Change
    STEPHEN H. SCHNEIDER AND WILLIAM W. KELLOGG

Chapter 6
The Carbon Dioxide Cycle: Reservoir Models to Depict the Exchange of Atmospheric Carbon Dioxide with the Oceans and Land Plants
   CHARLES D. KEELING

PART I. FORMULATION AND MATHEMATICAL SOLUTION OF THE MODEL EQUATIONS

PART II. CHEMICAL SPECIFICATION AND NUMERICAL RESULTS

Chapter 1

THE ROLE OF NATURAL AND ANTHROPOGENIC POLLUTANTS IN CLOUD AND PRECIPITATION FORMATION

Hans R. Pruppacher

Associate Professor, Department of Meteorology
University of California, Los Angeles

1. INTRODUCTION

Since the resurgence of interest in cloud and precipitation physics, stimulated by the demonstration in 1946 of the possibility of cloud modification by artificial nucleation, an ever-increasing amount of research has been reported in the literature on the mechanisms which underlie the formation of clouds and precipitation in the atmosphere. This research has essentially been carried out along two lines of study: studies of the cloud microphysical processes including the phase change of water substance in the atmosphere and the growth of the formed cloud particles to precipitation particles, and studies of the cloud macrophysical processes including all those dynamic processes which govern the formation, growth, dimensions, shape, organization, lifetime, and vigor of the cloud as a whole. In recent years, it has been realized that the cloud dynamical and cloud microphysical processes have to be viewed in a combined manner for a complete understanding of the formation of clouds and precipitation. Thus, much work is currently being done with this aim in mind.

In the field of cloud microphysics, emphasis has been placed in the past on the study of the role of pollutants from natural sources. Recently, however, scientists have seriously become concerned about the rising level of anthropogenic pollutants in the atmosphere, as indicated for instance, by the rising world-wide level of atmospheric CO_2 [1] and the steady decrease in the electrical conductivity of the atmosphere over the North Atlantic [2] during the last 100 years, and concern has been expressed regarding possible effects of these pollutants on cloud and precipitation formation. It is the purpose of this chapter to review recent studies in cloud and precipitation physics, in particular as these are related to the role of both natural and anthropogenic pollutants. In order to do this, a few preliminary remarks are in order.

The term "air" can most generally be defined as the mixture of gases and solid and liquid particulate matter which comprises the earth's atmosphere. Some of the gaseous constituents of air, such as N_2, O_2, He, Ne, A, Kr, Xe, and H_2, have extremely long residence times [3]. Others, such as CO_2, O_3, N_2O, and CH_4, have residence times of the order of a few to many years, while gases such as H_2O vapor, NO_2, NO, NH_3, SO_2, H_2S, CO, HCl, and I_2 are very variable in time and space and have residence times as short as a few days or weeks [3]. Some gases, particularly CO_2, CO, O_3, NO_2, NO, NH_3, SO_2, and H_2S, have anthropogenic as well as natural sources and may therefore appear in much higher concentrations in air over industrialized cities than in air over the countryside [3,4].

A suspension of solid and liquid particulate matter in a gaseous medium is defined as an aerosol. Typical constituents of the atmospheric aerosol are: silicates; salts such as NaCl, $MgCl_2$, $MgSO_4$, Na_2SO_4, $NaNO_3$, $(NH_4)_2SO_4$, NH_4Cl, NH_4NO_3; acids such as H_2SO_4; metal oxides; organic combustion products; biological material; volcanic material and extraterrestrial material of metallic and stony compounds [3]. Atmospheric aerosol particles cover a wide range of sizes, from a radius of about 6×10^{-8} cm for the small ions which consist of a few neutral air molecules clustered around an ion, to a radius of more than 150 μm (1.5×10^{-2} cm) [5] for the largest particles.

The electrically charged particles, or ions, can be divided into four groups according to their electric mobility B [6]: the small ions have $1.0 > B \geq 10^{-2}$ cm^2 V^{-1} sec^{-1} ($6.6 < r < 78 \times 10^{-8}$ cm); the large ions, $10^{-2} > B \geq 10^{-3}$ cm^2 V^{-1} sec^{-1} ($78 < r < 250 \times 10^{-8}$ cm); the Langevin ions, $10^{-3} > B \geq 2.5 \times 10^{-4}$ cm^2 V^{-1} sec^{-1} ($250 < r < 570 \times 10^{-8}$ cm); and the ultralarge ions, $B < 2.5 \times 10^{-4}$ cm^2 V^{-1} sec^{-1} ($r > 570 \times 10^{-8}$ cm). The electrically neutral particles are customarily divided into three size groups: the Aitken particles ($r < 10^{-5}$ cm), the large particles ($10^{-5} < r < 10^{-4}$ cm), and the giant particles ($r > 10^{-4}$ cm). The number concentrations of the particles in the atmosphere also cover an enormous range. The concentration of small ions near the earth's surface ranges typically from 100 to 1000 cm^{-3} and that of large ions from 1000 up to 80,000 cm^{-3} in highly polluted air [6,7]. The concentration of the Aitken nuclei near the earth's surface ranges typically from a few hundred per cm^3 over the ocean [8,9] to 10^3–10^5 cm^3 over the land, often exceeding 10^6 cm^{-3} in highly polluted air of an industrial city [3,10,11]. The concentration of aerosol particles decreases with increasing size and over the land is typically 10^3 cm^{-3}, 1 cm^{-3}, 1 liter^{-1}, 1 m^{-3} for particles larger than 10^{-5}; 10^{-4}, 10^{-3}, 10^{-2} cm radius, respectively [3,5]. In air over the ocean, with winds just strong enough to produce white caps, the concentration of all hygroscopic particles including sea salt is typically 10^2 cm^{-3}, 10 cm^{-3}, 1 cm^{-3}, 1 liter^{-1} for particles with masses larger than 10^{-16}, 10^{-14}, 10^{-12}, 10^{-9} g, respectively [12]. The upper size limit of the aerosol particles is controlled by sedimentation. Particles larger than 20 μm in radius remain airborne only for very short times and their occurrence is therefore restricted to the vicinity of their source [3]. The lower end of the size spectrum is controlled by coagulation causing particles smaller than about 5×10^{-7} cm to become rapidly attached to larger particles. Thus, small ions may exist only a few minutes in relatively clean air and only for a few seconds in heavily polluted air. The lifetime of particles of intermediate sizes is controlled in the troposphere by the combined action of sedimentation, coagulation with other aerosol particles and with cloud drops, and rainout and washout. In the lower troposphere, their lifetime ranges typically from two days to two weeks (the shorter lifetimes applying more likely to midlatitudes), while in the upper troposphere, the residence time is typically between two and four weeks [3,13].

From the point of view of cloud formation, water vapor is the most significant gaseous constituent of air. Atmospheric air is rarely, if ever, perfectly dry. In the troposphere, water vapor is present in highly varying amounts and is continuously transported, by undergoing phase changes, between its two temporary storage locations: the earth's surface and the clouds. The average residence time of water vapor in the troposphere varies from about eight days at low latitudes to about 15 days at high latitudes [3]. The average residence time of liquid water in the troposphere is about 11 hours [3]. This is longer than the average lifetime of tropospheric clouds, which is about 1 hour, and indicates that most clouds undergo several evaporation–condensation cycles before precipitating. Highest mixing ratios of water vapor are found at low altitudes near the earth's surface, where the mixing ratio may become as high as 30×10^{-3} g g^{-1}. The water vapor mixing ratio normally decreases toward the poles and with altitude. It is thought that about 90 % of the atmospheric moisture lies below 6 km and probably less than 1 % above the tropopause, thus limiting cloud formation to the troposphere (with the exception of the occasional formation of nacreous clouds in the lower stratosphere). The variation of water vapor with altitude has been studied extensively in the recent past. Newer measurements using improved techniques [14-16] agree reasonably well with each other and show that the water vapor mixing ratio decreases with altitude throughout the troposphere to a minimum at about 2 km above the local tropopause and remains practically constant thereafter in the stratosphere, at least up to 31 km. The water vapor mixing ratio in the lower stratosphere ranges typically between 1.5×10^{-6} and 3.5×10^{-6} g g^{-1} and undergoes characteristic seasonal changes which reflect the removal capability of moisture by the tropical tropopause and allow the conclusion that the level of stratospheric moisture is regulated by the temperature of the tropical tropopause.

2. IF THERE WERE NO AEROSOLS IN THE ATMOSPHERE

Under the condition that air is completely free of particulate matter and ions, the phase change of water vapor to water droplets entirely depends on chance collisions between water molecules. Such collisions lead to the formation of small molecular aggregates or "embryos" which are formed and disrupted continuously because of microscopic thermal and density fluctuations in the vapor. If these aggregates surpass a critical size, the size of a "germ," they will survive and continue to grow to a macroscopic drop. According to the classical theory of homogeneous nucleation based on the work of Volmer, Becker, and Döring [see Refs. 17, 18], the free energy for formation of a germ is

$$\Delta F_g = \tfrac{1}{3}\sigma_g \Omega_g \tag{1}$$

where Ω_g is the surface area and σ_g the surface energy of the germ. The rate J at which germs are created within the vapor is an exponential function of ΔF_g and can be expressed by the relation

$$J = C_1 w (\Omega_g/3g)(\sigma_g \Omega_g/\pi kT)^{1/2} \exp[-(\Delta F_g/kT)] \tag{2}$$

where T is the temperature, k the Boltzmann constant, g the number of molecules in the germ, w the collision rate of water molecules per unit surface area of the condensed phase, and C_1 the number of molecules per cubic centimeter in the vapor. If kinetic gas theory is applied to water vapor and if the small molecular aggregates of the condensed phase are treated as well-defined spherical droplets having thermodynamic and physical properties of bulk liquid (which is an unsatisfactory feature of the classical theory), the radius r_g of the germ, ΔF_g, and J are given by the relations

$$r_{g,} = 2\sigma_{wv} M_w/\rho_w RT \ln S \tag{3}$$

$$\Delta F_g = 16\pi\sigma_{wv}^3 M_w^2/3\rho_w^2 R^2 T^2 (\ln S)^2 \tag{4}$$

$$J = \frac{\alpha_w}{\rho_w} \left(\frac{2N_A{}^3 \sigma_{wv} M_w}{\pi} \right)^{1/2} \left(\frac{Sp_\infty}{RT} \right)^2 \exp\left[-\frac{16\pi N_A M_w{}^2 \sigma_{wv}^3}{3(\ln S)^2 \rho_w{}^2 R^3 T^3} \right] \tag{5}$$

where M_w is the molecular weight of water, ρ_w the density of water, R the universal gas constant, σ_{wv} the surface tension of water, N_A the Avogadro number, $S = p/p_\infty$ the saturation ratio, p the pressure of the supersaturated vapor, p_∞ the equilibrium water vapor pressure over a flat surface at temperature T, and α_w the condensation coefficient. The rate of nucleation J is very strongly dependent on S. As S increases, J remains essentially zero until a critical saturation ratio S_c is reached at which J becomes very large over a very short range of S. Usually one defines S_c as the saturation ratio at which J = one drop cm^{-3} sec^{-1}. Numerical evaluation of Eq. (5) shows that for $T = 265°K$, $S_c \approx 5.0$. The value of S_c increases with decreasing temperature and at $+30°C$ has a value of about three and at $-30°C$ a value of about seven. Thus, by defining $(S - 1) \times 100$ as the supersaturation in per cent of the vapor, we can conclude that for the formation of drops under homogeneous conditions, supersaturations of several hundred per cent are needed. In most experiments which were designed to verify the classical theory, expansion chambers were used [for a review and criticism, see Refs. 17,19]. Recently, however, diffusion chambers have been employed. These gave results for the n-alkanes, methanol, ethanol, and water which were in good agreement with the classical theory [20,21]. Despite this agreement, the classical theory has recently come under heavy criticism, first, because macroscopic quantities are used to describe σ_g and Ω_g, and second, because the equilibrium distribution of embryos of the condensed phase is obtained by assuming that the free energy ΔF_i of an embryo of i molecules is equal to the free energy of the same number of molecules in the bulk liquid plus the free energy due to the surface of the embryo:

$$\Delta F_i = \tfrac{4}{3}\pi r_i^3\,\Delta F_v + 4\pi r_i^2 \sigma_i \tag{6}$$

where ΔF_v is the free-energy difference per unit volume of liquid between molecules in the liquid and vapor states and is negative for supersaturated vapor. Lothe and Pound [22] pointed out that the above expression refers to droplets at rest, whereas, they argue, the tiny embryos in a condensing vapor are really moving rotationally and translationally like macromolecules in a gas. The free energy of an embryo should therefore contain contributions from its translation and rotation. However, in a recent paper Reiss [23] showed on the basis of rigorous statistical mechanics that the criticism of Lothe and Pound is not justified.

In analogy to homogeneous nucleation of water droplets in supersaturated vapor, we would expect that at high enough supersaturations and low enough temperatures, ice crystals will be nucleated in homogeneous vapor. Following the classical treatment of homogeneous nucleation [18], we can express the nucleation rate of ice crystals from homogeneous vapor by the relation

$$J = C_1 w(\Omega_g/3g)[(\sum_{\beta} \sigma_g{}^\beta \Omega_g{}^\beta)/\pi kT]^{1/2} \exp[-(\Delta F_g/kT)] \tag{7}$$

where $\sum_\beta \sigma_g{}^\beta \Omega_g{}^\beta$ represents the sum of the surface-energy contributions of all faces β of surface area Ω^β of the crystalline germ, Ω_g is the total surface area of the germ, and

$$\Delta F_g = \tfrac{1}{3}s\sigma_g r_g{}^2 \tag{8}$$

if the ice germ is assumed to be a hexagonal prism compatible with Wulff's relations [18] and s is a shape factor equal to 4π for a sphere. If one attributes to the ice germ physical properties of bulk ice, one finds

$$r_g = 2M_w\sigma_{IV}/\rho_I RT \ln S \tag{9}$$

$$J = \frac{\alpha_I}{\rho_I}\left(\frac{sN_A{}^3\sigma_{IV}M_w}{2\pi^2}\right)^{1/2}\left(\frac{Sp_\infty}{RT}\right)^2 \exp\left[-\frac{4sN_A M_w{}^2\sigma_{IV}^3}{3(\ln S)^2\rho_I{}^2R^3T^3}\right] \tag{10}$$

where α_I is the condensation coefficient for ice, σ_{IV} is the interface free energy between ice and vapor, and ρ_I is the density of ice. A numerical analysis of Eq. (10) for $T = 261°K (-12°C)$ shows that even at a saturation ratio as high as $S = 6$, the nucleation rate is still as small as $J = 3.4 \times 10^{-52}$ ice crystals $cm^{-3} sec^{-1}$. From a comparison between Eqs. (5) and (10) in terms of the temperature variation of the isonucleating ($J = 1$) partial pressures of water vapor expressed as $\ln p$ versus $1/T$, Dufour and Defay [18] demonstrated that down to a temperature well below $-100°C$, drop formation is energetically easier and therefore more likely than ice crystals formation. Under homogeneous conditions, ice formation will therefore always occur in two stages: (1) appearance of supercooled water drops and (2) freezing of the supercooled water drops by homogeneous nucleation of ice within the drops. This result shows that, despite earlier evidence which cannot withstand criticism [18], Ostwald's *Stufenregel* (law of stages) holds for water substance at least to a temperature of $-100°C$. Experimental verification of this result has been given by Maybank and Mason [24], who showed that in completely clean, moist air, ice crystals always appear as a result of homogeneous drop formation followed by homogeneous freezing of drops rather than by direct ice crystal formation from the vapor.

Contrary to homogeneous nucleation in water vapor, which essentially can be treated as an ideal gas of single molecules, homogeneous nucleation of ice in supercooled liquid water must take into account the viscous nature of water where the motion of the molecules are hindered by the bonds formed between neighboring molecules. Following Ref. 18, the rate of formation of ice germs $cm^{-3} sec^{-1}$ can be written as

$$J = C_1(kT/h)b(\Omega_g/3g)\left[\left(\sum_\beta \sigma_g{}^\beta\Omega_g{}^\beta\right)/\pi kT\right]^{1/2} \exp[-(\Delta F^{\ddagger} + \Delta F_g)/kT] \tag{11}$$

where h is the Planck constant, b is the number of molecules of liquid in contact with unit area of ice surface, $\Delta F^{\ddagger}$ is approximately equal to the activation energy for self-diffusion of molecules in the liquid, C_1 is the number of water molecules per cubic centimeter, and where the free energy for the formation of a germ is given by Eq. (8) with

$$r_g = 2\sigma_{Iw}M_w/\rho_I \int_T^{T_0}\frac{L_s(T) - L_v(T)}{T}dT \tag{12}$$

where L_s is the latent heat of sublimation, L_v is the latent heat of evaporation, σ_{IW} is the interface free energy between water and ice, and $T_0 = 273°K$. It can be shown that $b(\Omega_g/g)(\sum_\beta\sigma_g{}^\beta\Omega_g{}^\beta)/\pi kT \approx 1$ for temperatures between

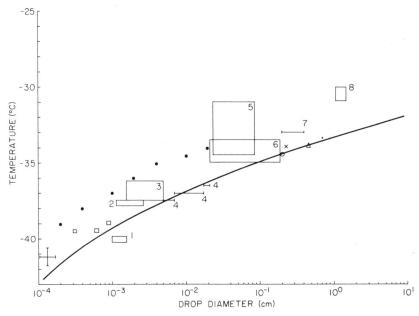

Fig. 1. Comparison of the freezing temperature of very pure water drops of various sizes with the freezing temperature predicted from homogeneous nucleation theory (based on data from Refs 25–37). (1) [37], (2) [30], (3) [36], (4) [34], (5) [31], (6) [35], (7) [32], (8) [33], (●) [28], (×) [26], (△) [25], (○) [27], (□) [29], (data bars) cloud chamber data.

273 and 233°K. Thus

$$J \approx C_1 kT/h \exp[-(\Delta F^{\ddagger} + \Delta F_g)/kT] \tag{13}$$

Many freezing experiments using highest-purity water have been reported in the literature. The results of some selected experiments [25–37] are plotted in Fig. 1. Since these experiments effectively measure the temperature at which the rate JV of nucleation of ice in a drop of volume V is of the order of 1 sec^{-1}, Eqs. (8), (12), and (13) were evaluated in the manner suggested by Fletcher [38] setting $JV = 1$, assuming $s = 4\pi$, and $L_s - L_v = L_f$, where L_f is the latent heat of fusion, and using the most recent values for $\sigma_{IW}(T)$, and $\Delta F^{\ddagger}(T)$. Figure 1 shows that the theoretical line constitutes a lower bound to all experimental results. This is somewhat expected considering the experimental difficulties in purifying water and considering the fact that in some experiments the cooling rate affected the experimental results. Most importantly, however, Fig. 1 shows that the rate of formation of ice germs in supercooled liquid water under homogeneous conditions is a function of the volume of the supercooled liquid and is the smaller the smaller the volume. Thus, drops of raindrop size (a few millimeters in diameter) supercool to temperatures between -34 and $-35°C$ before they freeze, while drops of cloud drop size (a few microns in diameter) supercool to about $-40°C$ before they freeze.

Let us now view these results in terms of atmospheric processes. Study of the motions in the atmosphere show that as air rises, it constantly mixes with its environment and consequently does not cool in an adiabatic manner. Under such conditions, the supersaturations of several hundred per cent required for homogeneous drop formation cannot be reached in atmospheric updrafts. Since under homogeneous conditions, the ice phase can only appear via the liquid phase, neither drops nor ice crystals can form in an atmosphere completely devoid of foreign particles. Thus, cloud drops and ice crystals must form in the atmosphere via a heterogeneous process involving the atmospheric aerosol. This conclusion can also be reached on the basis of direct observations in the atmosphere.

Hoffer [39] and recently Warner [40] made measurements of the relative humidity and supersaturation in atmospheric clouds. These observations, summarized in Fig. 2, showed that in atmospheric clouds, the supersaturation rarely if ever surpassed 2% and that the median value of the measured supersaturations was about 0.1%. Only in 2% of the observations was the supersaturation larger than 0.5%, and only in 1% of the observations the air was supersaturated by more than 1%. On the other hand, certain portions of the cloud were found to be subsaturated by as much as 2% and a considerable percentage of samples were in air subsaturated by 1%.

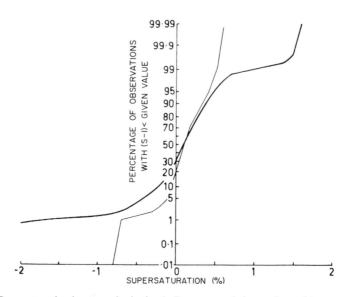

Fig. 2. Supersaturation in atmospheric clouds. Percentage of observations with supersaturation less than a given value for all samples (heavy line) and for samples taken within 300 m of cloud base (light line) [40]. (By courtesy of *J. Rech. Atmos.*)

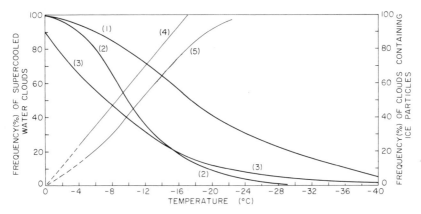

Fig. 3. Variation of frequency of supercooled clouds and clouds containing ice crystals. Curves 1–3 pertain to ordinate at left; curves 4 and 5 pertain to ordinate at right. (1) European territory of USSR, mixed clouds [42]. (2) Germany, all-water clouds [41]. (3) European territory of USSR, all-water clouds [42]. (4) Tasmania (Australia), mixed clouds [43] (5) Minnesota (USA), mixed clouds [44]. (By courtesy of *Quart. J. Roy. Meteorol. Soc.*, Morris and Braham, and the Israel Program for Scientific Translations.)

Selected aircraft observations on the formation of the ice phase in the atmosphere made by Peppler [41], Borovikov [42], Mossop *et al.* [43], and Morris and Braham [44] are summarized in Fig. 3. This figure shows that the frequency with which supercooled clouds were observed is reasonably consistent with the frequency of clouds containing ice crystals, despite the rather different locations at which these observations were made. It is seen that, in general, most atmospheric clouds consist of supercooled cloud drops if their temperature is warmer than $-12°C$ and mostly of ice particles if their temperature has fallen below $-22°C$. However, not all clouds follow this general trend. Some clouds contain ice particles in significant numbers at temperatures much warmer than $-12°C$, while other clouds may stay supercooled down to temperatures considerably colder than $-22°C$. However, $-40.6°C$ represents the lowest temperature ever measured at which cloud drops were still in supercooled form [42].

Comparing these observations with the requirements for homogeneous ice formation, we are led to conclude that cloud drop formation as well as ice formation in the atmosphere are heterogeneous processes involving the presence of the atmospheric aerosol.

Further evidence which supports this conclusion is derived from a direct analysis of ice particles and water drops sampled in atmospheric clouds and fogs. Numerous studies of the residue of sublimated snow crystals have been carried out by means of an electron microscope or by means of electron diffraction techniques [45–51]. In more than 95% of the cases, one solid nucleus was found in the central portion of a snow crystal. These center nuclei had diameters between 0.1 and 15 μm, with a mode between 0.4 and

TABLE 1. Type of Center Particles in Snow Crystals Collected at Different Locations

	Hokkaido (Japan)		Honshu (Japan)		Michigan (U.S.A.)		Missouri (U.S.A.)		Thule (Greenland)	
Estimated substances	Number	%	Number	%	Number	%	Number	%	Number	%
Mineral particle	176	57	46	88	235	87	70	28	302	84
Microscopic particle	57	19	0	0	2	1	5	2	2	1
Combustion product	26	8	2	4	6	2	7	3	0	0
Microorganism	3	1	0	0	0	0	3	1	0	0
Unknown material	30	10	4	8	25	9	100	40	39	11
Not observed	15	5	0	0	3	1	65	26	13	4
Total	307	100	52	100	271	100	250	100	356	100
Reference	[51]		[51]		[51]		[54]		[52]	

*Based on data from [51,52,54].

1 μm, and consisted mainly of silicates, most of which were identified as clay particles as seen from Tables 1 and 2. Besides silicates, hygroscopic particles and combustion products could be identified. Soulage [55,56] dissected the residue of snow crystals under the microscope and found that the larger particles consisted both of water-soluble and water-insoluble components, thus corroborating the suggestion of Junge [57] that most

TABLE 2. Comparison of Snow Crystal Forms and Center Particle Substance for Snow Crystals Sampled at Houghton, Michigan (USA)([5])*

Substances	Hexagonal plate	Dendritic crystal	Column	Pyramid and bullet	Needle	Total
Clay minerals:						
Kaolin minerals	76	35	6	13	8	138
Montmorillonite minerals	24	3	3	5	1	36
Illite minerals	16	6	6	5	0	33
Attapulgite or Sepiolite	6	3	2	1	0	12
Related minerals	8	4	3	0	1	16
Other substances:						
Potassium chloride	0	2	0	0	0	2
Carbon particles	4	1	0	1	0	6
Unknown material	13	8	1	2	1	25
Not observed	3	0	0	0	0	3
Total	150	62	21	27	11	271

*By courtesy of *J. Atmos. Sci.* and the authors.

aerosol particles are "mixed particles." About 2% of the center nuclei of the ice crystals or ice fog particles analyzed consisted of peculiar spherical particles (spherules) of diameter between 0.6 and 6 μm. Some of these spherules could be identified as extraterrestrial material [53,58]. The fact that in portions other than the central part of an ice crystal numerous smaller particles of diameters 0.15–0.05 μm were found led to the supposition that the large central particles must have been responsible for the initiation of the crystal, while the smaller ones in the outer portion of the crystal probably were captured by the crystal subsequent to its formation. These investigations demonstrate the active involvement of the atmospheric aerosol particles in the initiation of the ice phase in the atmosphere. In contrast, the involvement of aerosols in the initiation of the liquid phase is less evident from an analysis of cloud and fog drops. This is due to the fact that one cannot clearly differentiate between those water-soluble or -insoluble particles that were responsible in initiating the drop formation and those that were captured subsequent to the drop formation. The water-soluble substances lose their identity since they become dissolved in the drop, and the water-insoluble particles swim around within the drop with no preferred location that would allow any one particle to be identified as that particle that was responsible for the drop formation. Nevertheless, from the numerous investigations of the residue of cloud and fog droplets, some important conclusions could be drawn [59–68]. It was found that sea salt and combustion particles constituted by far the largest portion of the cloud and fog drop residue. Sea salt was more abundant in clouds and fogs over or near the ocean, while combustion material was more abundant in clouds and fogs formed farther inland and near cities. Analyses of cloud drop residues are summarized in Table 3. These contrast the analyses of ice crystal residues summarized in Table 2. Since the differences in the type of residue could hardly be attributed to differences in captured

TABLE 3. Type of Particles Found in Residue of Cloud and Fog Drops*

Substance	Number			Per cent		
	[59,60]	[63,64]	[66,67]	[59,60]	[63,64]	[66,67]
Sea salt particles	5	11	37	13	16	54
Mixed particles with:						
combustion particles	20	25	15	51	36	22
soil particles	11	16	10	28	23	14
Unknown or no observed particles	3	17	7	8	25	10
Total	39	69	69	100	100	100

*Based on data from [59,60,63,64,66,67].

particles, the conclusion was inevitable that sea salt and combustion particles are the primary particles involved in cloud drop formation, while soil particles are the primary particles involved in ice crystal formation. In the salt residues, NaCl was most abundant, but $MgCl_2$ and $MgSO_4$ were also present. The particles most frequently found were Junge-type mixed particles consisting of water-insoluble and water-soluble hygroscopic material. These investigations demonstrate the important role of the atmospheric aerosol particles in serving as catalyzing agents to bring about the phase change of water substance in the atmosphere.

While from a microphysical point of view, atmospheric clouds are the result of the action of nuclei, from a macrophysical point of view, clouds are the result of the cooling of moist air caused by the expansion of air during forced vertical motion, by contact with colder air masses, or by contact with colder land or water surfaces. Since the amount of water vapor that can be contained in a given mass of air decreases with decreasing temperature, the temperature of a parcel of moist air may eventually decrease to the point where the water vapor contained in it is sufficient to saturate or slightly supersaturate the air with respect to a flat, liquid water surface or ice surface. Further cooling, depending on the aerosol present, may result in the formation of a visible cloud consisting of water droplets, or ice crystals if the temperature is low enough. It is therefore the air motions in and around a cloud in combination with the aerosol which act as nuclei that determine the concentration, initial size distribution, and physical nature of the cloud particles. As soon as these have formed, the microphysical processes begin to broaden the initial cloud particle size spectrum until particles of precipitation size are formed. Because the air motions govern the dimension, water content, and duration of a cloud, they decisively influence the rates of these growth processes and also the period over which they operate and thus the maximum size which the precipitation particles attain. If precipitation should result, it is again the air motion which decisively influences its distribution intensity and duration. Conversely, however, the growth and evaporation of the cloud and precipitation particles, accompanied by changes of phase and water concentration, provide sources and sinks of heat that can profoundly influence the air motion by adding positive or negative buoyancy which promotes and sustains updrafts or promotes the formation of downdrafts.

3. THE ROLE OF AEROSOLS IN NUCLEATING CLOUD DROPS

3.1. The Role of Ions

The phase change of water vapor to water drops is aided by the presence of ions. This can be shown by considering a water drop as a dielectric sphere of dielectric constant ε formed around an ion of electric charge q. Based on the classical theory of nucleation and assuming that the dielectric constant of air is unity and that macroscopic concepts can be applied to small droplets,

Tohmfor and Volmer [69] derived the following relationship between the drop size and the supersaturation:

$$\ln \frac{p}{p_\infty} = \frac{2\sigma_{wv} M_w}{\rho_w RTr} - \frac{q^2 M_w}{8\pi \rho_w RTr^4}\left(1 - \frac{1}{\varepsilon}\right) \tag{14}$$

Equation (14) is numerically evaluated in Fig. 4 for $T = 273°\text{K}$, $q = 4.8 \times 10^{-10}$ esu, and $\varepsilon = \infty$, which means that the drop is assumed to be a conductor, and for $\varepsilon = 1.85$. The curve for $\varepsilon = 88$ (the dielectric constant of bulk water at 0°C) is practically identical with the curve for $\varepsilon = \infty$. The critical saturation ratio at which nucleation in an atmosphere containing ions will occur is not given by the maximum of these equilibrium growth curves. To determine the critical threshold for nucleation, the theory of statistical fluctuations has to be applied. Tohmfor and Volmer found that the rate J of germ formation can be expressed by the relation

$$J = Z_1 w \frac{\Omega_g}{3g}\left\{\frac{\sigma_g \Omega_g - (q^2/r_g)[1 - (1/\varepsilon)]}{\pi kT}\right\}^{1/2} \exp -\left(\frac{\Delta F_g}{kT}\right) \tag{15}$$

where

$$\Delta F_g = \frac{4}{3}\pi\sigma_g(r_g^{\ 2} - r_a^{\ 2}) - \frac{2q^2}{3}\left(1 - \frac{1}{\varepsilon}\right)\left(\frac{1}{r_a} - \frac{1}{r_g}\right) \tag{16}$$

and Z_1 is the number of ions per cubic centimeter. The quantity r_a (see Fig. 4) represents the radius of the drop in metastable equilibrium with an environment that is characterized by a saturation ratio S_c at which fluctuations in the vapor can form a germ r_g at a rate $J = 1 \text{ cm}^{-3} \text{ sec}^{-1}$. For the case that

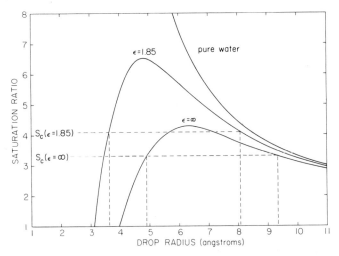

Fig. 4. Variation of the equilibrium saturation ratio with size of electrically charged and uncharged water drops (for 0°C).

$\varepsilon = 90$, $T = 265°K$, and $Z_1 = 10^3 \, cm^{-3}$, evaluation of Eqs. (15) and (16) gives $S_c = 3.3$. This value for S_c is considerably below $S_c = 4.1$, the experimentally found value (for review of experimental work, see Ref. 12). However, it can easily be shown that, due to dielectric polarization of the water by the ion, the dielectric constant of water is not that of bulk water but is considerably smaller. Thus, ε for crystal water ranges between 1 and 8, water bound to metal oxides behaves as a liquid with $1.1 \leq \varepsilon \leq 6$, and water bound to clay has values of ε ranging between 4 and 31. Experiment and theory come into agreement if it is assumed that $\varepsilon = 1.85$ [69]. Experiments carried out recently [70] verified the theory given above for the case of negative ions. Positive ions require larger saturation ratios. This probably is due to steric effects. In the surface of water, the water dipoles are preferentially oriented with their negative end outward. Thus, a positive ion would induce an opposite orientation in the water structure which is associated with a higher energy barrier ΔF_g for drop formation.

A comparison between the saturation ratio necessary for drop formation under homogeneous conditions and drop formation in the presence of ions shows that, at least for negative ions, $S_c(ions) < S_c(homogeneous)$. However, the supersaturation is still of the order of several hundred per cent, which will never be reached by the air motions typical for the atmosphere.

3.2. The Role of Water-Soluble, Hygroscopic Particles

An important characteristic of water-soluble particles is their hygroscopic nature which causes their size to vary with relative humidity. The growth of such particles normally starts at relative humidities considerably below 100%. At a fixed temperature, a particle consisting of a certain chemical compound has one characteristic relative humidity at which the dry particle changes into a saturated solution droplet. In the case of a single chemical compound, application of the phase rule shows that at a fixed temperature, a triple point exists (humidity being taken as the remaining variable) at which solid, liquid, and vapor are in equilibrium. This triple-point humidity is the humidity at which the phase change from solid to liquid takes place and is, in fact, the humidity in equilibrium with the saturated solution at the given temperature. The characteristic humidity $(RH)_c$ computed for various salts from their solubility in water and Raoult's law [71,72] is listed in Table 4. These values are in reasonably good agreement with the experimental values for temperatures above 0°C [71,73] and below 0°C [74].

The growth of a saturated solution droplet has been studied experimentally and theoretically by many investigators (for a review, see Refs. 3, 12, 18, 75, 76). Following a recent theoretical treatment [72,77], the variation of the radius r of the solution drop can be written

$$\ln(p_r'/p_\infty) = (2M_w \sigma'/\rho' RTr) + \ln a_w, \tag{17}$$

TABLE 4. Critical Relative Humidity to Form a Saturated Solution from a Given Dry Salt*

Substance	LiCl	CaCl$_2$ ·4H$_2$O	MgCl$_2$ ·6H$_2$O	NH$_4$NO$_3$	NaNO$_3$	NaCl	NH$_4$Cl	(NH$_4$)$_2$SO$_4$	KCl	Na$_2$SO$_4$	MgSO$_4$
$(RH)_c$ %	13	18	33	62	74	76	77	80	85	86	91

*Based on data from [73].

where p_r' is the vapor pressure over the solution drops, p_∞ is the saturation vapor pressure over a pure, flat water surface, σ' is the surface tension of the solution, ρ' is the density of the solution, and a_w is the activity of the water in the solution. Introducing the mean ionic activity coefficient $\gamma_\pm$ of the solute in solution, we can write

$$-\ln a_w = \nu \mathcal{M} W + \nu W \int_1^{\gamma_\pm} \mathcal{M} \, d(\ln \gamma_\pm) \qquad (18)$$

where $\mathcal{M}$ is the molarity defined as the number of moles of solute per 1000 g of water, ν is the number of moles of ions per moles of solute, and $W = M_w/1000$. Introducing the practical osmotic coefficient ϕ defined by Bjerrum, we can write [78]

$$-\ln a_w = (x_s/x_w)\phi = \nu \mathcal{M} W \phi = \nu \phi m_s M_w / M_s (\tfrac{4}{3}\pi r^3 \rho' - m_s), \qquad (19)$$

where x_w and x_s are the mole fractions for the salt and the water in solution, m_s is the mass of the salt, and M_s is the molecular weight of the salt. The osmotic coefficient in terms of $\gamma_\pm$ is then given by $\phi = 1 + (1/\mathcal{M})\gamma_\pm(\mathcal{M})$, with $\gamma_\pm(\mathcal{M}) = \int_1^{\gamma_\pm} \mathcal{M} \, d(\ln \gamma_\pm)$. Equations (17) and (19) which describe the equilibrium growth of a solution drop were recently solved numerically by Low [79] for a large number of different salts. The results for NaCl and (NH$_4$)$_2$SO$_4$ and typical masses of atmospheric salt particles are plotted in Fig. 5. It is seen from this figure that the equilibrium value for the saturation ratio reaches a characteristic maximum S_c which varies strongly with the type of salt and the mass of the salt particle. It is this characteristic saturation ratio which has to be reached or surpassed in an atmospheric air parcel if salt particles contained in it are to be activated to water drops that will subsequently freely grow further by diffusion from the vapor. Values for S_c of a variety of masses of NaCl particles are tabulated in Table 5 and plotted for $T = 273°K$ in Fig. 6. From Fig. 6, the effectiveness of NaCl particles as cloud-drop-forming nuclei (commonly termed cloud condensation nuclei and henceforth abbreviated CCN) is clearly evident. At supersaturations smaller than about 1% (which is, as discussed, a typical upper bound of supersaturation reached in atmospheric clouds), all NaCl particles with a radius larger than about 1.3×10^{-6} cm can act as CCN.

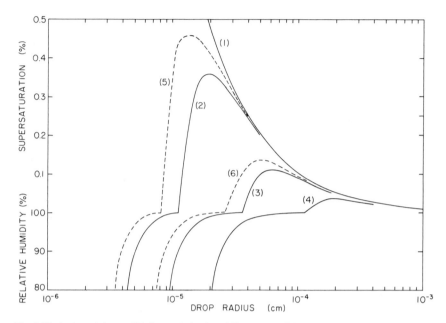

Fig. 5. Variation of the equilibrium relative humidity and equilibrium supersaturation with size of drops of various aqueous solutions. (Based on data from Ref. 79.) (1) Pure water drop; (2) solution drop with 10^{-16} g NaCl; (3) with 10^{-15} g NaCl; (4) with 10^{-14} g NaCl; (5) with 10^{-16} g $(NH_4)_2SO_4$; (6) with 10^{-15} g $(NH_4)_2SO_4$.

Recently, quantitative experiments were carried out by Winkler [80,82] to determine the growth characteristics of known salt aerosols and unknown maritime and continental aerosols. These aerosols were impacted on foils and the mass of water taken up by the deposit in equilibrium with an environment of known relative humidity was measured and compared to the original dry mass by a weighing method. The equilibrium growth of an aerosol deposit of a known salt under this condition is given by Raoult's law

$$p'/p_\infty = \gamma_w x_w = \gamma_w[n_w/(n_w + n_s)]$$ (20)

TABLE 5. Critical Supersaturation S_c for NaCl Particles of Mass m_N and Dry Radius r_N for Activation to Freely Growing Drops*

m_N (g)	10^{-17}	10^{-16}	10^{-15}	10^{-14}	10^{-13}	10^{-12}	10^{-11}	10^{-10}	10^{-9}	10^{-8}
S_c (%), 20°C	1.11	0.36	0.11	3.59×10^{-2}	1.11×10^{-2}	3.59×10^{-3}	1.11×10^{-3}	3.59×10^{-4}	1.11×10^{-4}	3.59×10^{-5}
S_c (%), 0°C	1.33	0.42	0.133	4.2×10^{-2}	1.33×10^{-2}	4.2×10^{-3}	1.33×10^{-3}	4.2×10^{-4}	1.33×10^{-4}	4.2×10^{-5}
r_N (cm)	1.02×10^{-6}	2.22×10^{-6}	4.8×10^{-6}	1.02×10^{-5}	2.22×10^{-5}	4.8×10^{-5}	1.02×10^{-4}	2.22×10^{-4}	4.8×10^{-4}	1.02×10^{-3}

*Based on data from [79].

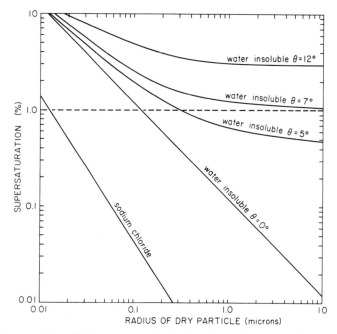

Fig. 6. Variation of the critical supersaturation for drop formation on spherical water-soluble particles and spherical water-insoluble particles of various wetting angles (for 0°C).

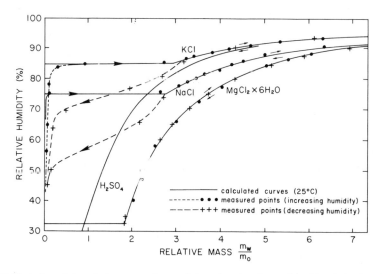

Fig. 7. Variation of the relative mass of aerosol deposits of various salts with relative humidity [81,82]. (By courtesy of *Ann. der Meteorologie* and Junge.)

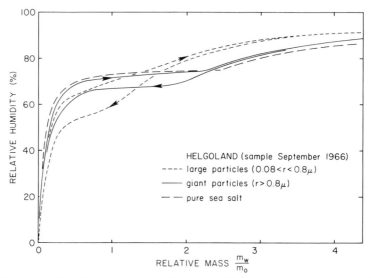

Fig. 8. Variation of the relative mass of a maritime aerosol deposit with relative humidity [80,81].
(By courtesy of *Ann. der Meteorologie* and Junge.)

where γ_w is the activity coefficient of the water in solution and n_w and n_s are, respectively, the number of moles of water and number of moles of salt in solution. For a given humidity and a given salt, the mass of water taken up under equilibrium conditions was theoretically computed from Eq. (20) and was compared by Winkler with experimental values for the case of three different salts and sulfuric acid in Fig. 7. The agreement between experiment and theory is good except for the fact that the phase transition from the dry salt to a saturated solution takes place at a smaller relative humidity than computed and that the equilibrium curve behaves differently during decreasing relative humidity than during increasing relative humidity. The latter effect was attributed to the formation of salt-supersaturated solutions. The difference in behavior of a maritime and a continental aerosol is demonstrated in Figs. 8 and 9. Both aerosols began to take up moisture at very low relative humidities. However, while the maritime aerosol sample started to take up moisture rapidly at relative humidities between 60% and 80%, analogous to a sea salt aerosol, approximated by an aerosol of 80% NaCl and 12% $MgCl_2$ [80], significant growth of the continental aerosol sample took place only after the relative humidity had exceeded 80%. A comparison of the results obtained for natural aerosols with the results obtained by laboratory experiments for mixed particles composed of known water-soluble and water-insoluble substances demonstrated that the growth curves of natural aerosols have to be explained in terms of Junge's mixed-particle concept [82].

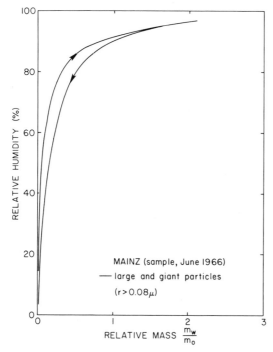

Fig. 9. Variation of the relative mass of a continental aerosol deposit with relative humidity [81]. (By courtesy of *Ann. der Meteorologie.*)

3.3. The Role of Water-Insoluble Particles

The theory of heterogeneous drop formation involving water-insoluble particles is due to Fletcher, who extended the classical theory of Volmer–Weber and Becker–Döring to include the effects of particle size (see Ref. 75). According to Fletcher's theory, the nucleation rate of water drops in a heterogeneous vapor containing spherical water-insoluble particles is given by the relation

$$J = 4\pi K r_N^2 \exp[-(\Delta F_g/kT)] \qquad (21)$$

where r_N is the radius of the aerosol particle, J is the nucleation rate per particle, and K is, in analogy to Eq. (2), given by $K = C_1' w(\sigma_g \Omega_g/\pi kT)^{1/2}$, where the surface area of a critical embryo on the water-insoluble substrate is approximately given by $\Omega_g \approx \pi r_g^2$, C_1' is the number of single water vapor molecules adsorbed per unit surface area of the nucleating surface, and w is the rate of impact of water molecules from the vapor per unit surface area. Near 0°C, the kinetic factor K varies between 10^{24} and 10^{27} depending on the particle's effectiveness for adsorbing water vapor. The prefactor of the exponent can approximately be taken to be $10^{25} 4\pi r_N^2$ [75]. Assuming the drop germ takes the shape of a spherical cap of radius r_g, an assumption recently justified by Barchet [83], which meets the surface of the solid, water-insoluble particle at a contact angle θ, Fletcher showed that

$$\Delta F_g = [16\pi\sigma_{wv}^3 M_w^2/3\rho_w^2 R^2 T^2 (\ln S)^2] f(m,x) \tag{22}$$

where

$$x = r_N/r_g \tag{23}$$

$$r_g = 2\sigma_{wv} M_w/\rho_w RT \ln S \tag{24}$$

$$f(m,x) = \tfrac{1}{2} + \tfrac{1}{2}[(1 - mx)/u]^3 + \tfrac{1}{2}x^3\{2 - 3[(x - m)/u] + [(x - m)/u]^3\}$$
$$+ \tfrac{3}{2}mx^2\{[(x - m)/u] - 1\} \tag{25}$$

$$u = (1 + x^2 - 2mx)^{1/2} \quad \text{and} \quad m = \cos\theta \tag{26}$$

For the case $r_N = 0$, $x = 0$, and $f(m,x) = 1$, ΔF_g has the same value as for homogeneous nucleation. For the case that the particle is completely wettable, $\theta = 0$, $m = 1$, and $\Delta F_g = 0$. In this case, $r_g = r_N$ and the critical saturation ratio necessary to nucleate a drop can simply be obtained from the Kelvin equation giving

$$\ln S_c = 2\sigma_{wv} M_w/\rho_w RT r_g \tag{27}$$

For particles which exhibit a finite wetting angle toward water, Eqs. (21)–(26) have to be solved simultaneously. Recently, McDonald [84] transformed this system of equations for the case $J = 1$ germ particle^{-1} sec^{-1} and $T = 273°K$ into a working equation applicable to atmospheric cloud conditions where small supersaturations prevail. According to MacDonald, we can write

$$\theta = \cos^{-1}\{1 - [(x - 1)/x][0.662 + 0.022 \ln r_N]^{1/2} \ln S_c\} \tag{28}$$

A numerical evaluation of Eq. (28) is given in Fig. 6, where S_c is plotted as a function of r_N for $\theta = 5°$, $7°$, and $12°$ and where comparison is also made with the case $\theta = 0°$ evaluated from Eq. (27). The results show that S_c strongly depends on the wetting angle of a particle as pointed out by McDonald [84]. Even particles which are water-insoluble but completely wettable hardly will get involved in the drop formation process in the atmosphere unless their radius is larger than about 1.2×10^{-5} cm. Insoluble particles with wetting angles larger than about $7°$ will not get involved at all in the condensation process no matter what their size is if it is assumed that 1% supersaturation is a typical upper limit in atmospheric clouds.

The results given in Fig. 6 have to be viewed with some caution. First, it is very unlikely that small particles of a hydrophobic substance have the same wetting angle as the same substance in bulk form. Recent experiments carried out in a diffusion chamber [85] showed that small particles of substances which in bulk had a wetting angle of more than $90°$ behaved as fully wettable particles. Second, we should recall that a large portion of the water-insoluble particles in the atmosphere consist of silicate particles from the soil of the earth's surface. It was pointed out by MacDonald [84] that these particles, even though fully wettable, are not spherical, but platelike (kaolinite) or needlelike (halloysite). Considering particle shapes different than spherical, McDonald predicted that fully wettable disks or oblong plates can be activated to drops if their "diameter" D satisfies the inequality $D > 2r_g$, where r_g is the Kelvin radius given by Eq. (3) and D is the greatest transverse dimension of the particle. However, even if most of the water-insoluble particles in the atmosphere seem to behave as fully wettable particles, it is evident from Fig. 6 that these particles will much less readily assume the role of CCN than salt particles. Figure 6 shows that a water-insoluble particle with the same size as a dry salt particle needs a supersaturation for activation to a drop which is more than one order of magnitude larger than that needed by a salt particle.

In summary, it can be stated that aerosol particles which are to serve as a CCN have to fulfill two major requirements, one with regard to their size and one with regard to their water affinity. The drop-forming capability of CCN is the better the larger the particles, the larger their water solubility, and, if water-insoluble, the smaller their wetting angle.

4. THE ROLE OF AEROSOLS IN THE FORMATION OF PRECIPITATION IN WARM CLOUDS

In order to link the atmospheric aerosol to cloud and precipitation formation, it is vital to determine the number of atmospheric aerosol particles which can serve as CCN. Measurements of the concentration of CCN on a worldwide basis have recently been made by Twomey and Wojciechowski [86]. These measurements, carried out during aircraft flights over Australia, Africa, the United States, and the Atlantic and the Pacific Oceans, are summarized in Figs. 10 and 11. The measurements indicated no systematic latitudinal or seasonal variation in the worldwide concentration of CCN. On the other hand, the results confirmed previous conclusions that continental air masses are systematically richer in CCN than maritime air masses. Within a particular air mass and at flight level distance from the earth's surface, the variation of the CCN spectra was surprisingly small [86]. On the other hand, measurements made at one particular location on the earth's surface [87–89] indicated that the concentration of CCN at 1% supersaturation varied considerably depending on the continental and maritime character of the air mass and depending on the proximity of the observation site to CCN sources, and ranged from $10 \, \text{cm}^{-3}$ or less up to $10^4 \, \text{cm}^{-3}$. In maritime or modified maritime air masses (which stayed less than two days over land), the CCN concentration was rarely larger than $100 \, \text{cm}^{-3}$. High concentrations in excess of $10^3 \, \text{cm}^{-3}$ ranging up to $10^4 \, \text{cm}^{-3}$ were only found in air which

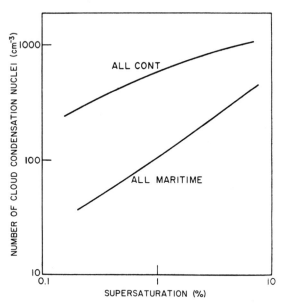

Fig. 10. Variation of the median concentration of cloud condensation nuclei with supersaturation for continental aerosols [86]. (By courtesy of *J. Atmos. Sci.* and the authors.)

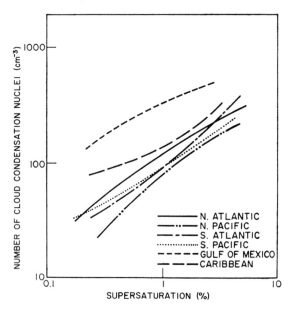

Fig. 11. Variation of the median concentration of cloud condensation nuclei with supersaturation for maritime aerosols [86]. (By courtesy of *J. Atmos. Sci.* and the authors.)

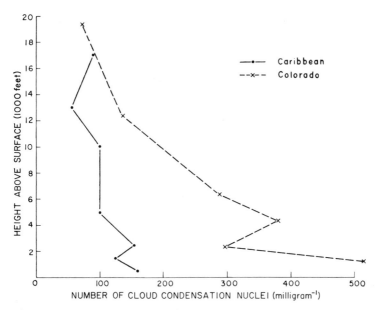

Fig. 12. Variation with height of the mean concentration per milligram of air of cloud condensation nuclei activated at 0.75% supersaturation [90]. Broken line: over the plains of Colorado (USA); full line: over the Caribbean Sea (1 mg air = 1 cm³ at 20°C and 840 mb). (By courtesy of *J. Atmos. Sci.* and the authors.)

had been over land for several days. Evidence that the contrast between the CCN concentration in continental and maritime air masses, which is very prominent near the earth's surface, decreases steadily with height was given by Squires and Twomey [90]. The results of their measurements of the CCN concentration at a supersaturation of 0.75% over Colorado (U. S.) and the Caribbean Sea are reproduced in Fig. 12.

Since one of the requirements for a good CCN is its hygroscopic nature, a relationship is expected between the number of CCN and the number of hygroscopic particles such as sea salt particles. The question whether sea salt particles alone can account for the number of drops found in clouds of maritime air masses has received wide attention. Based on earlier evidence, Mason [12] concluded that the bubble burst mechanism at the surface of the ocean is insufficient by a factor of 5–10 in order for sea salt particles to account for the loss of nuclei by cloud formation. Recent measurements confirm these earlier speculations. Twomey [91,92] and Dinger *et al.* [93], during investigations of CCN over the ocean, noticed that most aerosol particles heated to about 300°C evaporated and behaved analogously to a heated $(NH_4)_2SO_4$ or NH_4Cl aerosol but unlike an NaCl aerosol, which withstands temperatures up to 500°C. Out of a typical population of 100 CCN cm^{-3} measured near the ocean at a supersaturation of 0.75%, as many as 50–90 CCN cm^{-3} were found to be volatile and thus did not consist

TABLE 6. Typical Cloud Drop Concentrations in Various Cloud Types ([96])*

	Cloud type		
Drop concentration	Hawaiian orographic cloud	Maritime cumulus	Continental cumulus
Median drop concentration (cm^{-3})	10	45	228
Maximum drop concentration (cm^{-3})	370	470	2800
Total number of observations	123	438	403

*By courtesy of *Tellus*.

of NaCl. An even larger percentage was volatile at higher elevations. Above about 10,000 ft, practically all the CCN were volatile. Assuming a concentration of nonvolatile CCN of about 10 cm^{-3} and a residence time for CCN over the ocean of about two to three days, Twomey [92] estimated the production rate of sea salt particles from the ocean to be about 50 cm^{-2} sec^{-1}. This value is in good agreement with the independent recent estimate of Blanchard [94] of 25–100 cm^{-2} sec^{-1}, which was based on realistic splashing experiments with ocean water and on observations along the shore of the island of Hawaii. The result is also in agreement with the earlier experiments of Twomey [95], who estimated a production rate of 30–240 CCN cm^{-2} sec^{-1} at a supersaturation of 1 % for an ocean surface where three bubbles cm^{-2} sec^{-1} of diameter 0.5–2 mm are produced. Comparing these results with some typical cloud drop concentrations given in Table 6 [96], Twomey

TABLE 7. Comparison between Number of Aitken Nuclei (Activated above 100 % Supersaturation) and the Number of Cloud Condensation Nuclei (Activated at 1 % Supersaturation)*

Location	Number of Aitken nuclei (cm^{-3})	Number of cloud condensation nuclei (cm^{-3})
Washington, D. C. [98]	78,000	2000
	68,000	2000
	57,000	5000
	50,000	7000
Long Island (N.Y.) [99]	51,000	220
	18,000	110
	6,500	150
	5,700	30
Yellowstone National Park (Wyoming) [100]	1,000	15

*Based on data from [98–100].

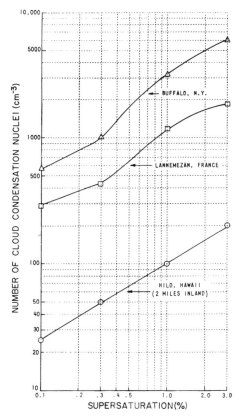

Fig. 13. Variation of the concentration of cloud condensation nuclei with supersaturation of various locations [97]. (By courtesy of *J. Rech. Atmos.*)

concluded that the total number of aerosol particles active at supersaturations lower than 1 % is sufficient to account for the number concentration of drops in maritime clouds, while the number of sea salt particles alone is not sufficient to do so. Comparing the number of CCN to the total number of nuclei over the ocean, which is typically between 500 and 900 cm^{-3}, it is evident that over the ocean only, one-fifth to one-tenth of the total nuclei population is able to act as CCN.

This fraction decreases to $\frac{1}{100}$ or less for continental aerosols and aerosols near cities. While the number of CCN may increase considerably as one passes from the ocean inland and close to a city, as shown in Fig. 13 [97], the number of Aitken nuclei increases even more rapidly. This fact is illustrated in Table 7, based on observations made by Allee [98], Twomey and Severynse [99], and Auer [100]. It is seen from this table that high Aitken nuclei counts (representative of the total number of nuclei present in the air) may not mean higher CCN counts, a result proposed earlier by Twomey [87]. The reason for this behavior can be found in the nucleus source itself, which may be largely devoid of particles meeting the requirements for CCN specified above, or else is due to a powerful poisoning mechanism resulting from coagulation of CCN with Aitken nuclei, rendering the CCN incapable of being activated at supersaturations of 1% or less. Nevertheless, it is well established from observations that city fogs are much denser, i.e., have a larger drop concentration, and are more persistent than fogs over the countryside. This is in part a direct consequence of the larger concentration of CCN usually found in city air as compared to air over the countryside. Larger CCN concentrations tend to be related to drops of smaller sizes since the given amount of moisture now has to be distributed over a larger number of nuclei. Thus, they tend to stay uniformly small and hence are quite incapable of growing by colliding with each other. Such fogs are therefore colloidally very stable. Furthermore, the stability of fogs in city air can drastically be increased by the presence of vapors such as SO_2, which in water becomes oxidized to H_2SO_4. Sulfuric acid droplets are very hygroscopic (see Fig. 7) and are capable of existing at relative humidities far below 100%. The stability of fogs in heavily polluted air may also be affected by organic substances which are surface active and form partial or complete monolayers on the surface of water drops. Such monolayers are capable of increasing the lifetime of a water drop drastically by reducing its rate of evaporation up to 20 times if the drop is several hundred microns in radius and up to a hundred times or more if the drop is several tens of microns in radius or smaller [101–103]. Molecules containing straight hydrocarbon chains were found to be most effective in reducing the rate of evaporation of water drops [103]. Recently, the rate of evaporation of 70-μm-radius drops polluted by gasoline exhausts has been studied in a wind tunnel by Hoffer and Mallen [104]. These investigators found that the rate of evaporation of the drops progressively decreased until at some finite size, evaporation did completely stop. Several investigators suggested that surface-active, organic substances are present in the surface of oceans and may become airborne by means of the bubble-bursting mechanism [105–108] and thus may stabilize sea fogs [109]. However, Garrett [103] pointed out that a substantial portion of the components recovered from the ocean surface consisted of nonlinear molecules or were only weakly surface-active and that consequently films of natural origin on water drops are not likely to reduce their evaporation rate significantly, in contrast to monolayers of anthropogenic pollutants. It has been suggested that surface films on water drops present in heavily polluted air also may alter the coalescence efficiency of drops (which is the fraction of colliding drops that actually coalesce). However, the coalescence efficiency is not only a function of surface-active material on the drops, but also is a complicated function of the impact velocity and impact angle between the drops, the electric charges on the drops, and the temperature and relative humidity in the air. The combined effect of these factors is difficult to predict. No quantitative measurements on this problem are available.

TABLE 8. Comparison of Natural and Anthropogenic Production Rates of Cloud Condensation Nuclei [110]*

	Area of region (cm^2)	Fuel consumption in region $(g\ sec^{-1})$	Mean regional production rate of anthropogenic nuclei, A (nuclei cm^{-2} sec^{-1})	Mean regional production rate of natural nuclei, N (nuclei cm^{-2} sec^{-1})	A/N %
USA	7.8×10^{16}	3.5×10^7	72	500	15
Northern hemisphere	2.55×10^{18}	1.5×10^8	9	250*	4

*The oceanic production rate [94] is included; by courtesy of *J. Rech. Atmos.*

A comparison between Fig. 10 and Table 6 suggests that the number of CCN available in continental atmospheric aerosols is sufficient to account for the number of drops found in continental clouds. This conclusion is analogous to the conclusion one can draw for the case of maritime aerosols and maritime clouds. Thus, cloud drop formation in the atmosphere can proceed without difficulty, in contrast to ice formation, as we shall see later.

In this context, it is instructive to ask in what percentage anthropogenic aerosols enter the local and global budgets of CCN. Based on current figures for fuel consumption and based on laboratory as well as field observations of CCN (active at 0.5 % supersaturation) produced by fuel combustion, Twomey [95] and recently Squires [110] attempted to roughly estimate the contribution of the anthropogenic CCN to the worldwide CCN budget. Squires's [110] estimate is summarized in Table 8. In agreement with Twomey [95], Squires concluded that at present the anthropogenic production rate of CCN is only a few per cent of the natural production. However, a continually increasing fuel consumption on a worldwide scale may change this figure upward. Over densely inhabited and heavily industrialized areas, the anthropogenic production of CCN contributes significantly to the total production. According to Squires's estimate, this contribution over the USA is as high as 15 %. Over smaller regions such as the Eastern USA or Western Europe, the anthropogenic CCN production may even approach that due to natural causes.

The question whether the concentration of CCN could affect the microstructure of atmospheric clouds has attracted much attention. According to the studies of Squires [111–114] there exist marked systematic differences in the microstructure of atmospheric clouds. Low drop concentrations are usually associated with broad drop size spectra and large maximum and average drop sizes. Conversely, high drop size concentrations are usually associated with narrow drop size spectra and small maximum and average drop sizes. Maritime clouds tend to belong to the former group, while

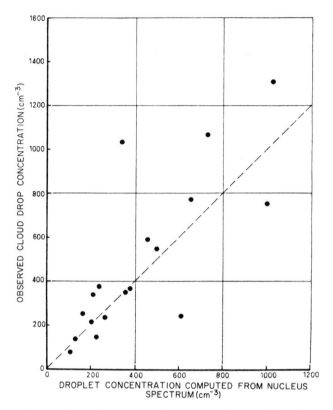

Fig. 14. Comparison of observed mean cloud drop concentrations with cloud drop concentrations computed from observed cloud condensation nucleus spectra for an updraft of 3 m sec^{-1}. The dashed line represents exact agreement between observed and computed values [116]. (By courtesy of *J. Atmos. Sci* and the authors.)

continental clouds tend to belong to the latter. Squires suggested that differences in the cloud microstructure of maritime and continental clouds have to be explained in terms of differences between the maritime and continental aerosol and only to a lesser extent in terms of differences in upcurrent, turbulence level, or total lifetime of parcels of cloud air. Twomey and Squires [115] and Twomey and Warner [116] computed the cloud drop concentration which would be expected from the observed CCN spectra and the observed updrafts in various clouds and compared these computed values with the observed cloud drop concentrations. This comparison is reproduced in Fig. 14. A regression analysis on these data gave a correlation coefficient in excess of 90% and showed that, at least in the early history, the condensation nucleus population of the atmosphere accounts for the initial concentration of cloud droplets. A problem of considerably greater difficulty arises when one attempts to infer the drop size distribution in a certain drop from a given aerosol population. Mordy [117], Neiburger and

Chien [118], and recently Kornfeld [119] and Chen [120,121] studied theoretically the change with time of various size distributions of aerosol particles as a function of the composition and initial sizes of the aerosol particles and as a function of the dynamic and thermodynamic cloud parameters. The results of these studies suggest that, given a hygroscopic aerosol, the initial size distribution of the aerosol particles seems to be a major factor that determines the size distribution of the drops in a cloud. Thus, for the development of a broad size spectrum (in the sense of its extent towards larger sizes), the presence of a broad size spectrum of aerosol particles seems to be necessary. The same conclusions have recently been reached by Junge and McLaren [122] from a study of the growth of "mixed" CCN composed in various proportions of water insoluble and water soluble substances. During the growth by diffusion of a population of hygroscopic aerosol particles of one particular chemical compound the originally single-modal size distribution changes progressively to an aerosol with a bimodal size distribution. The gap in the size spectrum develops as a result of differences in the critical supersaturation needed for activation of aerosol particles to visible drops. As a parcel of air rises the saturation ratio increases. Once saturation is surpassed, condensation begins on the most efficient CCN, which are those that have the lowest values of critical supersaturation. As the supersaturation continues to rise, more and more particles become activated to drops. However, eventually the consumption of water vapor by the growing drops prevents the supersaturation from rising further. All these aerosol particles whose critical supersaturation is smaller than the maximum supersaturation reached by the air parcel essentially stop growing at this supersaturation and remain inactivated while all particles with a smaller critical supersaturation keep on growing rapidly to visible cloud drops. The development of a bimodal aerosol size spectrum and the associated gap between those aerosol particles that were left inactivated and those that were activated to cloud drops is illustrated in Fig. 15a,b [120,121]. While earlier theoretical computations modeled the condensation process by means of air parcels which cool adiabatically as they rise in the atmosphere [117–119], the most recent computations [120,121] are based on a thermodynamic and dynamic model which includes entrainment. In Fig. 15 the effect of the initial size spectrum of the dry, hygroscopic nuclei on the drop size spectrum is clearly reflected in the different times needed for aerosol particles in equilibrium with 99% relative humidity (assumed to be the level of visible cloud) at time $t = 0$ to grow to drops of 20 and 30 μm radius. After a time of about 7 min, 80 drops liter^{-1} with radii larger than 20 μm and 6 drops liter^{-1} with radii larger than 30 μm were present in a cloud which originated on an aerosol with the dry particle size distribution given in Fig. 15(a). After the same time, only 370 drops m^{-3} with radii larger than 20 μm and 7 drops m^{-3} larger than 30 μm were present in a cloud which originated on an aerosol with the much narrower size distribution given in Fig. 15(b). Figure 15 shows further that

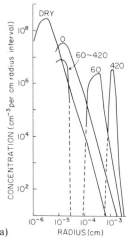 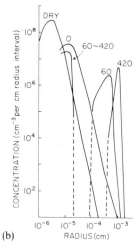

(a)　　　　　　　　　　　　　　　　　(b)

Fig. 15. Differential cloud drop size distribution after various elapsed times for the case of drop formation on two different cloud condensation nucleus spectra [120]. Curves are labeled in seconds. (By courtesy of the author.)

growth by condensation results in the formation of a population of cloud drops with a single-mode size distribution.

Quite frequently the observed drop size distributions in atmospheric clouds are significantly broader than those predicted by computations of droplet growth on salt nuclei in a closed homogeneous parcel of air rising in a constant updraft. Even if entrainment is included in the computational growth model, the drop size spectrum is not significantly broader than the spectrum computed from a closed parcel model [121]. Kornfeld [119] showed that the presence of water insoluble particles or "mixed" particles may contribute to the broadening of the drop size spectrum in particular if neighboring air parcels contain different numbers of water soluble and water insoluble particles on which condensation takes place. According to Paluch [123] broadening of a cloud drop size distribution may come about as a result of small fluctuations in relative humidity coupled with droplet sedimentation. During the early stages of condensation, small local variations in relative humidity can produce significant variations in droplet sizes. The resulting nonuniformities in droplet sedimentation rate create inhomogeneities in droplet concentration which in time cause variations in vapor pressure and differences in droplet growth rates. Numerical computations of Paluch showed that the initial conditions for this model to produce a broad drop size spectrum are sufficiently general to be of common occurrence.

Clouds with drops of radii larger than 20 to 30 μm tend to be colloidally unstable due to the rapidly increasing efficiency with which drops larger than this size are able to collide and coalesce with other cloud drops. The large number of field studies carried out by various investigators in different parts of the world demonstrates that the colloidal instability of a cloud is well correlated to a broad cloud drop size spectrum and a large average drop size. Both of these characteristics are less likely to develop in continental air masses whose aerosols are characterized mainly by a narrow nucleus spectrum, high nucleus concentrations, and water insoluble or mixed particles than in maritime air masses whose aerosols are characterized mainly by a wide nucleus spectrum, lower nucleus concentrations, and mostly hygroscopic particles [112–114,124–127]. Thus, from the point of view of growth of cloud drops by collision and coalescence, it is understandable that continental clouds, with their initially narrow drop size spectrum and small mean drop radius, will have to reach larger heights and widths than maritime clouds before precipitation can develop.

Detailed studies of the condensation and collision–coalescence mechanism show that precipitation in all-water clouds is the result of a continued selection of the most qualified particles. A small, fortunate number of nuclei at the large end of the nucleus size spectrum essentially decides whether drops are formed that are capable of initiating the collision–coalescence process, and again a statistically small, but fortunate number of drops at the large end of the drop size spectrum provides the precipitation by growing faster than the rest of the drops. This is exemplified by the fact that in clouds, the concentration of drops of precipitation size (drops larger than about 200 μm diameter) is typically 100–1000 m^{-3} [126], while the total drop concentration is typically 100–1000 cm^{-3} or a factor of about 10^6 times larger. Our theoretical knowledge of the stochastic growth mechanism of cloud drops is due to a number of investigators [128–131]. Recently, Chin [132] studied the stochastic collision–coalescence growth of cloud drops which follow a Khrgian–Mazin size spectrum given by $n(r) = Ar^2 \exp(-Br)$, where r is the drop size, $A = (1.45/R_{av}^6)L$, $B = 3/R_{av}$, L is the liquid water content, and R_{av} is the average radius. He found that the 100th largest drop m^{-3} had a radius of 125 μm after 20 min and a radius of 187 μm after 30 min for the cloud drop size spectrum given in Fig. 16(a), while the 100th largest drop m^{-3} had a size of only 39 and 53 μm after 20 and 30 min, respectively, for the drop size spectrum given in Fig. 16(b).

The pronounced correlation found between the colloidal stability, or precipitation tendency, of clouds and the concentration, size distribution, and chemical constitution of the aerosol motivated several investigators to search for correlations between anthropogenic nuclei and precipitation from warm clouds. While significant local increases in the number of CCN were found as a result of burning of sugar cane leaves during harvesting season of sugar cane in Australia and Hawaii [133–135], as a result of forest fires [136], and as a result of certain industrial sources such as pulp and paper mills in the state of Washington in the USA [137], the increases in CCN had no clear-cut effects on precipitation. In Australia, a definite increase in the colloidal stability of clouds and an associated decrease in rainfall amounts downwind

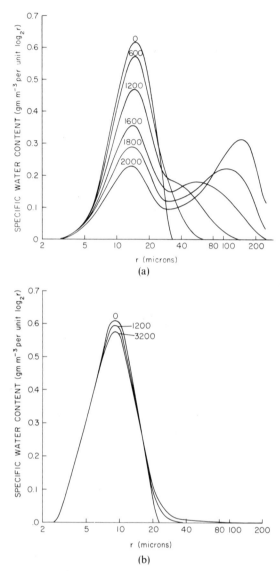

Fig. 16. Evolution with time of two spectra of cloud drops growing by collision and coalescence [132]. Curves are labeled in seconds. (By courtesy of the author.)

from the CCN sources was found [133,134], but no such effect could be detected in Hawaii [135]. An opposite situation was found in the state of Washington, where pulp and paper mills added significant numbers of large and giant hygroscopic nuclei into the atmosphere which destabilized the clouds and caused a precipitation increase downstream of the sources [137].

5. THE ROLE OF AEROSOLS IN NUCLEATING ICE CRYSTALS

Ice formation in the atmosphere is a comparatively rare cloud physical event. This fact is illustrated in Fig. 3, which shows that ice formation at 0°C, the equilibrium freezing temperature of bulk water, is extremely unlikely in atmospheric clouds. Most clouds consist entirely of supercooled drops as long as the cloud summit temperature is warmer than -10°C. Even at temperatures between -10 and -20°C, a large proportion of atmospheric clouds are either supercooled, all-water clouds, or mixed clouds containing

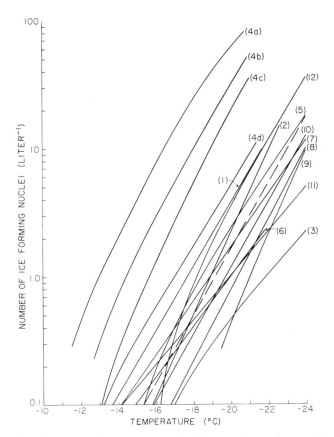

Fig. 17. Variation of the number of ice-forming nuclei with temperature and geographical location (based on data from Refs. 43 and 139–148). (1) Bracknell (England), 51°N, 0°W [139]; (2) Clermont–Ferrand (France), 46°N, 3°E [140]; (3) Corvallis, Oregon (USA), 44°N, 123°W [141]; (4) Tokyo (Japan), 36°N, 140°E [142], (a) fresh arrival of Loess, (b) explosion of volcano Mihara, (c) air mass from N. China, (d) air mass from S. China; (5) Tucson, Arizona (USA), 32°N, 111°W [141]; (6) Jerusalem (Israel), 32°N, 35°E [143]; (7) Palm Beach, Florida (USA), 27°N, 80°W [144]; (8) Hawaii (USA) 20°N, 158°W [145]; (9) Swakopmund (S. Africa) 34°S, 14°E [146]; (10) Sidney (Australia) 34°S, 151°E [147]; (11) Tasmania (Australia), 43°S, 147°E [43]; (12) Antarctica 78°S [148]. The dashed line represents $\Delta T = 10^{-5} \exp(0.6 \, \Delta T)$.

supercooled drops and ice crystals. It is only as clouds reach up to levels in the atmosphere where the temperature is $-20°C$ or below that a major proportion of them becomes completely glaciated.

This observation suggests that only a small fraction of the aerosol particles in the atmosphere serve as ice-forming nuclei (henceforth abbreviated IFN) and of those, only a small percentage seem to be capable of initiating the ice phase in the atmosphere at temperatures warmer than $-10°C$. A large number of observations made by many investigators and at different parts of the earth substantiates this conclusion (for a summary of earlier work, see Refs. 12, 75, 138). Aerosol particles may initiate the ice phase (1) directly from the vapor by adsorbing water vapor from a surrounding humid atmosphere, (2) from the supercooled liquid water by coming in contact with supercooled cloud drops and freezing those from the three-phase boundary, or (3) from the supercooled liquid water by freezing cloud drops from within. None of the presently available instruments are capable of exposing the natural aerosols to all three ice-forming mechanisms together and at the same time also simulate the time scale and low supersaturations which are typical in the atmosphere. The available results of IFN measurements therefore have to be interpreted with caution. Despite the insufficiencies of the present instrumentation, some useful conclusions can be drawn regarding the ice-forming capability of the atmospheric aerosol.

In Fig. 17, a selected number of observations is given of the mean IFN concentration for various locations on the earth surface and for various temperatures at which the aerosols were activated [139-148]. This figure shows that the lower the temperature, the larger is the number of aerosol particles which can serve as IFN. Between -10 and $-30°C$, the IFN concentration seems to vary about logarithmically with temperature, changing by a factor of about ten for a temperature change of $4°C$. For comparison with the observations, the relation $n = n_0 \exp(\beta\Delta T)$ proposed by Fletcher [75] is given for $\beta = 0.6$ and $n_0 = 10^{-5}$ liter^{-1}, where n is the number of IFN active at a temperature warmer than T and $\Delta T = T_0 - T$, with $T_0 = 273°K$. While there is sufficient evidence that the IFN concentration is a quantity which, at a given temperature, strongly fluctuates both in time and with location, varying at $-20°C$ from zero to as high as several thousand per liter near IFN sources, the time mean concentration far from a source seems to vary surprisingly little at a given temperature and shows no systematic variation with geographical latitude or longitude, or between the northern and southern hemisphere as pointed out by Bigg [147] (curves 1–3, 4d, 5–12 in Fig. 17). Bigg and Miles [149] suggested that this behavior of the IFN concentration at the earth's surface may be explained on the basis of a uniform, worldwide influx of IFN through the upper atmosphere from an extra-terrestrial source. As a basis of their suggestion, they invoked Bowen's [150] hypothesis that particulate matter resulting from meteorites entering the earth's atmosphere could account for pronounced peaks in the average daily rainfall summed over a worldwide network of observation stations if a time lag of about 30 days between the annually reoccurring showers and the peak in the rainfall rates is assumed. A full account of earlier arguments for and against the Bowen theory is given by Fletcher [75] and Mossop [138]. The more recent evidence seems to be as controversial as the earlier evidence. Bigg and Giutronich [151] felt that the earlier laboratory experiments of Mason and Maybank [152] and Qureshi and Maybank [153] with ground material of stony and iron meteorites sampled at the earth's surface were unrealistic and showed that freshly condensed particles from meteorites which were evaporated under conditions similar to those in the high atmosphere acted as an abundant source of IFN active at $-10°C$. On the other hand, recent samples of extraterrestrial material collected by Gokhale and Goold [154] at 80 km height in the atmosphere showed that these particles had little ice-forming capability even at $-20°C$. A number of investigators measured the vertical variation of IFN through the troposphere up into the lower stratosphere [149,155–160]. These measurements showed that the IFN concentrations were significantly higher in the upper troposphere and lower stratosphere than those at the ground, as indicated in Fig. 18, and were closely correlated to the upper air circulation. These results were interpreted in terms of an extraterrestrial source of IFN which reach the ground in variable numbers through vertical mixing associated with hydrodynamic instabilities of the air, thus producing IFN "storms" which are often observed at the ground. In support of this interpretation are the recent observations of Maruyama and Kitagawa [161], who found a close correlation between the occurrence of meteorite showers and IFN concentration, and the findings of Bigg [162] and Bigg and Miles [159], who found a close correlation between the moon phase and the IFN concentration, suggesting a modulation of the extraterrestrial influx of IFN by the moon. However, a number of investigators did not see any need for invoking an extraterrestrial source to explain the behavior of the IFN in the atmosphere. Some observers did not find any correlation between meteorite showers and the concentration of IFN [143,163,164]. Others found that IFN "storms" could be attributed to evaporating cloud and precipitation particles [165–167]. Higuchi and Wushiki [168] recently sampled aerosol particles at Barrow, Alaska, and in Japan on Mt. Fuji (3770 m) at temperatures below 0°C. Those aerosol samples that were never heated above 0°C often showed IFN concentrations as high as 10 liter^{-1} (at $-17°C$), while those aerosol samples that were warmed above $+20°C$ consistently showed IFN concentrations below 0.5 liter^{-1} (at $-17°C$). These investigators concluded that at sufficiently cold temperatures, as are found at the higher levels of the atmosphere or in polar air masses, terrestrial aerosol particles may become "activated" to high-quality IFN. Further evidence in favor of a terrestrial source of IFN comes from a careful analysis of the central portions of snow crystals which, as discussed in an earlier section, in most cases contain a silicate particle, often identifiable as clay, from the soil of the

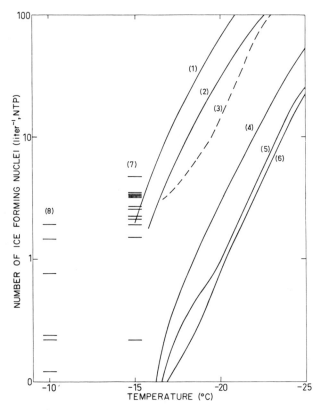

Fig. 18. Variation of the concentration of ice-forming nuclei with temperature and height in atmosphere [149] (1) at 13 km over Australia; (2) at 7.5 km over Australia; (3) highest ground concentration in Australia and Antarctica; (4) ground concentration, Antarctica, 78°S; (5) ground concentration, Australia, 27°S; (7,8) at 13–27 km over Australia. (By courtesy of *Tellus*.)

earth's surface [45–54]. Ice-forming nuclei counts and trajectory analysis of air masses showed further that in the northern hemisphere, the arid regions of North China and Mongolia constitute an abundant source of clay particles which, once airborne by the action of local storms, are transported via Japan eastward in the air masses associated with these storms and are carried in the jet stream further eastward to the North American continent [142,163,169].

Many experiments have been conducted to determine the ice-forming capability of particles from the surface soil. Soils from different parts of the world as well as pure silicates, in particular clays, were tested. Examining soil particles from different parts of the USA, Schaefer [170] found that many soils were capable of forming ice crystals from the vapor with threshold temperatures for ice formation (one particle in about 10^4 particles produces an ice crystal) between -10 and $-20°C$. Some of the soils had a threshold as warm as $-9°C$. On the other hand, soils from different parts of Australia were

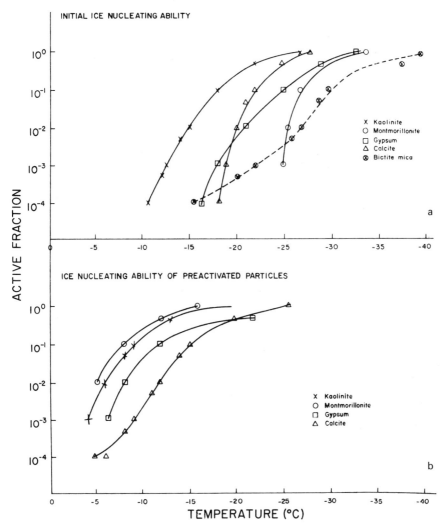

Fig. 19. Variation of the ice nucleating ability of mineral dusts with temperature [174]. Active fraction = number of ice-forming nuclei per total number of particles introduced. (By courtesy of *Quart. J. Roy. Meteorol. Soc.*)

distinctively less active, with threshold temperatures ranging between -18 and $-22°C$ [171]. Similar experiments with numerous pure clays gave threshold temperatures as warm as -9 to $-10°C$ for kaolinite, illite, metabentonite, and anauxite, and threshold temperatures between -10 and $-20°C$ for most other clays [152,172–174]. Figure 19(a) shows that at temperatures colder than the threshold temperature, the ice-forming ability of clays increases with decreasing temperature, the active fraction approaching one, i.e., each particle acts as an IFN, if the temperature falls below $-25°C$. However, clays, as well as many other natural and artificial substances, are capable of undergoing "preactivation," achieving during this process a substantially improved capability to initiate the ice phase. Particles may become "pre-activated" once they have been involved in ice crystal formation, after which the temperature did not rise above $0°C$ and the relative humidity did not drop too far below ice saturation, although it was low enough for the macroscopic crystal to disappear [152,174–177]. Particles may also become "pre-activated" once they have been cooled to temperatures below $-40°C$ at relative humidities below ice saturation but not lower than 50% [168,178]. Experiments [174] demonstrated that by "preactivation" clays exhibit threshold temperatures as warm as $-3°C$ and approach an active fraction of one if the temperature falls below $-15°C$ (Fig. 19b).

There is evidence that clay particles submerged in water drops have a considerably lower capability of initiating the ice phase than if they would act directly from the vapor. Vali [179] found that soil particles were considerably better IFN than pure clay particles, a fact which he attributed to some unknown organic substances in the soil. Hoffer [34] examined the freezing temperature of 100–120 μm diameter drops of aqueous suspensions of various clays and found that none of the drops froze at temperatures warmer than $-13.5°C$ and none of the median freezing temperatures of a population of drops were warmer than $-24°C$, in agreement with some later results of Gokhale [180]. A surprising result was found when clays were tested as contact nuclei. Experimenting with millimeter-size drops, Gokhale and Goold [181] showed that drops of this size would freeze at temperatures between -5 and $-8°C$ on contact with clay particles, while the median freezing temperature of a population of millimeter-size drops with clays submerged in the drops was about $10°C$ lower.

Active volcanoes represent another source of IFN in the atmosphere. Even though not of great importance to the worldwide IFN budget due to the limited number of active volcanoes at any one time, active volcanoes may cause strong local effects. Significant peaks in the concentration of IFN due to active volcanoes in Japan were reported by Isano and co-workers [142,182, 183], who also showed by laboratory experiments that volcanic ashes produced significant numbers of ice crystals from the vapor at temperatures as warm as $-8°C$. In contrast to these findings, Price and Pales [184] found no significant increase in the IFN concentration above the background concentration during the eruption of volcanoes on the Hawaiian Islands.

Recent studies showed that man's activities constitute another source of IFN which, although not proven to be of a worldwide significance, still are of considerable importance to the local IFN budget. Evidence for the presence of anthropogenic IFN in the atmosphere has been given by several recent studies carried out in France [140,185,186], the USA [187,188], and Australia [189]. These studies showed that certain industries, in particular steel mills, aluminum works, sulfide works, and some power plants, release considerable amounts of IFN into the atmosphere. Near these sources, the concentration of IFN at $-20°C$ was found to be 100–1000 times as large as the background concentration. The pollutants from electric steel furnaces in France were found to be of particular high ice-forming capability, with an activation threshold of $-9°C$, thus being of similar quality as clay particles. With these findings in mind, it is not difficult to understand why the concentration of IFN is rather high all across the heavily industrialized European continent (Table 9, [190–192]). Chemically, these industrial pollutants seem in part to consist of metal oxides. Metal oxides tested in the laboratory were found to exhibit a wide range of threshold temperatures. The compounds Ag_2O, Cu_2O, NiO, CoO, Al_2O_3, CdO, Mn_3O_4, and MgO had threshold temperatures between -5 and $-12°C$, CuO, MnO_2, SnO, ZnO, SnO_2, and Fe_3O_4 had threshold temperatures between -12 and $-20°C$, and FeO, Fe_2O_3, TiO_2, PbO_2, HgO, Co_2O_3, and PbO had threshold temperatures below $-20°C$ [193–198]. Despite the high threshold temperatures of some metal oxides, the activity ratio of particles of these substances rarely exceeded 1 IFN per 100 particles. Furthermore, most metal oxides exhibited a much poorer ice-forming capability when they were submerged in the supercooled water drops than when they initiated the ice phase directly from the vapor.

Recently, numerous investigators concerned themselves with the question whether anthropogenic combustion products could serve as efficient IFN. It has been known for some time that lead iodide (PbI_2) particles act as excellent IFN with a threshold temperature of about $-6°C$. It also has been known for some time that aerosols of urban areas contain considerable amounts of lead. The lead content of the aerosols over major cities in the USA was found to range typically between 1 and 9 $\mu g\ m^{-3}$ [199].

TABLE 9. Concentration of Ice-Forming Nuclei (Activated at $-20°C$)
for Different Sites in Europe*

Location	Reference	Mean number of ice nuclei (liter^{-1})
Budapest (Hungary)	[190]	2.6
Kl. Feldberg, 800 m (Germany)	[191]	14.1
Vienna (Austria)	[192]	8.4
Lorient (France)	[192]	53.7
Clermont-Ferrand (France)	[192]	39.9
Mezilhac, 1130 m (France)	[192]	6.8
Valdelore, 1050 m (France)	[192]	20.8
Toulouse (France)	[192]	15.0
Valencia (Spain)	[192]	19.1
Campo Catino, 1787 m (Italy)	[192]	10.8

*Based on data from [190–192].

The presence of lead in the air can be attributed to the combustion of gasoline, which may contain lead up to about 0.1 % by weight. Schaefer [200] suggested that lead particles from automobile exhausts constitute powerful "latent" IFN. Indeed, laboratory and field experiments demonstrated that automobile exhausts and polluted air from urban areas contained large numbers of IFN if the gas samples were "activated" by injection of iodine vapor [201–207]. Furthermore, Schaefer [202] found that large numbers of IFN were formed if automobile exhausts and smoke from certain woods were mixed together. Based on these results, Schaefer suggested that the concentration of iodine in the atmosphere may be sufficient to produce lead iodide (PbI_2) at the surface of the lead particles in the automobile exhausts and in polluted air of urban areas. Acceptance of such a suggestion obviously depends on whether one can establish that the concentration of iodine in the atmosphere is indeed large enough for the reaction $Pb + I_2 = PbI_2$ to take place. The values reported in literature for the concentration of iodine in air show considerable scatter. Even relatively recent values range from a few nanograms to several micrograms per cubic meter of air [208–210]. Moyers et al. [211] attributed the differences among the earlier and the more recent measurements in part to unreliable sampling and analyzing methods, to contamination from industries engaged in the burning of sea weed, and to the fact that little distinction was made between gaseous and particulate iodine. The first reliable measurements of gaseous iodine in maritime air have recently been made by Moyers [212] in Hawaii. He found an average concentration of 8 ± 3.5 ng m^{-3} of gaseous iodine in an unpolluted maritime atmosphere. Recently, Moyers et al. [211] reported on their simultaneous measurements of the concentration of gaseous and particulate iodine and lead in the polluted air of the Boston–Cambridge area. The concentration of gaseous iodine varied from 10 to 18 ng m^{-3}, that of particulate iodine from 2 to 15 ng m^{-3}, and that of lead from 0.4 to 3.7 μg m^{-3}. Although the concentration of gaseous iodine was not as high as had previously been expected for a polluted atmosphere, Moyers et al. estimated from their measurements that the concentration of gaseous iodine in a polluted atmosphere is sufficient to activate all the lead-containing aerosol particles to IFN. However, this conclusion sharply contradicts the finding that in polluted air of urban areas, in exhausts from airplanes and cars, and in smoke of burning wood and coal, no significantly elevated IFN concentrations are found in the presence of iodine concentrations typical for the atmosphere [170,202,213–216]. Moyers et al. suggested the following reasons for this discrepancy: (1) Experiments carried out by Clough et al. [217] suggest that the adsorption of gaseous iodine on aerosols produced by combustion processes is largely a reversible process. No experiments are available which would demonstrate whether adsorbed layers of iodine are capable of causing ice nucleation. (2) It is highly questionable whether the surface of lead particles contained in automobile exhausts have a surface which consists of metallic lead. Rather, it is expected that right from the very beginning of their lifetime, the surface of these particles consists of lead oxide or lead bromide layers which inhibit the reaction that forms lead iodide. (3) The lead particles emitted from automobile exhausts are very small and thus are expected to coagulate rapidly with other particles which may "shield" the lead from reacting with iodine. (4) Even if the circumstances would allow lead iodide to be formed at the surface of a lead particle, the reaction would only proceed very slowly at atmospheric iodine concentrations, which are several orders of magnitude smaller than those which must have prevailed during the laboratory experiments of Schaefer and others.

Combustion particles may constitute an important source of IFN if the temperature is very low. This is exemplified, for instance, by the formation of "ice fog" which forms in stable air over cities with abundant sources of water vapor and combustion particles if the temperature of the air drops below about $-35°C$ [218]. At these low temperatures, injection of water vapor into air containing an abundance of combustion particles first causes the formation of water drops, which quickly cool to the ambient temperature, at which point they freeze by the ice-forming action of the particles within the supercooled drops. The resulting ice particles are small (several microns in diameter) and appear in large concentrations (several hundred particles per cubic centimeter), thus causing a severe visibility hazard. Contrail formation from aircraft is another example of the effect of an abundant source of combustion particles coupled with a water vapor source in an environment of low temperature. The conditions for contrail formation have been studied by Appleman [219], Pilié and Jiusto [220], and others. As in the case of ice fog, small drops are formed initially which contain substantial amounts of soluble materials and combustion particles. The latter subsequently nucleate the supercooled drops to ice particles. If one compares the variation of the mixing ratio of water vapor with the variation of the air temperature with height in the atmosphere, one finds that the proximity to saturation of water vapor tends to increase with height in the troposphere. This, together with the fact that the residence time of gases and particulate matter increases with height, reaching about six months to three years near the tropopause and in lower stratosphere [13], makes these levels in the atmosphere very "vulnerable" to cloud formation, as may result from the injection of particles and water vapor contained in the exhausts of jet aircrafts [221]. Such contrail "clouds" may, if they occur with sufficient frequency, affect the heat budget of the atmosphere and serve as a source of ice crystals capable of "seeding" supercooled clouds at lower levels.

Since a major fraction of the atmospheric aerosol consists of water-soluble, hygroscopic particles, it seems obvious to pose the question whether these particles can serve as IFN. In recent years, this question has been answered by field measurements and by laboratory studies. Field observations demonstrated clearly that maritime air masses are consistently correlated to low concentrations of IFN [142,146,222–226]. In agreement with this finding, laboratory tests with numerous types of salts showed that water-soluble particles and salts in particular are poor IFN [227]. Recent laboratory studies with populations of drops of aqueous solutions indicated that salts and most other soluble substances, including soluble gases and organic macromolecules, either hinder the ice nucleation process in supercooled water or have no effect on it at all. This result holds provided that no interaction, chemical or otherwise, takes place between the solute and the insoluble particles present in the water that could change their surface characteristics and improve their ice nucleability [27,34,228–232]. It was found that as long as the solute concentration in the solution drops is smaller than about 10^{-3} moles liter^{-1}, the freezing temperature of the drops is not affected by the solute. At higher concentrations, drops freeze at a temperature colder than that of pure water drops [34,228]. By means of Table 10, it is shown that solution drops growing by diffusion from the vapor on atmospheric salt particles dilute very quickly during their growth. Drops forming on CCN of NaCl with masses larger than about 10^{-14} g have, after growth to their activation size, salt concentrations smaller than 10^{-3} moles liter^{-1} and thus beyond that size will not be affected in their freezing behavior by the dissolved salt. Drops forming on sodium chloride particles of $m_N < 10^{-4}$ g must grow beyond their actuation size in order to achieve dilutions at which their freezing temperature is unaffected by the solute, a fact pointed out already by Junge [233]. Since atmospheric clouds and fogs have drops of radii larger than 1 μm, it is not expected that the freezing temperature of these drops will be affected by the solute in the drops unless large amounts of soluble material are being picked up by the drops subsequent to their formation by scavenging pro-

TABLE 10. Variation of the Dilution of a NaCl Solution Drop Contained in Drops Growing with Relative Humidity*

	Mass of salt in drop (g)	
	10^{-16}	10^{-14}
Relative dilution at RH $\equiv 80\%$	1.0	1.0
Relative dilution at RH $\equiv 90\%$	1.7	1.9
Relative dilution at RH $= 100\%$	19.3	187
Relative dilution at S_c	110	1060
Drop size at S_c (cm)	2.5×10^{-5}	2×10^{-4}
Salt concentration in drop at S_c (moles liter^{-1})	5.2×10^{-2}	5.2×10^{-3}
Drop size at which salt concentration in drops is 1×10^{-3} moles per liter (cm)	7.5×10^{-5}	3.5×10^4

*Based on data from [79].

cesses in heavily polluted air, or unless the drops are in portions of a cloud or fog where the relative humidity has dropped below 100%. The first condition is rarely realized since the salt concentrations in cloud and rain water typically range from 0 to about 10^{-5} mole of salt liter^{-1} (Ref. 3, and numerous recent articles). Recent measurements show that even in cloud water from low-level stratus over urban areas the total solute concentration may approach, but rarely will exceed, 10^{-3} moles liter^{-1} [233]. On the other hand, relative humidities below 100% are an often observed phenomenon in fogs [234,235] and in clouds [39,40]. This suggests that ice formation by freezing of cloud drops will most readily start in those cloud portions where the humidity is highest [39].

Not only water-soluble materials, but also water-insoluble particles accumulate in cloud drops by the action of scavenging mechanisms. The larger the number of water-insoluble particles which become accumulated in a drop, the larger is the probability that a good IFN has been collected and thus the larger is the probability that the drop freezes at a warmer temperature. In support of this thought, Vali [239] estimated that the concentration of insoluble particles larger than 0.01 μm radius is about 10^{10}–10^{11} cm^{-3} of cloud water, a concentration which he felt to be reasonable in the light of the measurements of Rosinski [236,237], who found that in cloud and rain drops the concentration of insoluble particles of radius larger than 1 μm ranged between 10^3 and 10^6 cm^{-3}. Vali [179,238,239] determined that drops of 0.01 cm^3 (1.34 mm radius) formed from the water of cloud and precipitation particles froze within a wide temperature interval but some of them at a temperature as warm as $-6°C$. In contrast to these experiments, Hoffer and Braham [240] observed that drops from melted ice pellets froze at a temperature considerably colder than the minimum possible temperature in the cloud from which they were collected. At least in the light of Vali's results, it seems to be justifiable to expect that in certain clouds, in which the drops have grown by collision coalescence to precipitation size and meanwhile by chance collected some good IFN, some drops may freeze at temperatures warmer than $-10°C$. Ice formation at such warm temperatures could also be expected by the action of contact nuclei or "preactivated nuclei." While freezing, these drops may shatter and splinter and thus in an avalanchelike manner glaciate the whole cloud. It was suggested by Braham [241] that the splintering mechanism could account for clouds which glaciated even though their cloud summit temperatures were as warm as -4 to $-10°C$ [241-247]. Measurements showed that in clouds with summit temperatures varying between -5 and $-25°C$, the ice crystal concentration n_{IC} ranges between 1 and 10^3 liter^{-1}, with most frequent concentrations ranging, in a practically temperature-independent manner, between 10 and 100 liter^{-1} [43,44,241,248-252]. These concentrations are far higher than those which are expected on the basis of the IFN concentration in the atmosphere, which is about 1–10 m^{-3} (10^{-3}–10^{-2} liter^{-1}) at $-5°C$ and 10–100 liter^{-1} at $-25°C$

(see Fig. 17). Thus, for temperatures between -5 and $-25°C$, n_{IC}/n_{IFN} ranges approximately between 10^4 and 1 [253-255].

There are two major objections against explaining the observed glaciation of relatively warm clouds by the splintering mechanism as envisioned above: (1) Mossop [43,256], in analyzing previous literature on the splintering mechanism, concluded that the number of splinters produced by freezing of single drops is far too small to account for the powerful ice multiplication mechanism which is needed for the glaciation of such relatively warm clouds. As an alternative, he proposed that the ice multiplication mechanism by splintering may involve the freezing of drops on growing graupel particles [257,258] rather than the freezing of isolated drops. However, this seems unlikely since in this case, freezing proceeds from the ice substrate outward and the stresses which are necessary for fragmentation of the freezing drop are unlikely to build up. (2) Glaciation of relatively warm clouds is quite frequently observed in maritime clouds [43,246] where it is not expected that large numbers of water-insoluble particles are present to be scavenged or to act as contact nuclei. From laboratory experiments, one further would deduce that the active fraction of preactivated nuclei is too low at a temperature of $-5°C$ to account for the number of ice particles needed for starting the splintering mechanism. Apart from the clue that ice formation in these clouds is probably associated with the presence of numerous large drops, which is certainly a characteristic feature of maritime clouds, it is still an open question how glaciation in these relatively warm clouds gets started.

Our inadequacy to explain ice formation in the specific case of relatively warm clouds reflects quite generally our lack of a precise understanding of the ice nucleation mechanism. Nevertheless, the large number of experimental ice nucleation studies reported in literature, in particular those with artificial substances, enable us at least to list some of the major characteristics which aerosol particles have to possess if they should be capable of initiating the ice phase.

As can be learned from heterogeneous drop formation, the rate-determining step of heterogeneous ice formation is the energy needed for the formation of an ice embryo large enough to become freely growing in the surrounding supersaturated or supercooled environment. Any foreign particle will tend to promote phase change by providing a stable surface on which such an embryo can grow and thus reduce the energy penalty involved in the formation of a critical embryo or germ. Consequently, water insolubility is the first requirement an IFN has to meet. If the nucleus is water-soluble, the mutual interaction between the polar water molecules and the molecules or ions of the underlying substrate causes disintegration of the crystal lattice of the substrate followed by diffusion of the freed molecules or ions of the substrate into the water. Such a condition is not conducive to the formation of an ice germ.

The second requirement an IFN has to meet pertains to its size. An IFN must have a size comparable to or larger than the critical ice embryo. Sano et al. [259] found from experiments that the ice-forming capability of a particle decreases strongly if its size becomes smaller than $0.2\ \mu m$. In agreement with this result, Georgii [222] and Georgii and Kleinjung [260] found a good correlation between the concentration of IFN and the concentration of large aerosol particles ($0.1 \leq r \leq 1\ \mu m$) present in the atmosphere and a similarly excellent correlation between the concentration of IFN and the concentration of giant aerosol particles ($r > 1\ \mu m$), while no correlation was found between the number of IFN and the number of Aitken particles ($r < 1\ \mu m$) in the atmosphere.

The third and broadest requirement of an IFN embraces all the characteristics of the interface between the nucleus and ice. The free energy associated with the interface between an IFN and ice must be as small as possible. This energy is affected by the following factors.

(a) The chemical nature of an IFN, characterized by the type of chemical bonds which are exhibited at its surface, clearly affects the interface energy. Considering the fact that the ice lattice is held together by hydrogen (O–H . . . O) bonds, it is desirable that an IFN exhibit at its surface hydrogen bonds of the same bond strength and thus of the same polar character as those prevailing between the water molecules in ice. It is also essential that the hydrogen-bonding molecules in the surface of an IFN have rotational symmetry. Asymmetric molecules tend to point their active H-bonding groups inward to achieve minimum surface free energy of the crystal, while molecules with rotational symmetry cannot avoid exposing their active H-bonding groups at the surface. Considering the bond requirement for nucleation, it is not surprising that numerous authors have reported that certain organic substances are excellent IFN (see, e.g., Refs. 261–263). Also, some of the silicates expose H bonds at their surface. Thus, kaolinite exposes at its cleavage surface $Al_2(OH)_4$ octahedra from which emanate H bonds from the hydroxyl groups.

(b) The crystallographic nature of an IFN also affects the interface energy between the nucleus and ice. The crystallographic nature is characterized by the geometrical arrangement of the chemical bonds exhibited at the surface of the nucleus, the bond lengths, and the bond angles. As water molecules are bonded to the surface of the IFN in preferred configurations, the ability of the water molecules to be bonded to each other will be affected. Clearly, the crystallographic nature of the particle surface should resemble as closely as possible the crystallographic nature of ice. An exposed face of the IFN should have at least one set of atomic sites which are close to the geometry of water molecules in some low-index plane of ice. In this manner, atomic matching across the interface may be achieved which leads to epitaxial or oriented overgrowth of ice on the substrate. If there are small crystallographic differences between ice and the substrate, either the lattice of ice or of the nucleus may elastically deform to join coherently to the other lattice. Thus, strain considerations suggest that the solid substrate have a relatively low shear modulus to minimize the elastic strain energy. If there are large crystallographic differences between ice and the substrate, dislocations at the interface will result. Dislocations at the ice–nucleus interface will raise the interface energy, and any elastic strain within the ice embryo will raise the bulk free energy of the ice molecules. Both effects reduce the ability of a particle to serve as an IFN. The apparent crystallographic misfit between the substrate nucleus and ice is usually defined by $\delta = (na_{0N} - ma_{0I})/ma_{0I}$, where a_{0N} is the lattice constant of a particular face of the nucleus, a_{0I} is the corresponding lattice constant in the ice lattice, and n and m are integers chosen such that δ becomes minimal. As an example, the apparent crystallographic misfits of three substances are given in Table 11 [198,264]. In order to compute the actual crystallographic misfit, it is necessary to allow for strain in the ice embryo. Assuming that the ice embryo can be strained by an amount $\varepsilon = (x_0 - a_{0I})/a_{0I}$, where a_{0I} and x_I are the lattice parameters of the ice embryo in the strain-free and strained conditions, respectively, the actual crystallographic misfit between the ice embryo and a particular face of the IFN is then given by $\delta - \varepsilon$, where δ is measured at $n = m = 1$, i.e., $\delta = (a_{0N} - a_{0I})/a_{0I}$. As long as the embryo can be strained by the full amount δ in two dimensions, $\delta = \varepsilon$, i.e., $\delta - \varepsilon = 0$. Under these conditions, the ice embryo fits a surface element of the IFN precisely and the ice embryo is then coherent with that surface element. If the embryo cannot by strained the full amount δ, then $\delta - \varepsilon > 0$, i.e., $\delta > \varepsilon > 0$. Under these conditions, the ice embryo is incoherent with the IFN and the boundary

TABLE 11. Crystallographic Matching between Ice and Selected Crystalline Substances*

Substance	Crystal symmetry	Lattice constants (Å)			Substrate plane	Misfit δ (%) between lattice plane and ice lattice			Threshold temperature (0°C)
		a_0	b_0	c_0		in [1210] direction	in [1010] direction	in [0001] direction	
Ice	Hexagonal	4.52	—	7.36	—	—	—	—	—
AgI	Hexagonal	4.58	—	7.49	(0001)	+1.3(1:1)†	+1.3(1:1)	+1.8(1:1)	−4
CuO	Monoclinic	4.56	3.14	5.11	(001)	+2.9(1:1)		−7.4(1:2)	−7
					(010)	+2.9(1:1)	−2.0(2:3)		
Kaolinite	Pseudo-hexagonal	2.98	—	7.08	(001)	−1.1(2:3)	−1.1(2:3)	−3.8(1:1)	−9

*Based on data from [198,264].
†If the distance of each first atomic row on a particular face of a crystalline nucleus ($n = 1$) is compared with the distance of each first atomic row on a particular face of ice ($m = 1$) the value of δ is attributed the symbol (1:1).

regions between the ice embryo and the surface of the IFN can be pictured as being made up of local regions of good fit bounded by line dislocations. Usually, epitaxial overgrowth of ice on a substrate can take place as long as $\delta - \varepsilon$ stays below a certain critical value $(\delta - \varepsilon)_{\text{crit}}$, i.e., as long as $0 \leq (\delta - \varepsilon) \leq (\delta - \varepsilon)_{\text{crit}}$.

An unequivocal proof of the necessity, but not the sufficiency, of the crystallographic similarities between an IFN and ice was given by Evans [265], who studied AgI as an IFN for ice-I and for the high-pressure phase ice-III. He showed that even under conditions where ice-III was stable, ice-I, which fits the crystallography of AgI much better than ice-III, was consistently nucleated by AgI.

(c) Fletcher [266] pointed out that the electrical nature of the surface of a particle is also of great importance for characterizing the interface energy between an IFN and ice. A set of atomic sites on which epitaxial growth is geometrically possible should contain approximately equal numbers of positively and negatively charged members. Such an electrical arrangement is energetically more conducive to the formation of ice, where the water dipoles have a random orientation, than a surface with atomic members of the same electric sign. If the polar water molecules would all be oriented parallel to one another by coming in contact with a polar surface of uniform electric sign, the entropy of any ice embryo on this surface would be reduced, the free energy of the ice embryo raised, and consequently the ice-forming capability of the surface reduced. In confirmation of this requirement, Edwards and Evans [267] showed that the nucleation efficiency of a substance is greatest near its isoelectric point.

It should be noticed that all the above-mentioned factors characterizing an IFN enter equally for all particles of similar size, habit, and of a given material. If these would be the only factors necessary to define an IFN, then all particles of a given size, shape, and well-defined material would exhibit the same ice-forming capability. Fletcher based his earlier work [75] on this assumption and theoretically treated ice nucleation as a function of the size and shape of an aerosol particle and as a function of its interface characteristics combined together in the parameter m. In the case of nucleation of a water drop by a solid, water-insoluble particle, the parameter m was defined in terms of the contact angle θ so that $m = \cos \theta$. In the case of ice nucleation, the contact angle has little meaning. However, by Young's relation, $\cos \theta$ can, for equilibrium conditions, be related to the interface energies. Thus, for the deposition of ice from the vapor, $m = (\sigma_{NV} - \sigma_{NI})/\sigma_{IV}$ and for ice formation from supercooled water, $m = (\sigma_{NW} - \sigma_{NI})/\sigma_{IW}$, where σ_{NV}, σ_{NW}, σ_{IV}, and σ_{IW} are the interface free energies between the nucleus and water vapor, the nucleus and supercooled water, ice and vapor, and ice and supercooled water, respectively. The parameter m ranges from $m = -1$ for an inactive IFN to $m = +1$ for a very good IFN. Thus, a good IFN should have a value of σ_{NI} which is as small as possible and values for σ_{NW} and σ_{NV} which are as large as possible. As pointed out above, this can best be achieved if there is crystallographic and chemical "matching" at the interface. In analogy to heterogeneous nucleation of water drops from the vapor, Fletcher [75] showed that for nucleation of ice from the vapor by a spherical particle of radius r_N, the nucleation rate J per particle is given by

$$J = K r_N^2 \exp[-(\Delta F_g'/kT)] \tag{29}$$

where $\Delta F_g' = f(m,x)\,\Delta F_g$, ΔF_g being given by Eqs. (8) and (9) with $s = 4\pi$, and where strain effects are neglected. The function $f(m,x)$ is given by Eq. (25) with m as defined above and x is given by Eq. (23), and the kinetic constant $K \approx 10^{26}$ cm^{-2} sec^{-1}.

For the rate of nucleation of ice in supercooled water by a spherical IFN, Fletcher [75] gave the relation

$$J = K r_N^2 \exp[-(\Delta F^*/kT)] \exp[-(\Delta F_g'/kT)] \tag{30}$$

where $K \approx 10^{28}$ cm^{-2} sec^{-1} and $\Delta F_g' = f(m,x)\,\Delta F_g$, ΔF_g being given by Eqs. (8) and (12) with $s = 4\pi$. The numerical evaluation of Eqs. (29) and (30) was carried out by Fletcher for $\sigma_{IV} \approx 100$ ergs cm^{-2} and $\sigma_{IW} \approx 20$ ergs cm^{-2} and for $J = 1$ sec^{-1} particle^{-1}. His results are displayed in Fig. 20 for the case of heterogeneous ice formation in a water-vapor-saturated environment and in Fig. 20(b) for the case of heterogeneous ice formation in supercooled water. It is seen from these two figures that in the case of ice formation from the vapor, the size of the nucleus is relatively unimportant for particles larger than about 0.1 μm radius, while below this size, the nucleation efficiency drops off sharply. In the case of ice formation in super-cooled water, the size effect becomes important only below a radius of about 0.03 μm. In both cases, the interface parameter m has a very large effect on ice nucleation. In order to predict by means of Fletcher's theory the nucleation ability of atmospheric aerosol particles, it would be necessary to specify the parameter m characteristic of these particles. Unfortunately, m is not known exactly for atmospheric aerosols.

It has to be kept in mind that these theoretical arguments represent quite a simplification of the actual ice nucleation process. Even Fletcher's improvements of the theory by taking into account particles of geometrical shapes other than spherical [268] left some serious gaps. Katz [269], Edwards et al. [270], and Mossop and Jayaweera [271] maintained on the basis of experimental evidence the view that the nucleation activity varies strongly even among particles of a nearly monodisperse aerosol and of a well-defined substance. A major part of this "activity spectrum" was attributed to the distribution of certain "active sites" upon the surfaces of the IFN. Two types of active sites have to be considered. Studies on the mechanism of adsorption of water vapor onto surfaces of a water-insoluble substance doped to various degrees with a hydrophilic impurity demonstrated that "active sites" for ice nucleation can be represented by hydrophilic sites located on an otherwise

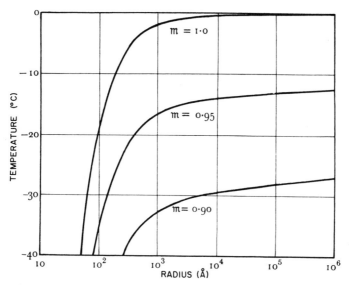

Fig. 20a. Temperature at which a spherical particle of given radius and interface parameter m will nucleate an ice crystal in 1 sec by deposition from an environment at water saturation. (From Fletcher [75], courtesy of Cambridge Univ. Press.)

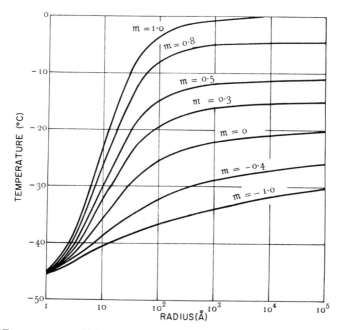

Fig. 20b. Temperature at which a spherical particle of given radius and interface parameter m will nucleate an ice crystal in 1 sec in supercooled water by freezing. (From Fletcher [75], courtesy of Cambridge Univ. Press.)

hydrophobic surface [272-279]. Other experimental studies showed that ice crystals preferentially are nucleated at the edges and on steps and cracks of a crystalline surface, indicating that topographical features such as crystallographic defects on a surface may serve as "active sites" for ice nucleation [198,264,280-284]. Such defects can be caused by screw or edge dislocations in a crystalline solid which emerge at the surface in form of a step. Recently, Fletcher [285] expanded his ice nucleation theory to include in the nucleation rate equation the surface topography of a particle, idealizing a surface defect by a reentrant conical cavity. Numerical analysis of Fletcher's theory showed, as was expected from experiments, that small surface defects have a pronounced effect on the ice nucleation behavior of a particle in the sense of enhancing its ice-forming capability as long as the particle has a radius which is larger than 10^{-2} μm.

We shall now discuss the mechanism of ice nucleation from the vapor and supercooled water in some more detail. An aerosol particle essentially has three modes of action available to initiate the ice phase.

A first mode of action is exhibited when an aerosol particle finds itself surrounded by moist air. Under this condition, the particle may adsorb onto its dry surface water molecules directly from the surrounding gas phase. This adsorption takes place preferentially on the "active sites" at the surface of the particle in the form of two-dimensional water patches as long as the vapor pressure in the surrounding atmosphere is still small. At higher vapor pressures, these patches may grow to multilayered, three-dimensional structures or clusters [279]. The larger these three-dimensional structures can become before competitive adsorption starts on less active sites of the surface, the higher is the nucleation efficiency of the particle. The earlier adsorption studies of Mason and van den Heuvel [198] and Bryant et al. [264] and the more recent adsorption studies of Sano et al. [194] with oxides and of Roberts and Hallett [174] with clays showed that the transformation of these clusters to ice requires an environment which is at water saturation as long as the temperature of the environment is warmer than a critical temperature. At temperatures colder than the critical temperature, transformation to ice occurs below water saturation at a characteristic ice supersaturation (see Fig. 21). Thus, in contrast to homogeneous ice nucleation, Ostwald's rule becomes inverted for heterogeneous ice nucleation at a temperature which is characteristic to each ice-forming substance.

A second mode of action by which aerosol particles initiate the ice phase is exhibited when the aerosol particle finds itself surrounded by a supercooled drop after having been captured by the drop or after having been involved in the formation of the drop by condensation. The molecules in supercooled water are not distributed randomly in space but exhibit a definite structural arrangement which becomes the more pronounced the lower the temperature (for a review of earlier and recent work on the structure of water and ice, see Refs. 38,286–288). This water "structure" is due to the presence of clusters

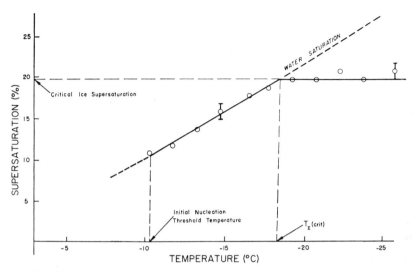

Fig. 21. Variation with temperature of the minimum supersaturation over ice required for ice nucleation by kaolinite. Each plotted point represents the lowest supersaturation necessary for an ice crystal to nucleate at that temperature [174]. (By courtesy of *Quart. J. Roy. Meteorol. Soc.*)

of water molecules whose size and lifetime increase with decreasing temperature, the latter being typically 10^{-11}–10^{-10} sec. Comparing this time with the much longer dielectric relaxation time of water molecules in ice ($-10°C$), which is about 10^{-5} sec, it seems reasonable to assume that the first stage of ice formation in a supercooled water drop at a temperature warmer than its homogeneous nucleation temperature consists in the stabilization of water clusters at the surface of an aerosol particle in the drop. As in the case of ice formation from the vapor, this may take place by the adsorption of water molecules on active sites at the surface of the particle. Edwards et al. [289] and Evans [290,291] proposed on the basis of their recent experimental work that, in supercooled water, ice formation hinges upon the development of patches of an absorbed monolayer of water molecules on a solid surface which is provided by the particle. If this solid surface has only a moderate affinity for water and is mostly hydrophobic with some active sites, the monolayer will be rather disordered with little geometrical arrangement and much freedom for the movement of the water molecules by surface diffusion as long as the temperature is warmer than a certain characteristic temperature. Below this characteristic temperature, the solid surface, assuming it meets the crystallographic requirements of an IFN, is capable of transforming progressively the disordered monolayer into a "crystalline" (in a two-dimensional sense) icelike ordered state (2D ice). The temperature at which the monolayer is well "crystallized" may be considered below the two-dimensional disordering temperature of the most stable, ordered monolayer patch, depening on the properties of the solid surface. This ordered, icelike state is stabilized predominantly by lateral hydrogen bonding and anchored on the surface by its active sites. Nucleation of bulk ice proceeds from this ordered state with no or very little further supercooling. If, on the other hand, the solid surface has a strong affinity for water due to the presence of strongly polar groups, an excessive number of hydrogen bonds, or strongly hydrating cations (as is the case for many clays), the adsorbed layer may adopt a strongly bonded structure which, although the epitaxial fit in plan may be perfect, is unsuitable for a nucleation of 2D ice since the orientation of the individual water dipoles is incompatible with the random orientation they have in 2D ice. In this case, a second adsorbed layer has to form which may assume a state that is sufficiently disordered so that the transition to 2D ice can take place. Thus, heterogeneous ice nucleation in supercooled liquid water is viewed as the result of a nucleation process which takes place within the adsorbed layer at the surface of the IFN depending both on the nature of the IFN and the statistical fluctuations within the adsorbed layer. The actual barrier to ice formation is the formation of this 2D ice layer on which subsequently, by fluctuation in the supercooled water, a critical cluster or germ is nucleated and bulk ice forms. This model is in contrast to the classical approach, which views ice nucleation in supercooled water as a process during which a critical ice embryo is formed directly at the surface of the IFN by statistical fluctuation of single molecules in the supercooled water.

The classical point of view was adopted by Bigg [36,292,293], Carte [294], and Dufour and Defay [18], who attempted to explain the freezing of water drops by assuming that at a given temperature all the ice embryos formed in a population of equal-size water drops have equal chance of reaching the size of a critical embryo or germ by random fluctuations among single water molecules in the water and that the nature of foreign particles present have negligible effect in this process (stochastic hypothesis). Under these conditions, drop freezing may be described by the relation [18]

$$\ln[1 - P(V,t)] = -(V/\alpha) \int_T^{T_0} J(T)\,dT \qquad (31)$$

which can be expressed in the form [36,295]

$$\ln[1 - P(V,t)] = -(BV\tau/\alpha)(e^{-\theta/\tau} - 1) \qquad (32)$$

or for $\theta < -5°C$, for which $e^{-\theta/\tau} \gg 1$, in the form

$$\ln(N_\theta/N_0) = -(BV\tau/\alpha)\,e^{-\theta/\tau} \qquad (33)$$

where $P(V,t)$ is the probability that freezing will be initiated in a drop of volume V maintained for t seconds at a supercooling $-\theta = \Delta T = T_0 - T$, $T_0 = 273°K$; $\alpha = d\theta/dT$ is the cooling rate; $J(T)$ is the rate of nucleus formation at temperature T; N_θ is the number of drops unfrozen when the temperature θ is reached; N_0 is the total number of drops in the population;

and B and τ are constants. It can be shown that Eq. (33) may be rewritten for the case of a freezing process at constant supercooling $\theta = \theta_c$ as

$$\ln(N_t/N_0) = -BV(e^{-\theta_c/\tau})t \tag{34}$$

where N_t is the number of drops still unfrozen after time t has elapsed. For a constant cooling rate $\alpha = \alpha_c$, or a constant temperature, Eqs. (33) and (34) are given in their differential forms by

$$\frac{1}{N}\left(\frac{dN}{d\theta}\right)_{\alpha=\alpha_c} = \frac{BV}{\alpha_c}e^{-\theta/\tau} \tag{35a}$$

$$\frac{1}{N}\left(\frac{dN}{dt}\right)_{\theta=\alpha_c} = -BVe^{-\theta_c} \tag{35b}$$

It is seen from Eqs. (35a) and (35b) that, according to the stochastic hypothesis, the fraction of drops frozen per temperature interval while cooled at a constant cooling rate α_c exponentially increases with increasing supercooling, and that the fraction of drops frozen per time interval while kept at a constant supercooling θ_c is a constant whose value only depends on the supercooling.

A different point of view for drop freezing has been taken by Levine [296] and Langham and Mason [35], who assumed that every nucleus contained in a drop has a characteristic temperature at which it is certain to initiate freezing. According to this hypothesis, the freezing temperature of a drop is determined by that particle in the drop that has the warmest characteristic temperature (singular hypothesis). Assuming that the number concentration of nuclei which become ice-forming between $0°C$ and θ is given by $n(\theta) = n_0e^{-\theta/\tau}$, the probability that a drop of volume V contains at least one effective nucleus on reaching the temperature θ is given by $1 - e^{-n(\theta)V}$, the freezing process can be described by the relation

$$\ln(N_\theta/N_0) = -n_0V\,e^{-\theta/\tau} \tag{36}$$

It is seen from Eq. (36) that according to the singular hypothesis, the number of frozen drops is independent of the cooling rate and stays constant if the supercooling is constant. From Eq. (36), we find

$$\frac{1}{N}\frac{dN}{d\theta} = \frac{n_0V}{\tau}e^{\,\theta/\tau} \tag{37}$$

Thus, for a constant cooling rate, the stochastic hypothesis expressed in Eq. (35a) and the singular hypothesis expressed in Eq. (37) predict the same freezing behavior. It can easily be shown that from both the stochastic and the singular hypotheses the relation

$$\ln V = A + a\theta_m = A - a(\Delta T)_m \tag{38}$$

can be derived, where $\theta_m = -(\Delta T)_m$ is the supercooling at which 50% of the drops in the population are frozen. Equation (38), first derived and experimentally verified by Bigg [36], shows that the singular and the stochastic hypotheses predict a linear relationship between the median freezing temperature of a population of uniform water drops and the logarithm of the drop volume. However, the stochastic and the singular hypotheses make completely different predictions for the freezing behavior as a function of cooling rate or at a constant supercooling. From a large number of experiments [297,298], Vali and Stansbury concluded that drop freezing neither has completely the features of the stochastic hypothesis nor completely the features of the singular hypothesis. However, the results can be understood if in a drop a population of nuclei is assumed each of which possesses a characteristic range of temperatures within which it can initiate ice nucleation, and if the ice nucleation probability associated with each nucleus is a function of the temperature of the water. This explanation seems to agree, at least qualitatively, with the freezing mechanism envisioned by Edwards and Evans, who proposed that heterogeneous freezing depends on the capability of a particle to form a diffuse and subsequently ordered adsorbed layer of water molecules by the specific action of its active sites and by the temperature-dependent chance fluctuations within the adsorbed layer and in the water outside of it. Thus, it appears that heterogeneous ice formation in the vapor as well as in supercooled water should be looked at from the point of view of the formation of critical ice embryos on patches of adsorbed layers at the surface of an IFN, rather than in terms of the formation of critical ice embryos directly as the surface of the IFN by chance fluctuations of single molecules.

A third mode of action by which an aerosol particle initiates ice formation is exhibited when an aerosol particle comes into contact by collision with a supercooled water drop. At first sight, it seems to be difficult to understand why a water-insoluble particle should have a better capability to initiate ice when it comes into contact with a supercooled water drop than when it is completely submerged within the drop. Fletcher [299] pointed out that the experimentally established difference between the two modes of action may lie in the partial solubility of any particle especially if it has a very small size. Particularly vulnerable are the active sites at the surface of a particle. These may be etched by the water preferentially in comparison with the rest of the particle surface so that they lose in part or completely their nucleation ability. Evans [300] suggested that contact nucleation, as any other form of "mechanical" nucleation, may be explained on the basis of his adsorption model. A particle immersed in supercooled water may have a strongly adsorbed layer unsuitable for transformation into a 2D ice ordered layer. This barrier to ice formation may be considerably reduced during the initial moments of contact between a dry particle and a supercooled water drop which, for a time which is of the order of the rotational period of a water molecule, an adsorbed layer of disordered state may form. Such a layer may transform more readily into a 2D ice layer than the strongly bonded layer which is likely to form at the surface of a particle completely immersed in the drop. Ice formation by a particle completely immersed in a drop may further be hindered by means of foreign substances dissolved in the drop which may preferentially "poison" the active sites on the particle and thus destroy their ice-forming ability. Although dissolved salts may not necessarily enter the adsorbed layer of a particle immersed in a drop, they will, in high enough concentrations, impede ice formation by their action on the water outside the adsorbed layer. It is well established that due to their size and their strong radial electric field, ions destroy the natural water structure by breaking hydrogen bonds and by forming hydration shells. Also, ice formation may be severely hindered by the adsorption of foreign gases and vapors onto the active sites at the surface of an IFN. This has been demonstrated by Georgii [301,302], who found that the effectiveness of IFN is substantially reduced in the presence of foreign gases such as SO_2, NH_3, or NO_2. The reduction was the more pronounced the larger the concentration of the gas present. Poisoning of active sites at the surface of IFN in air may also be caused by coagulation of the IFN with Aitken particles.

While certain processes were found to poison and "deactivate" IFN, other processes were found to activate IFN, making them capable of initiating ice at temperatures warmer than before the activation procedure. As mentioned earlier, an IFN may become activated and in this state may survive relative humidities considerably below ice saturation after it had been involved once in ice formation or after it had become exposed once to a temperature below $-40°C$. An earlier explanation of the phenomenon of preactivation was based on a retention of ice embryos in small cavities or capillaries where they survive evaporation of the macroscopic ice because of the curvature effect. However, through the work of Roberts and Hallett [174] and Edwards et al. [289], it became evident that the observed facts are better explained in terms of the retention of patches of an ordered, icelike layer adsorbed at the surface of the IFN.

6. THE ROLE OF AEROSOLS IN THE FORMATION OF PRECIPITATION IN SUPERCOOLED CLOUDS

An atmospheric aerosol particle may serve as an IFN as well as a CCN. Those aerosol particles that consist of only water-soluble substances will, even at temperatures below 0°C, only serve as CCN to form drops and not get involved in ice formation. Unless the drops capture insoluble particles subsequent to their formation, they will stay supercooled and grow further by collision and coalescence. Those aerosol particles that are "mixed particles," consisting of water-soluble, hygroscopic substances and water-insoluble substances, may serve as both CCN and IFN. After their initial action as CCN, the water-insoluble portion of the particle will have become immersed in the formed cloud drop from where, after the drop has become sufficiently diluted and has become sufficiently supercooled, it may initiate

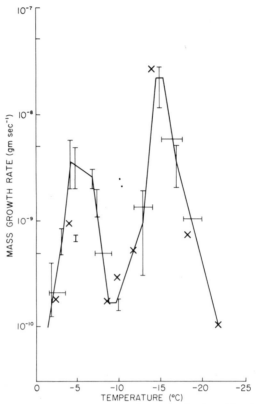

Fig. 22. Variation of the growth rate of ice crystals growing by deposition from the vapor as a function of temperature (based on data from Refs. 304–306; vertical bars, data from Ref. 304; horizontal bars, data from Ref. 305; ($\times$) data from Ref. 306.

freezing. Those aerosol particles that consist of water-insoluble substances only have little chance to serve as a CCN and thus may serve as IFN, initiating the ice phase by an adsorption process from the vapor or by contact with supercooled cloud drops.

Thus, we see that the growth processes of cloud particles to precipitation particles in supercooled clouds may involve all three phases of water: vapor, supercooled drops, and ice crystals. Such "mixed" clouds are colloidally unstable since the water vapor pressure over supercooled water is larger than over ice at the same temperature. Thus, ice crystals in such a cloud will grow at the expense of the water drops as long as water drops are there or until the ice crystals have fallen out of the cloud (Bergeron–Findeisen mechanism). The growth rate of an ice crystal by diffusion is a pronounced function of the habit of the crystal [303]. This fact is illustrated by Fig. 22, which is based on the results of Ono [304], Fukuta [305], and Neumann et al. [306]. This figure shows that the growth rate of ice crystals in the atmosphere has two maxima, one near -5 to $-6°C$ and one near -15 to $-16°C$. Solution of the classical diffusion equation for the case of growth from water vapor of an ice crystal of one particular shape, which may be idealized by a sphere, a circular disk, or an oblate or prolate spheroid, yields an asymmetric curve with one broad maximum at $-14.25°C$ or $-16.75°C$ depending on whether the atmospheric pressure is 1000 or 500 mb [76]. This suggests that the variation with temperature of the observed growth rate is due to a variation of the ice crystal shape with temperature. Indeed, laboratory experiments and observations in natural clouds have shown that an ice crystal whose basic habit is a hexagonal prism has a preferred growth along the crystallographic c axis or a axis depending on temperature. Growth is predominantly along the a axis at temperatures between 0 and $-3°C$, along the c axis at temperatures between -3 and $-8°C$, along the a axis at temperatures between -8 and $-25°C$, and along the c axis at temperatures below $-25°C$. Thus, the first growth rate maximum falls into the temperature interval of pronounced c-axis growth, where the formation of ice needles is common, while the second growth rate maximum falls into the temperature interval of pronounced a-axis growth, where the formation of ice dendrites is common. The growth minimum near 0°C is expected from thermodynamic reasons, while the second and third growth rate minima fall into the temperature intervals where the transition from c- to a-axis growth and vice versa takes place. Even though a complete theoretical description of this habit variation with temperature is not yet available, the phenomenon has quantitatively been related to the temperature variation of the propagation velocity of ice layers on the prism and basal faces of an ice crystal which is controlled by the mean migration distance of a water molecule on these faces [282].

Recently, Jayaweera [307] computed the growth rates and masses of ventilated ice crystals falling at terminal velocity in a water saturated atmosphere of temperature between 0°C and $-20°C$. These computations were based on a semi-empirical model that uses the electrostatic analogy, and assumes that the crystals follow the experimentally observed growth modes and keep axis ratios that follow the observed relations of Ono [304]. Jayaweera's calculations show that at $-18°C$, about 600 mb, and after 300 sec (5 min) a crystal reaches a mass which is equivalent to a water drop of about 62 μm radius, and after 1000 sec (about $16\frac{1}{2}$ min) reaches a mass equivalent to a water drop of about 134 μm radius. At $-15°C$ and about 640 mb the corresponding masses are equivalent to drop sizes of 78 and 416 μm radius. If one would like to translate these data into precipitation rates it is to be remembered that the precipitation rate is not only dependent on the masses of the precipitation particles but also on their terminal fall velocity, which is considerably smaller for ice crystals [308] than for water drops [309], and on the number concentration of ice crystals, which according to Fig. 17 is only about 100 cm^{-3} at $-15°C$ and about 600 cm^{-3} at $-18°C$.

Little is known quantitatively on the growth rate of ice crystals by collision with other ice crystals or with supercooled water drops, principally because it is difficult to make experimental measurements that realistically simulate conditions in atmospheric clouds and secondly because no realistic values for the collection efficiency of ice crystals are available. Recent laboratory experiments [310] indicated that the collection efficiency of ice crystals colliding with ice crystals is quite low due to the low sticking efficiency of these crystals, except at temperatures close to 0°C or in the presence of an external electric field. Contrary to this, ice crystals colliding with supercooled water drops have a freeze-on efficiency which is practically 100%. Assuming this, and assuming collision efficiencies given by Langmuir, Braham [311] theoretically followed the growth of ice crystals in a cloud for the case of crystals growing by diffusion and collision with drops. Braham assumed that the drop size distribution in the cloud was that typical for the arid Southwest of the USA in summer, that the updraft speed in the cloud was 5 m sec^{-1}, that the in-cloud lapse rate was saturation adiabatic with an equivalent potential temperature of 352.8°K, and that the ice crystal growth rate from the vapor as a function of temperature was given by observed values as exemplified in Fig. 22. The results of Braham's computations are reproduced in Fig. 23 for the case of a cloud with liquid water content (LWC) of 0.3 and 1.0 g m^{-3} and for ice crystals nucleated at $-5°$ and $-10°C$. It is seen from this figure that for a LWC of 0.3 g m^{-3}, a 0.5 mm graupel particle (equivalent

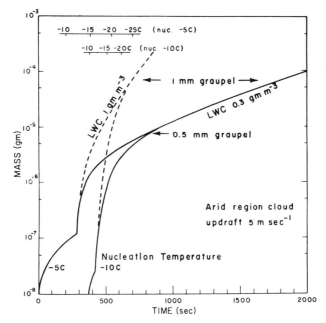

Fig. 23. Variation of the mass of an ice particle growing by vapor deposition followed by riming [311]. (By courtesy of *Bull. Am. Meteorol. Soc.* and the author.)

to a water drop of 120 μm radius) is formed in 14 min and a 1-mm graupel particle (equivalent to a water drop to 254 μm radius) in 30 min. In a cloud of LWC of 1 g m^{-3}, a 1-mm graupel particle is formed in 12 min and a graupel equivalent to a water drop of 623 μm radius in 17 min. From his results Braham drew the interesting and significant conclusion that in cumulous clouds the rapid diffusional growth regime at temperatures between -12 and $-17°C$ (see Fig. 22) so completely dominates ice crystal growth by vapor diffusion that crystals nucleated at a temperature warmer than $-10°C$ will end up with virtually identical masses by the time they are carried up to the $-20°C$ level in the cloud. If we recall now that the concentration of ice crystals formed by IFN of the atmospheric aerosol at temperatures warmer than $-10°C$ is typically less than 0.01 liter^{-1} (10 m^{-3}), we easily may convince ourselves that the concentration of ice particles formed by natural IFN at temperatures warmer than $-10°C$ is too small to account for significant precipitation even though the ice particles may grow to millimeter-size graupel within a reasonably short time. Thus for precipitation to occur by the action of IFN, a cloud has to reach levels in the atmosphere where the temperature is -15 to $-20°C$ unless an efficient ice multiplication mechanism is going on in the cloud or nuclei in addition to those normally present

TABLE 12. Effect of Urban Areas on Precipitation [314]*

	Urban–rural difference (increase) expressed as a per cent of rural value						
	Chicago	La Porte	St. Louis	C.-U.†	Tulsa	Wash., D.C.	New York
Precipitation							
Annual	5	31	7	5	8	7	16
Warmer half-year	4	30	**	4	5	6	12
Colder half-year	6	33	**	8	11	9	20
All rain days							
Annual	6	0	**	7	**	**	**
Warmer half-year	8	0	**	3	**	**	**
Colder half-year	4	0	**	10	**	**	**
Moderate-heavy rain days							
Annual	10	34	**	5	**	**	**
Warmer half-year	15	54	**	9	**	**	**
Colder half-year	0	5	**	0	**	**	**
Thunderstorm days							
Annual	6	38	11	7	**	**	**
Summer	7	63	20	17	**	**	**

*By courtesy of Bull. Am. Meteorol. Soc. and the author.
†C.-U. = Champaign-Urbana.
**Results unavailable or data insufficient to make a comparison.

in the atmosphere are penetrating the cloud. Additional particles may possibly come from local anthropogenic sources. Indeed, Changnon [312-315] has found increasing evidence that urban areas experience more precipitation than their adjoining rural environment, a finding which he attributed to the additional nuclei provided by these areas and the additional convection stimulated by heat and frictional effects (see Table 12).

In summary, it should be stressed that in our discussion we did not attempt to give a complete quantitative account of all the processes leading to the formation of atmospheric clouds and precipitation. As a goal, we merely wanted to demonstrate the profound role which the atmospheric aerosol plays by its chemical and physical characteristics in determining the microstructure of atmospheric clouds and in bringing about precipitation. While we mostly concerned ourselves with the role of pollutants from natural sources, we have given evidence that the effects of anthropogenic pollutants can by no means be disregarded. Even though anthropogenic effects are at present still confined to a local scale, the worldwide steady rise of the global level of anthropogenic pollutants makes it imperative to closely monitor the anthropogenic pollutants on a local as well as worldwide scale and attempt to learn more about the effect of these pollutants on cloud physical processes.

REFERENCES

1. Pales, J., and Keeling, C. D., *J. Geophys. Res.* **70**, 6053 (1965).
2. Cobb, W., and Wells, H. J., *J. Atmos. Sci.* **27**, 814 (1970).
3. Junge, C. E., "Air Chemistry and Radioactivity," Academic Press, New York, 1963.
4. Georgii, H. W., *Bull. World Health Organiz.* **40**, 624 (1969).
5. Jaenike, R., and Junge, C. E., *Beitr. Phys. Atmosphäre* **40**, 129 (1967).
6. Israel, H., "Atmospheric Electricity," Vol. I (translated from German), Israel Program Sci. Transl., 1970.
7. Chalmers, J. A., "Atmospheric Electricity," 2nd ed., Pergamon Press, New York, 1967.
8. O'Connor, T. C., *J. Rech. Atmos.* **2** (2–3), 181 (1966).
9. Hogan, A. W., *J. Rech. Atmos.* **3** (1–2), 53 (1968).
10. Landsberg, H., The Climate of Towns, in "Man's Role in Changing the Face of the Earth," W. L. Thomas, ed., Chicago Univ. Press, 1956, p. 584.
11. Landsberg, H., Air Pollution and Urban Climate, in "Biometeorology," Volume 2, Pergamon Press, Oxford, 1966, p. 648.
12. Mason, B. J., "The Physics of Clouds," Oxford Press, London, 1971.
13. Martell, E. A., Advances in Chemistry No. 93, Am. Chem. Soc., Vol. 9, p. 138, 1970.
14. Mastenbrook, H. J., *J. Atmos. Sci.* **25**, 299 (1968).
15. Murcray, D. G., Kyle, T. S., and Williams, W. J., *J. Geophys. Res.* **74**, 5369 (1969).
16. McKinnan, D., and Morewood, H. W., *J. Atmos. Sci* **27**, 483 (1970).
17. Hirth, J. P., and Pound, G. M., "Condensation and Evaporation," Macmillan, New York, 1963.
18. Dufour, L., and Defay, R., "Thermodynamics of Clouds," Academic Press, New York, 1963.
19. Allen, L., and Kassner, J., *J. Colloid Interface Sci.* **30**, 81 (1969).
20. Katz, J., and Ostermier, B. J., *J. Chem. Phys.* **47**, 478 (1967).
21. Katz, J., *J. Chem. Phys.* **52**, 4733 (1970).

22. Lothe, J., and Pound, G. M., *J. Chem. Phys.* **36**, 2080 (1962).
23. Reiss, H., *J. Statistical Phys.* **2**, 83 (1970).
24. Maybank, J., and Mason, B. J., *Proc. Phys. Soc. London* **74**, 11 (1959).
25. Bayardelle, M., *Compt. Rend.* **239**, 988 (1954).
26. Roulleau, M., *J. Sci. Meteorol.* **9**, (36), 127 (1957).
27. Pruppacher, H., *J. Chem. Phys.* **39**, 1586 (1963).
28. Jacobi, W., *J. Meteorol.* **12**, 408 (1955).
29. Day, J., *J. Meteorol.* **15**, 226 (1958).
30. Carte, A. E., *Proc. Phys. Soc. London* **69B**, 1028 (1956).
31. Mossop, S. C., *Proc. Phys. Soc. London* **68B**, 193 (1955).
32. Meyer, J., and Pfaff, W., *Z. anorg. Chem.* **244**, 305 (1935).
33. Wylie, R. G., *Proc. Phys. Soc.* **66B**, 241 (1953).
34. Hoffer, T., *J. Meteorol.* **18**, 766 (1961).
35. Langham, E. J., and Mason, B. J., *Proc. Roy. Soc. A* **247**, 493 (1958).
36. Bigg, E. K., *Proc. Phys. Soc.* **66B**, 688 (1953).
37. Pound, G. M., Madonna, L. A., and Peake, S. L., Sci. Rept. No. 1, Metals Res. Lab. Carnegie Inst. of Tech., Pittsburgh, Contract No. AF19(122)–185, 1952.
38. Fletcher, N. L., "The Chemical Physics of Ice," Cambridge University Press, London, 1970.
39. Hoffer, T., University of Chicago Dept. Meteorol. TN. No. 22, p. 63, 1960.
40. Warner, J., *J. Rech. Atmos.* **3** (3), 233 (1968).
41. Peppler, W., *Forschg. u. Erfahrung. Reichsamt f. Wetterd.* **1** (1940).
42. Borovikov, A. M., in "Cloud Physics," A. Kh. Khrgian, editor. Translated by Israel Program for Scientific Translations. U.S. Dept. of Commerce and U.S. National Science Foundation 1963, p. 65.
43. Mossop, S. C., Ono, A., and Wishart, E. R., *Quart. J. Roy. Meteorol. Soc.* **96**, 487 (1970).
44. Morris, T. R., and Braham, R. R., in "Proc. Inter. Conf. Weather. Mod.," Albany, New York, 1968, p. 306.
45. Kumai, M., *J. Meteorol.* **8**, 151 (1951).
46. Aufm Kampe, H. J., Weickman, H. K., and Kedesdy, H. H., *J. Meteorol.* **4**, 374 (1952).
47. Isono, K., *J. Meteorol.* **12**, 456 (1955).
48. Nakaya, U., in "Proc. 1st Conf. Phys. of Clouds, Woods Hole," Pergamon Press, 1955, p. 35.
49. Kumai, M., *Geofys. pura e appl.* **36**, 169 (1957).
50. Isono, K., Tanaka, T., and Iwai, K., *J. Rech. Atmos.* **3**, (2–3), 341 (1966).
51. Kumai, M., *J. Meteorol.* **18**, 139 (1961).
52. Kumai, M., and Francis, M., *J. Atmos. Sci.* **19**, 434 (1962).
53. Kumai, M., *J. Geophys. Res.* **71**, 3397 (1966).
54. Rucklidge, J., *J. Atmos Sci.* **22**, 301 (1965).
55. Soulage, *Arch. Met. Geophys. Bioklimatol.* **8**, 211 (1955).
56. Soulage, *Ann. Géophys.* **13**, 103 (1957).
57. Junge, C. E., *Ann. Meteorol.* **5**, Beiheft (1952).
58. Kumai, M., CRREL Res. Rept. No. 245, 1969.
59. Kuroiwa, D., *J. Meteorol.* **8**, 157 (1951).
60. Kuroiwa, D., Hokkaido University Rept., p. 349, 1953.
61. Ogiwara, S., and Okita, T., *Tellus* **4**, 233 (1952).
62. Kuroiwa, D., and Tadano, B., *Low Temp. Sci.* **7**, 51 (1951).
63. Yamamoto, G., and Ohtake, T., *Sci. Rept. Tohoku Univ.*, Ser. 5 (*Geophys.*) **4**, 141 (1953).
64. Yamamoto, G., and Ohtake, T., *Sci. Rept. Tohoku Univ.*, Ser. 5 (*Geophys.*) **7**, 10 (1955).
65. Kumai, M., in "Proc. Int. Conf. Cloud Phys.," Tokyo, 1965, p. 52.
66. Isono, K., *Geofys. pura e applicata* **36**, 156 (1957).
67. Isono, K., *Japan. J. Geophys.* **2**, (2) (1959).
68. Dessens, H., Lafargue, C., and Stahl, F., *Ann. Géophys.* **8**, 21 (1952).

69. Tohmfor, G., and Volmer, M., *Ann. Phys. Folge.* **33**, 109 (1938).
70. Rathje, W., and Stranski, I. N., in "Proc. Conf. on Interfacial Phenomena and Nucleation," H. Reiss, ed., Geophys. Res. Papers No. 37, Geophys. Res. Dir., Air Force Cambridge Res. Center, Cambridge, Mass., 1955, Vol. I, p. 1.
71. Twomey, S., *J. Appl. Phys.* **24**, 1099 (1953).
72. Low, R. D., *J. Atmos. Sci.* **26**, 608 (1969).
73. Twomey, S., *J. Meteorol.* **11** 334 (1954).
74. Roussel, J. C., *J. Rech. Atmos.* **3** (3), 253 (1968).
75. Fletcher, N. H., "The Physics of Rain Clouds," Cambridge, Univ. Press, 1962.
76. Byers, H. R., "Elements of Cloud Physics," Univ. of Chicago Press, Chicago, 1965.
77. Low, R. D., *J. Rech. Atmos.* **4** (2), 65 (1969).
78. Lewis, G. N., and Randall, M., "Thermodynamics," McGraw-Hill, New York, 1961, p. 263.
79. Low, R. D., US Army Electronics Command, Report ECOM-5249, AD 691 709, Atom. Sci. Lab. White Sands Missile Range, New Mexico, 1969.
80. Winkler, Diplom Thesis, Meteorol.–Geophys. Inst. Univ. of Mainz, Mainz, Germany, 1967.
81. Winkler, *Ann. Meteorol. Neue Ser.* **4**, 134 (1968).
82. Abel, N., Winkler, P., and Junge, C. E., Final Rept. Contract AF 61(052)-965, Max Planck Inst. f. Chemie, Mainz, Germany, 1969.
83. Barchet, W. R., *J. Atmos. Sci.* **26**, 112 (1969).
84. McDonald, J. E., *J. Atmos. Sci.* **21**, 109 (1964).
85. Jiusto, J. E., and Kocmond, W. C., *J. Rech. Atmos.* **3** (1–2), 19 (1968).
86. Twomey, S., and Wojciechowski, T. A., *J. Atmos. Sci.* **26**, 684 (1969).
87. Twomey, S., *Geophys. pura e appl.* **43**, 227 (1959).
88. Jiusto, E., *J. Rech. Atmos.* **2** (2–3), 245 (1966).
89. Radke, L. F., and Hobbs, P. V., *J. Atmos. Sci.* **26**, 281 (1969).
90. Squires, P., and Twomey, S., *J. Atmos. Sci.* **23**, 401 (1966).
91. Twomey, S., *J. Rech. Atmos.* **3** (4), 281 (1968).
92. Twomey, S., *J. Rech. Atmos.* **4** (4), 179 (1969).
93. Dinger, J. E., Howell, H. B., and Wojciechowski, T. A., *J. Atmos. Sci.* **27**, 791 (1970).
94. Blanchard, D. C., *J. Rech. Atmos.* **4** (1), 1 (1969).
95. Twomey, S., *Bull. Obs. Puy de Dome* **1960** (1), 1.
96. Squires, P., *Tellus* **10**, 262 (1958).
97. Jiusto, J. E., and Kocmond, W. C., *J. Rech. Atmos.* **3** (1–2), 101 (1968).
98. Allee, P. A., in "Proc. 2nd Nat. Conf. on Weather Modific," Santa Barbara, 1970, p. 244.
99. Twomey, S., and Severynse, G. T., *J. Rech. Atmos.* **1** (2), 81 (1964).
100. Auer, H., *J. Rech. Atm.* **3** (2–3), 289 (1966).
101. Eisner, H. S., Quince, B. W., and Slack, C., *Disc. Faraday Soc.* **30**, 86 (1960).
102. Snead, C. C., and Zung, J. T., *J. Coll. Interface Sci.* **27**, 25 (1968).
103. Garrett, W. D., in "Proc. Cloud Phys. Conf. 1970, Fort Collins, Colorado," 1970, p. 131.
104. Hoffer, T., and Mallen, S. C., *J. Atmos. Sci.* **27**, 914 (1970).
105. Wilson, A. T., *Nature* **184**, 99 (1959).
106. Blanchard, D., *Science* **146**, 396 (1964).
107. Garrett, W. D., *Deep Sea Res.* **14**, 221 (1967).
108. Blanchard, D., in "Proc. Int. Conf. Cloud Physics," Toronto, 1968, p. 25.
109. Goetz, A., in "Proc. Cloud Phys. Conf.," Tokyo, Japan, 1965, p. 42.
110. Squires, P., *J. Rech. Atmos.* **2**, 297 (1966).
111. Squires, P., *Tellus* **8**, 443 (1956).
112. Squires, P., *Tellus* **10**, 256 (1958).
113. Squires, P., *Tellus* **10**, 262 (1958).
114. Squires, P., *Tellus* **10**, 272 (1958).

115. Twomey, S., and Squires, P., *Tellus* 11, 408 (1959).
116. Twomey, S., and Warner, J., *J. Atmos. Sci.* 24, 702 (1967).
117. Mordy, W., *Tellus* 11, 16 (1959).
118. Neiburger and Chien, C. W., Monograph No. 5, Am. Geophys. Union, 1960, p. 191.
119. Kornfeld, P., *J. Atmos. Sci.* 27, 256 (1970).
120. Chen, C. S., in "Proc. Conf. Cloud Phys., Fort Collins," 1970, p. 113.
121. Chen, C. S., Ph.D. Thesis, Dept. Meteor., UCLA, 1971.
122. Junge, C., and McLaren, E., *J. Atmos. Sci.*, 28, 382 (1971).
123. Paluch, I. R., *J. Atmos Sci.* 28, 629 (1971).
124. Battan, L. J., and Braham, R. R., *J. Meteorol.* 13, 587 (1956).
125. Morris, T. R., *J. Meteorol.* 14, 281 (1957).
126. Battan, L. J., *J. Appl. Meteorol.* 2, 333 (1963).
127. Brown, E. N., and Braham, R. R., *J. Atmos. Sci.* 20, 23 (1963).
128. Telford, J. W., *J. Meteorol.* 12, 436 (1955).
129. Twomey, S., *J. Atmos. Sci.* 23, 405 (1966).
130. Berry, E. X., *J. Atmos. Sci.* 24, 680 (1967).
131. Warshaw, M., *J. Atmos. Sci.* 24, 278 (1967).
132. Chin, H. C., Ph.D. Thesis, Dept. Meteorology, University of California, Los Angeles, 1970.
133. Warner, J., and Twomey, S., *J. Atmos. Sci.* 24, 704 (1967).
134. Warner, J., *J. Appl. Meteorol.* 7, 247 (1968).
135. Woodcock, A. H., and Jones, R. H., *J. Appl. Meteorol.* 9, 690 (1970).
136. Hobbs, P., and Radke, L. F., *Science* 163, 279 (1969).
137. Hobbs, P. V., Radke, L. F., and Shumway, S. E., *J. Atmos. Sci.* 27, 81 (1970).
138. Mossop, S. C., *Z. angew. math. Phys.* 14, 456 (1963).
139. Stevenson, C. M., *Quart. J. Roy. Meteorol. Soc.* 94, 35 (1968).
140. Soulage, G., *Nubila* 6, 43 (1964).
141. Kline, D. W., *Monthly Weather Review* 91, 681 (1963).
142. Isono, K., Komabayasi, M., and Ono, A., *J. Meteorol. Soc. Japan, Ser II* 37, 211 (1959).
143. Gagin, A., in "Proc. Int. Conf. Cloud Phys.," Tokyo, 1965, p. 155.
144. Heffernan, K. J., and Bracewell, R. N., *J. Meteorol.* 16, 337 (1959).
145. Droessler, E. G., and Heffernan, K. J., *J. Appl. Meteorol.* 4, 442 (1965).
146. Carte, A. E., and Mossop, S. C., *Bull. Obs. Puy de Dome*, 1960 (4), 137.
147. Bigg, K., in "Proc. Int. Conf. Cloud Phys.," Tokyo, 1965, p. 137.
148. Bigg, E. K., and Hopwood, S. C., *J. Atmos. Sci.* 20, 185 (1963).
149. Bigg, E. K., and Miles, G. T., *Tellus* 15, 162 (1963).
150. Bowen, E. G., *Austr. J. Physics* 60, 490 (1953).
151. Bigg, E. K., and Giutronich, J., *J. Atmos. Sci.* 24, 46 (1967).
152. Mason, B. J., and Maybank, J., *Quart. J. Roy. Meteorol. Soc.* 84, 235 (1958).
153. Qureshi, M. M., and Maybank, J., *Nature* 211, 508 (1966).
154. Gokhale, N., and Goold, J., *J. Geophys. Res.* 74, 5374 (1969).
155. Telford, J., *J. Meteorol.* 17, 86 (1960).
156. Bigg. K., Miles, G. T., Heffernan, K. J,, *J Meteorol.* 18, 804 (1961).
157 Droessler, E. G., *J. Atmos. Sci.* 21, 701 (1964).
158. Kassander, A. R., Sims, L. L., and MacDonald, J. E., Sci. Rept. No. 3, Univ. of Arizona, Inst. of Atm. Physics, 1956.
159. Bigg, E. K., and Miles, G. T., *J. Atmos. Soc.* 21, 396 (1964).
160. Bigg, E. K., *J. Atmos. Sci.* 24, 226 (1967).
161. Maruyama, M., and Kitagawa, T., *J. Meteorol. Soc. Japan, Ser. II* 65, 126 (1967).
162. Bigg, E. K., *Nature* 197, 172 (1963).
163. Isono, K., Komabayasi, K., Takeda, T., Tanaka, T., and Iwai, K., NWAP, Rept. 70–1, Water Res. Lab. Nag. Univ., Japan, 1970; Fujiwara, M., Met. Res. Inst., Tokyo, 1970.

164. Georgii, H. W., *Geofys. pura e appl.* **44**, 249 (1959).
165. Isono, K., and Tanaka, T., *J. Meteorol. Soc. Japan Ser. II* **44**, 255 (1966).
166. Ryan, B. F., and Scott, W. D., *J. Atmos. Sci.* **26**, 611 (1969).
167. Georgii, H. W., Monograph No. 5, Am. Geophys. Union, 1960, p. 233.
168. Higuchi, K., and Wushiki, H., *J. Meteorol. Soc. Japan* **48**, 248 (1970).
169. Rosinski, J., *J. Atm. Terr. Phys.* **29**, 1201 (1967).
170. Schaefer, V. J., Gen. Electric Res. Lab. Project Cirrus, Occas. Rept. No. 20, 1950.
171. Paterson, M. P., and Spillane, K. T., *J. Atmos. Sci.* **24**, 50 (1967).
172. Pruppacher, H. R., and Sänger, R., *Z. angew. math. Phys.* **6**, 407 (1955).
173. Mason, B. J., *Quart. J. Roy. Meteorol. Soc.* **86**, 552 (1960).
174. Roberts, P., and Hallet, J., *Quart. J. Roy. Meteorol. Soc.* **94**, 25 (1968).
175. Fukuta, N., *J. Atmos. Sci.* **23**, 741 (1966).
176. Mossop, S. C., *Proc. Phys. Soc.* **69**, 161 (1956).
177. Day, G. A., *Proc. Phys. Soc.* **72**, 296 (1958).
178. Higuchi, K., and Fukuta, N., *J. Atmos. Sci.* **23**, 187 (1966).
179. Vali, G., in "Proc. Inter. Conf. Cloud Phys.," Toronto, 1968, p. 302.
180. Gokhale, N., in "Proc. Inter. Conf. Cloud Phys.," Tokyo, 1965, p. 176.
181. Gokhale, N., and Goold, J., *J. Appl. Meteorol.* **7**, 870 (1968).
182. Isono, K., and Komabayasi, M., *J. Meteorol. Soc. Japan Ser II* **32**, 29 (1954).
183. Isono, K., *Nature* **183**, 317 (1959).
184. Price, S., and Pales, J., *Archiv. Meteorol. Geophys. Biocl. Ser. A* **13**, 398 (1963).
185. Soulage, G., *Bull. Obs. Puy de Dome* **1958** (4), 121.
186. Admirat, P., *Bull. Obs. Puy de Dome* **1962** (2), 87.
187. Langer, G., in "Proc. 1st Conf. Weather Mod., Albany," 1968, p. 220.
188. Langer, G., and Rosinski, J., *J. Appl. Meteorol.* **6**, 114 (1967).
189. Telford, J., *J. Meteorol.* **17**, 676 (1960).
190. Wirth, E., *J. Rech. Atmos.* **2** (1), 1 (1966).
191. Georgii, H., W., and Kleinjung, E., *J. Rech. Atmos.* **3** (4), 145 (1967).
192. Soulage, G., *J. Rech. Atmos.* **2** (2–3), 219 (1966).
193. Katz, U., *Z. Angew. Math. Phys.* **11**, 237 (1960).
194. Sano, I., Fukuta, N., Kojinia, Y., and Murai, T., *J. Meteorol. Soc. Japan* **41**, 189 (1963).
195. Serpolay, R., *Bull. Obs. Puy de Dome* **1958** (3), 81.
196. Serpolay, R., *Bull. Obs. Puy de Dome* **1959** (3), 81.
197. Fukuta, N., *J. Meteorol.* **15**, 17 (1958).
198. Mason, B. J., and van den Heuvel, A. P., *Proc. Phys. Soc.* **74**, 744 (1959).
199. Robinson, E., and Ludwig, F. L., *J. Air Poll. Control Assoc.* **17**, 664 (1967).
200. Schaefer, V. J., *Science* **154**, 1555 (1966).
201. Schaefer, V. J., *J. Rech. Atmos.* **3**, 181 (1968).
202. Schaefer, V. J., *J. Appl. Meteorol.* **7**, 148 (1968).
203. Schaefer, V. J., *Bull. Am. Meteorol. Soc.* **50**, 199 (1969).
204. Morgan, G. M., and Allee, P. A., *J. Appl. Meteorol.* **7**, 241 (1968).
205. Hogan, A. W., *Science* **158**, 800 (1967).
206. Morgan, G. M., *Nature* **213**, 58 (1967).
207. Parungo, F. P., and Rhea, J. O., *J. Appl. Meteorol.* **9**, 468 (1970).
208. Winchester, J. W., and Duce, R. A., *Tellus* **18**, 281 (1966).
209. Lininger, R. L., Duce, R. A., Winchester, J. W., and Matson, W. R., *J. Geophys. Res.* **71**, 2457 (1966).
210. Paslawska, S., and Ostrowski, S., *Acta Geophys. Polonica* **16**, 181 (1968).
211. Moyers, J. L., Zoller, W. H., and Duce, R. A., *J. Atmos. Sci.* **28**, 95 (1971).
212. Moyers, J. L., Ph.D. Thesis, University of Hawaii, Honolulu, 1970.
213. MacCredy, P. B., Smith, T. B., Todd, C. J., and Bessmer, K., Final Rept. of the Advisory Committee on Weather Control, 1957, Vol. II, p. 172.

214. Poppoff, I. G., Robbins, R. C., and Goettelman, R. C., *Bull. Am. Meteorol. Soc.* **39**, 144 (1958).
215. Soulage, G., *Nubila* **4**, 43 (1961).
216. Soulage, G., *Compt. Rend.* **240**, 2168 (1955).
217. Clough, W. S., Cousins, L. B., and Eggleton, A. E. J., *Int. J. Air Poll.* **9**, 769 (1965).
218. Ohtake, T., in "Proc. Int. Conf. on Physics of Snow and Ice, 1966, Sapporo, Japan," 1967, Vol. 1, p. 105.
219. Appleman, H., *Bull. Am. Meteorol. Soc.* **34**, 14 (1953).
220. Pilié, R., and Jiusto, J. E., *J. Meteorol.* **15**, 151 (1958).
221. Panel on Weather and Climate Modification "Weather and Climate Modification," Vol. II, Nat. Acad. Sci., Nat. Res. Council, Publication No. 1350, Washington D.C., 1966.
222. Georgii, H. W., *Ber. d. Deutsch. Wetter d. US. Zone*, **1959** (58).
223. Georgii, H. W., and Metnieks, A. L., *Geophys. pura e appl.* **41**, 159 (1958).
224. Mossop, S. C., *Proc. Phys. Soc.*, *B* **69**, 161 (1956).
225. Murty, Bh. V. R., *J. Meteorol. Soc. Japan* **47**, 219 (1969).
226. Hobbs, P. V., and Locatelli, J. D., *J. Atmos. Sci.* **27**, 90 (1970).
227. Fukuta, N., *J. Meteorol.* **15**, 17 (1958).
228. Pruppacher, H., *J. Atmos. Sci.* **20**, 376 (1963).
229. Pena, J., and Pena, R. G. de, *J. Geophys. Res.* **75**, 2831 (1970).
230. Parungo, F. P., and Wood, J., *J. Atmos. Sci.* **25**, 154 (1968).
231. Kuhns, I. E., *J. Atmos. Sci.* **25**, 878 (1968).
232. Kuhns, I. E., and Mason, B. J., *Proc. Roy. Soc.* (*London*) **A 302**, 437 (1968).
233. Junge, C. E., *Archiv. Met. Geophys. Biol.* **5**, 44 (1952).
234. Neiburger, M., and Wurtele, M. G., *Chemical Reviews* **44**, 21 (1949).
235. Pick, W. H., *Quart. J. Roy. Meteorol. Soc.* **57**, 288 (1931).
236. Rosinski, J., and Kerrigan, T. C., *J. Atmos. Sci.* **26**, 695 (1969).
237. Rosinski, J., *J. Appl. Meteorol.* **6**, 1066 (1967).
238. Stansbury, E. J., and Vali, G., Sci. Rept. MW-46, Stormy Weather Group, McGill University, 1965.
239. Vali, G., in "Proc. Conf. Severe Local Storms, St. Louis," 1967, p. 154.
240. Hoffer, T., and Braham, R. R., *J. Atmos. Sci* **19**, 232 (1962).
241. Braham, R. R., *J. Atmos. Sci.* **21**, 640 (1964).
242. Koenig, R., *J. Atmos. Sci.* **20**, 29 (1963).
243. Stewart, J. B., *Meteorol. Magaz.* **96**, 23 (1967).
244. Murgatroyed, R. J., and Garrod, M. P., *Quart. J. Roy. Meteorol. Soc.* **86**, 167 (1960).
245. Flohn, H., *Ber. d. Deutsch. Wetter d. US. Zone* **7** (51) (1959).
246. Coons, R. D., *Bull. Am. Meteorol. Soc.* **30**, 289 (1949).
247. MacCready, P. B., and Takeushi, D. M., *J. Appl. Meteorol.* **7**, 591 (1968).
248. Mossop, S. C., Ruskin, R. E., Heffernan, K. J., *J. Atmos. Sci.* **25**, 889 (1968).
249. Mossop, S. C., in "Proc. 7th Int. Conf. Condensation and Ice Nuclei," Vienna–Prague, 1969, p. 407.
250. Mossop, S. C., *J. Rech. Atmos.* **3**, 119 (1968).
251. Mossop, S. C., and Ono, A., *J. Atmos. Sci.* **26**, 130 (1969)
252. Mossop, S. C., Ono, A., and Heffernan, K. J., *J. Rech. Atmos.* **3**, 45 (1967).
253. Hobbs, P. V., *J. Atmos. Sci.* **26**, 315 (1969).
254. Auer, A. H., Veal, D. L., and Marwitz, J. D., *J. Atmos. Sci.* **26**, 1342 (1969).
255. Grant, L. O., in "Proc. Int. Conf. Cloud Phys.," Toronto, 1968, p. 305.
256. Mossop, S. C., *Bull. Am. Meteorol. Soc.* **51**, 474 (1970).
257. Latham, J., and Mason, B. J., *Proc. Roy. Soc. Ser. A* 260, 537 (1961).
258. Brownscombe, J. L., and Hallet, J., *Quart. J. Roy. Meteorol. Soc.* **93**, 455 (1967).
259. Sano, I., Fujtani, Y., and Maena, Y., *Memoirs Kobe Marine Observatory* **14**, 1 (1960).
260. Georgii, H. W., and Kleinjung, E., *Pure and Appl. Geophys.* **71**, 181 (1968).

261. Fukuta, N., *J. Atmos. Sci.* **23**, 191 (1966).
262. Parungo, F. P., and Lodge, J. P., *J. Atmos. Sci.* **22**, 309 (1965).
263. Parungo, F. P., and Lodge, J. P., *J. Chem. Phys.* **30**, 1476 (1967).
264. Bryant, G. W., Hallet, J., and Mason, B. J., *J. Phys. Chem. Solids* **12**, 189 (1959).
265. Evans, L. F., *Nature* **206**, 822 (1965).
266. Fletcher, N., *J. Chem. Phys.* **30**, 1476 (1959).
267. Edwards, L. F., and Evans, L. F., *Trans. Faraday Soc.* **58**, 1649 (1962).
268. Fletcher, N. H., *J. Chem. Phys.* **38**, 237 (1963).
269. Katz, U., *Z. angew. math. Phys.* **13**, 333 (1962).
270. Edwards, L. F., Evans, L. F., and La Mer, V. K., *J. Colloid Sci.* **17**, 749 (1962).
271. Mossop, S., and Jayaweera, K. O. L. F., *J. Appl. Meteorol.* **8**, 241 (1969).
272. Zettlemoyer, A. C., Tcheurekdjian, N., and Hosler, C. L., *Z. Angew. Math. Phys.* **14**, 496 (1963).
273. Tcheurekdjian, N., Zettlemoyer, A. C., and Chessik, J. J., *J. Phys. Chem.* **68**, 773 (1964).
274. Zettlemoyer, A. C., Tcheurekdjian, N., and Chessik, J. J., *Nature* **192**, 653 (1961).
275. Hamilton, W. C., Katsanis, E. P., and Zettlemoyer, A. C., in "Proc. 1st Nat. Weather Mod. Conf. Albany," 1968 p. 336.
276. Corrin, M. L., Moulik, S. P., and Cooley, B., *J. Atmos. Sci.* **24**, 530 (1967).
277. Corrin, M. L., and Nelson, J., *J. Phys. Chem.* **72**, 643 (1968).
278. Gravenhorst, G., and Corrin, M. L., in "Proc. Int. Conf. Nucleation, Prague–Vienna," 1969, p. 206.
279. Federer, B., *Z. angew. math. Phys.* **19**, 637 (1968).
280. Turnbull, D., "Artificial Stimulation of Rain," Pergamon Press, New York, 1955, p. 345.
281. Fletcher, N. H., *Austr. J. Phys.* **13**, 1408 (1960).
282. Hallet, J., *Phil. Mag.* **6**, 1073 (1961).
283. Fukuta, N., and Mason, B. J., *J. Phys. Chem. Solids* **24**, 715 (1963).
284. Kobayashi, T., Contrib. Inst. Low Temp. Sci., Ser. A, No. 20, p. 1 Hokhaido Univ. Sapporo, Japan, 1965.
285. Fletcher, N. H., *J. Atmos. Sci.* **26**, 1266 (1969).
286. Samoilov, O. Ya., "Structure of Aqueous Electrolyte Solutions," (transl. from Russian), Consultants Bureau, New York, 1965.
287. Kavanau, J. L., "Water and Solute–Water Interactions," Holden-Day, San Francisco, 1964.
288. Eisenberg, D., and Kauzmann, W. K., "The Structure and Properties of Water," Oxford University Press, 1969.
289. Edwards, G. R., Evans, L. F., and Zipper, A. F., *Trans. Faraday Soc.* **66** (565), 220 (1970).
290. Evans, L. F., *Nature* **213**, 384 (1967).
291. Evans, L. F., *Trans. Faraday Soc.* **63** (540), 1 (1967).
292. Bigg., E. K., *Quart. J. Roy. Meteorol. Soc.* **79**, 510 (1953).
293. Bigg, E. K., *Quart. J. Roy. Meteorol. Soc.* **81**, 478 (1955).
294. Carte, A. E., *Proc. Phys. Soc.* **73**, 324 (1959).
295. Vali, G., and Stansbury, E. J., *Canad. J. Phys.* **44**, 477 (1966).
296. Levine, J., NACA TN 2234, 1950.
297. Vali, G., and Stansbury, E. J., Sci. Rept. MW-41, Stormy Weather Group, McGill Univ., 1965.
298. Gokhale, N. R., *J. Atmos. Sci.* **22**, 212 (1965).
299. Fletcher, N. H., *J. Atmos. Sci.* **27**, 1098 (1970).
300. Evans, L. F., in "Proc. Conf. on Cloud Phys., Fort Collins," 1970, p. 14.
301. Georgii, H. W., *Z. angew. math. Phys.* **14**, 503 (1963).
302. Georgii, H. W., and Kleinjung, E., *J. Rech. Atmos.* **3**, 145 (1967).
303. Hallett, J., *J. Atmos. Sci.* **22**, 64 (1965).
304. Ono, A., *J. Atmos. Sci.* **27**, 649 (1970).

305. Fukuta, N., *J. Atmos. Sci.* **26**, 522 (1969).
306. Neumann, J., Gabriel, K. R., and Gagin, A., Paper presented at Int. Conf. on Water for Peace, May 23–31, 1967, Washington, D.C., 1967.
307. Jayaweera, K. O. L. F., *J. Atmos. Sci.* **28**, 728 (1971).
308. Jayaweera, K. O. L. F., *Quart. J. Roy. Meteorol. Soc.* **98**, 193 (1972).
309. Beard, K., and Pruppacher, H. R., *J. Atmos. Sci.* **26**, 1066 (1969).
310. Latham, J., and Saunders, C. P. R., *Quart. J. Roy. Meteorol. Soc.* **96**, 257 (1970).
311. Braham, R. R., *Bull. Am. Meteorol. Soc.* **49**, 343 (1968).
312. Changnon, S. A., "Air over Cities," Public Health Service, R. A. Taft Sanitary Eng. Center, SEC Tech. Rept. A62-5, 1961, p. 37.
313. Changnon, S. A., *Bull. Am. Meteorol. Soc.* **49**, 4 (1968).
314. Changnon, S. A., *Bull. Am. Meteorol. Soc.* **50**, 411 (1969).
315. Huff, F. A., and Changnon, S. A., in "Proc. 2nd Conf. Weather Modification, Santa Barbara," 1970, p. 215.

Chapter 2

PARTICULATE MATTER IN THE LOWER ATMOSPHERE

Richard D. Cadle

National Center for Atmospheric Research,* Boulder, Colorado

1. INTRODUCTION

Many scientific disciplines have contributed to our knowledge of the nature and concentration of fine particles in the atmosphere. These disciplines include theoretical and experimental physics, meteorology, astronomy, and chemistry. This chapter describes results obtained by such disciplines, discusses in detail many of the aspects of more than average interest, and suggests several problems that remain to be solved.

Both natural and man-produced particles are considered; in fact, it is rather arbitrary to distinguish between the two in the open or ambient atmosphere away from obvious sources of contamination. Although such particles are very small, and sometimes because they are so small, they can have considerable influence on our lives. For example, they affect the behavior of the atmosphere in many ways, such as by acting as condensation or freezing nuclei, thereby aiding in the formation of cloud droplets or snowflakes. By scattering and absorbing solar radiation, they can influence the earth's climate. Many kinds of particles found in the atmosphere are toxic, at least at sufficiently high concentrations. The potential hazards from radioactive fallout are very well documented. Suspensions of particles in smog and other

*The National Center for Atmospheric Research is sponsored by the National Science Foundation.

forms of air pollution may greatly decrease visibility; in fact, one of the most objectional aspects of smog is decreased visibility.

Scientists are interested in particles in the atmosphere for at least two reasons in addition to their direct or indirect effect on our lives. Knowledge of the nature and behavior of such particles often provides information concerning other aspects of our atmosphere; for example, certain types of particles such as radioactive fallout, pollens, and bacteria can serve as tracers for the movement of air masses and provide information concerning the dynamics of the atmosphere. There is also interest in the particles themselves as geophysical and geochemical phenomena.

Some definitions are in order. Obviously, we do not want to include molecules or birds when considering atmospheric particles. For purposes of this chapter, the range has been arbitrarily chosen to be 0.5–5 μm in radius (5×10^{-6} to 5×10^{-4} cm). The lower end somewhat overlaps the colloid range and the upper end is very roughly the upper size limit for particles that remain suspended in the atmosphere for fairly long periods of time. Thus, larger particles such as hailstones and snowflakes are not considered. Also for the purposes of this chapter both solids and liquids are included in the definition of the term particle.

There are many advantages to defining and naming particle size ranges. A system suggested by Junge [1] has become generally accepted for particles suspended in the atmosphere and defines three size ranges. Particles with radii less than 0.1 μm are called Aitken nuclei or Aitken particles. Those having radii between 0.1 and 1 μm are "large particles" and those with radii greater than 1 μm are "giant particles." The term "Aitken" resulted from the use of the Aitken nuclei counter [2], which is an expansion-type cloud chamber in which water condenses on the particles in the air drawn into the chamber. The droplets produced are counted to obtain a particle concentration and most of the particles have radii smaller than 0.1 μm. Large particles are responsible for much of the decrease in visibility produced by natural haze and by smog.

The terms particle and particle size may be very ambiguous unless they are properly defined. Thus it is necessary to distinguish between aggregates and the "ultimate" particles constituting the aggregates. The term particle size has been used to mean both radius and diameter, and must be further defined when applied to irregular particles. Fortunately, various definitions of particle size have been developed for irregular particles to avoid such uncertainties. These definitions fall into two classes. One class involves the definition of size as a statistical property based on individual measurements of some dimension of a large number of particles in a powder or suspension. The other class is based on measurements of some property of the suspension itself. One such property is the rate at which the particles settle in the suspension. Another is the nature of the light scattered from a beam of light falling on the suspension.

2. THE TROPOSPHERE

The troposphere is the lowest region of the atmosphere and extends from the earth's surface up through the region of generally decreasing temperature to the tropopause. The latter is defined as the altitude where the temperature ceases to decrease or may even increase. Large storms and turbulence characterize the troposphere, which receives much of its heat from the ground and from the condensation of water vapor rather than by the absorption of direct radiation from the sun. These characteristics of the troposphere play an important part in the behavior of the particles which it bears. Immediately above the tropopause is the stratosphere, which will be described later.

2.1. Sources of Particles

Much more of the earth is covered by oceans than by exposed land and therefore it is not surprising that particles consisting largely of sea salt should be an important fraction of all the particles in our atmosphere, especially over the oceans. These particles are widespread over both the oceans and continents and play an important role in cloud formation. Although the concentrations of salt particles over the oceans may be as high as 100 particles cm^{-3}, a concentration of 1 cm^{-3} seems to be more common. It is important to realize, however, that even in the atmosphere above the oceans, there are many different types of particles. This is partially because particles arising from the continents can be carried long distances by wind. Furthermore, salt particles undergo chemical reactions involving trace gases in the atmosphere and there is a continuous influx of extraterrestrial particles.

Sodium chloride crystals are hygroscopic. They form droplets consisting of solutions of sodium chloride when the relative humidity rises above about 75%. However, it is generally necessary for the relative humidity to decrease considerably below that figure before solid particles again separate from the droplets due to supersaturation. When humidities are between 75 and 100%, water vapor condenses on or evaporates from a droplet containing sodium chloride until the vapor pressure of the droplet becomes equal to the partial pressure of water vapor in the air, at least if conditions are such that equilibrium can be achieved. In our constantly changing atmosphere, however, equilibrium is seldom actually reached.

Obviously, particles of sea salt started life as droplets of ocean water. How the droplets are formed, however, is not so obvious. One mechanism is the blowing of spray from the tops of breaking waves. Another and now well-established mechanism is that the droplets are formed mainly by the breaking of very large numbers of sea bubbles as they reach the surface of the oceans. The bubbles are largely produced by the breaking of small waves. Droplets may also be produced in a number of other ways; for example, by rain or snow falling on the surface of the water.

The droplets that are formed by the blowing of spray from the tops of breaking waves are very large, and the sea salt particles that are produced when they evaporate are in the size range of Junge's "giant particles." These giant particles do not remain in the air for long periods of time and are often responsible for the heavy haze that may be observed along the coastline when the surf is severe and during an on-shore breeze. Even along the coast, however, the bulk of the airborne salt particles has usually been produced by the breaking of bubbles. For example, Randall [3] measured the airborne salt along the coast of Barbados and found that the trade winds and not the adjacent waves were the major source of salt particles, a finding which is consistent with the above statement.

Airborne salt particles cause considerable damage to crops and cause corrosion in coastal areas. Fallout of salt on oceanic islands and coastal areas in which the sea salt fallout from all processes is high ranges from 25 to 300 lb per acre per year with extremes as high as 3000 to 4000 lb per acre per year. On intermediate, humid coastal and inland areas, the fallout rate is 3–25 lb per acre per year [4].

The bursting of bubbles in the ocean seems to produce two size ranges of droplets by two different mechanisms. The first mechanism, which produces relatively large droplets, involves a very small jet of sea water that rises from the bottom of the bubble immediately after the bubble bursts. This jet quickly breaks into a number of droplets that are all nearly the same size and about one-tenth the size of the bubble. Therefore, the size of the droplets is largely determined by the size of the bubbles themselves. Since the smallest bubbles formed by small breaking waves are usually about 100 μm in diameter, the smallest droplets are about 10 μm in diameter. The concentration of salt in sea water is usually 3.3–3.6% by weight, so the evaporated droplets produce particles that are about 2.5 μm in diameter [5]. Particles produced in this way, as are those produced from spray, are much larger than most of the particles found over the oceans. The latter particles are produced from bursting bubbles by the second mechanism. Immediately before the breaking of the bubble, a thin film exists between the air of the bubble and the air of the open atmosphere. When this film breaks, a large number of very small droplets are produced. Mason [6] found that the bursting of each bubble produces 100–200 small droplets which evaporate to form particles of sea salt with maximum diameters of about 0.3 μm [6,7]. A noteworthy aspect of this process is that the droplets produced seem to have different concentrations of various ions relative to the sodium ion than does bulk sea water. It has been suggested that this may result from adsorption of various ions at the surface of sea water, but it is also possible that surface films of organic material may be partially responsible [8].

Eriksson [9] has calculated that under steady-state conditions, about 0.3% of the ocean surface must be covered with white caps. On the basis of this estimate, Junge [1] has estimated an order-of-magnitude production rate of one salt particle per square centimeter per second. He also has suggested that the average residence time of sea salt aerosol particles over the ocean is probably one to three days.

The major constituent of sea salt particles is, of course, sodium chloride, but smaller amounts of other inorganic substances are present such as carbonates, sulfates, potassium, magnesium, and calcium. Sea water also contains numerous organic substances, part of which are dissolved and part of which are suspended. Numerous studies have demonstrated that the ocean is covered with a layer of insoluble organic material that in most places is only a few molecules thick. All of this organic material does, of course, become incorporated to some extent with the sea salt particles. Such organic material may affect not only the composition of the particles, but also the mechanism of production. For example, Garrett [10] investigated the influence of monomolecular surface films on the production of condensation nuclei from bubbled sea water. He found that the addition of pure and mixed monomolecular organic films to the sea water surface increased the concentration of salt condensation nuclei by as much as threefold. Sources of organic material are numerous in the oceans: Sea life itself, both living and dead, is one source; natural oil seepage is another; and petroleum products that enter the ocean as a result of man's activities is a third. It is interesting to speculate that such organic material which was initially on the surface of sea water may eventually occur on the surface of cloud droplets formed about sea salt particles as condensation nuclei. Such adsorbed organic material may considerably influence both the electrical and the coalescence properties of cloud droplets; but little or nothing is known concerning such effects.

Another important source of particles in the troposphere is volcanoes, but their contribution varies tremendously with time and space. The reason for this variation is obvious: A single violent eruption may produce more particles than all of the subsequent eruptions for several years. Furthermore, there have been tremendous variations in the intensity of volcanism throughout geological time. During tertiary times, this intensity was perhaps two orders of·magnitude greater than that at present or during much of the previous geological history of the earth.

One of the most famous eruptions, and one of the most deadly during historical times, was that of Krakatoa in 1883 in the East Indies. It produced eruption clouds 18 miles high and turned day into night in Batavia about 100 miles away. A recent very violent eruption was by the volcano Gunung Agung in Bali in 1963. This eruption, like that at Krakatoa, injected particulate material into the stratosphere as well as into the troposphere, producing spectacular sunsets throughout the world. Meinel and Meinel [11] studied these sunsets and found that the appearance and intensity of the sunset glow changed from day to day. They suggested that a study of the appearance of the glow and measurements of the height over an extended period of time and from many places might provide important information concerning atmospheric circulation.

Newell [12] found that the presence of the aerosol from the Agung eruption caused increases of the stratospheric temperature of about 5°C. The increase was largest in the 60–80-mb region corresponding to the reported heights of the dust layer. Individual monthly increases in temperature ranged up to 8°C. Presumably, the particles directly absorbed solar radiation and transferred the resulting heat to the stratospheric air. Recent eruptions in the Galapagos Islands and in Iceland have also injected tremendous quantities of particles into the stratosphere and troposphere.

Most of our knowledge concerning the nature of particles formed by volcanoes has been obtained during the last few years. Some of the particles are merely finely divided lava, whereas others are produced by attrition of the crater walls. Still other particles are droplets of sulfuric acid. These contain dissolved crystalline material consisting largely of inorganic sulfates and halides that have probably sublimed from the magma. In 1965, McClaine and his co-workers [13] collected particles on filters from the eruption cloud of the volcano Syrtlingur off the coast of Iceland. Among a variety of particle compositions observed, they found that nickel and chromium were strongly evident, although these elements occur only in trace quantities in lava. They concluded that the particles were formed by selective vaporization from the magma and subsequent condensation from the eruption cloud.

Particles were collected by the Air Weather Service of the U.S. Air Force from an eruption near Hekla Volcano in Iceland in 1970. Analyses of these samples presently underway at the National Center for Atmospheric Research (NCAR) suggest that the bulk of the particles consist of fragments of lava (Fig. 1), but crystalline material such as calcium and ammonium sulfate and small spheres are also present.

During the last few years, the author and his associates at NCAR have made a number of collections of particles in the fume from the volcano Kilauea on the Island of Hawaii, both during and between eruptions. The Kilauea eruptions are relatively gentle and although lava particles are emitted into the air, they are sufficiently large that they settle out rapidly. The particles that remain in suspension are largely droplets of submicron size and are composed of sulfuric acid-containing dissolved salts. Following collection, crystals of various salts, such as calcium and ammonium sulfate, may separate from the droplets [14–16]. Figure 2 is an electron micrograph of particles collected by impaction from the fume of the 1967–68 eruption. The particles have partially evaporated, giving rise to the forms consisting of nuclei surrounded by satellite droplets, a pattern characteristic of sulfuric acid. Inorganic substances which have crystallized from the droplets also appear.

Particles collected from the fume from Halemaumau Crater of Kilauea Volcano before and after eruptions had a much higher soluble salt content

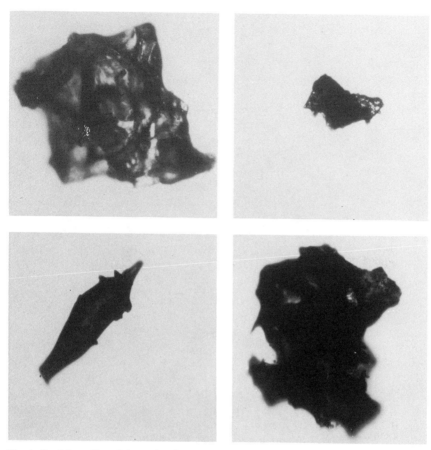

Fig. 1. Particles collected from the cloud produced by the eruption near Hekla Volcano in Iceland, 1970. These particles were a few hundred microns in size.

than those collected from the fume from the top of lava fountains (compare Figs. 2 and 3). Electron microprobe and electron diffraction analyses, however, indicated the nature of the salts was similar for the two types of situation. A possible explanation is that during the eruption, sulfer dioxide in the fume mixes at very high temperatures with oxygen from the air and is oxidized to sulfur trioxide.

Hydration of the sulfur trioxide will produce sulfuric acid droplets which dilute the metal salts. Some sulfur trioxide exists in the magma, but thermodynamic considerations of the equilibria existing in magmas suggest that the gaseous sulfur compounds must be largely in the form of hydrogen sulfide and sulfur dioxide [17]. Recently, the author and his associates have determined ratios of sulfur dioxide to sulfate in Kilauea fume by drawing the fume first through a filter and then through a solution which collects the sulfur dioxide. The filter, of course, removes particulate sulfate. The filter and

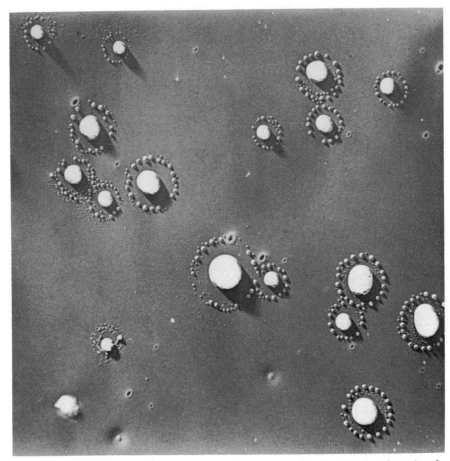

Fig. 2. Electron micrograph of particles collected from the fume directly off the lava fountains of the 1967–68 eruption of Kilauea. Distance across micrograph is 23 μm.

solution were then analyzed for sulfate and sulfur dioxide, respectively. The results have shown that in most primary fumaroles, the ratio of sulfur dioxide to sulfate is very high, often nearly 100 to 1 by weight. However, the fume emitted by one very hot fumarole which consisted of a very long, but only partially fuming crack contained almost equal amounts of sulfur dioxide and sulfate. This may have been a case of the oxidation of sulfur dioxide to sulfate by atmospheric oxygen. If this process also occurs in very explosive eruptions, such eruptions may inject tremendous amounts of sulfate directly into the atmosphere.

Recent studies have also been made of particles from the eruption clouds produced by the highly explosive volcanoes Mayon in the Philippines and Arenal in Costa Rica. The two volcanoes are very similar to each other with

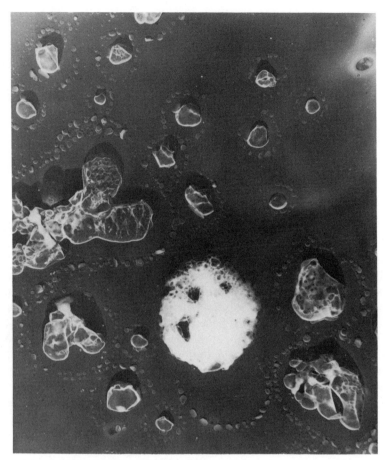

Fig. 3. Electron micrograph of particles collected from the fume from the east rift zone eruption of Kilauea in 1969 between eruptive phases but when tremendous amounts of fume were emitted. Distance across micrograph is 10 μm.

regard to type of eruption and of lava emitted, but quite different in these respects from Kilauea. The finely divided lava particles (the so-called "ash") were largely irregular and angular in shape, but a small percentage of spheres was also present. As at Kilauea, however, much of the fume consisted of droplets of sulfuric acid containing dissolved inorganic salts, although the percentage of sulfate was lower than at the Hawaiian volcano. Table 1 is an analysis based on aqueous extracts of filters containing fume particles from Arenal.

An estimate can be made of the mean annual emission of sulfur compounds by volcanoes over the last few hundred years, although the assumptions required for such estimates may be far from correct. Rittmann [18] has quoted Sapper concerning the amount of volcanic materials emitted between

TABLE 1. Weights of Various Constituents Collected on Polystyrene Fiber Filters from Arenal Eruption Fume

Sample	Si, μg	NH_4^+, μg	SO_4^{2-}, μg	Na^+, μg	Cl^-, μg	Ca^{2+}, μg	Mg^{2+}, μg	K^+, μg	Approximate volume sampled,* m^3
431	26,500	41	1630	184	387	590	220	60	8.3×10^2
451	23,000	10	1160	163	401	470	200	57	4.2×10^2
459	33,400	26	2300	189	480	810	330	89	1.2×10^3

*Values corrected to sea level.

the years 1500 and 1914. The amount of lava erupted was estimated to be 3.9×10^{17} cm^3 or, if we assume a density of three, 1.2×10^{18} g. MacDonald [19] estimated that the weight of gas evolved during eruptions of Kilauea was approximately 0.5% by weight of the lava evolved. Analyses by Shepherd [20] suggest that about 10% by weight of the gas from Kilauea is sulfur dioxide and this was the major form of sulfur. Thus the total weight of sulfur dioxide discharged over the approximately 400 years, assuming all volcanoes behaved as does Kilauea, must have been about 6×10^{14} g, or 1.5×10^{12} g yr^{-1}. This value can be compared with the total annual sulfur dioxide liberation to the atmosphere as pollution of 1.3×10^{14} g based on estimates by Robinson and Robbins [21]. If this evaluation gives the right order of magnitude for the emission of sulfur compounds calculated as sulfur dioxide, volcanoes contribute orders of magnitude less sulfur and its compounds to the atmosphere than do man's activities, except locally or immediately following large eruptions.

An important portion of the particle loading of the atmosphere comes from soil and rocks. This fact is dramatically emphasized by sand storms over deserts and dust storms over arid regions such as those that occurred over the "dust bowl" regions of the United States during the 1930's. The Sahara Desert produces airborne dust on a worldwide basis. An estimated ten million tons of red dust from Africa were deposited on England in 1903. More recently, dust from the African deserts has been collected along the coast of Barbados.

The effects of airborne dust and sand are dramatically demonstrated by the weirdly carved rock forms that are found in many arid regions such as the southwestern United States. Even weak winds raise particles of minerals which appear to be a constant part of the particulate loading of the atmosphere. For the most part, the sources are local. Such material probably settles from the atmosphere or is washed out by rain and then reentrained into the atmosphere a multitude of times before it is washed into the ocean to become part of the ocean sediments. Very little is known concerning the rate at which mineral matter and soil particles in general are introduced into the atmosphere and it would be very difficult to obtain meaningful information of this type because of the tremendous worldwide variations in the types of terrain and the intensity of the wind.

Grass, brush, and forest fires constitute another important source of tropospheric particles. Mason [22] has estimated that of the condensation nuclei responsible for the production of cloud droplets, about one-tenth are sea salt and the remaining are mixed nuclei and products of natural or man-made fires. Few studies have been made of the nature of the particles from natural fires. The "ash" is probably largely inorganic and consists of the minerals originally present in the vegetation. Carbon particles and partially burned, tarry material are also present.

Neuberger [23] estimated that an average grass fire extending over one acre produces about 20,000 billion billion (2×10^{22}) fine particles. Most of the particles are of the very small Aitken size defined earlier. As with mineral particles, it is very difficult to make meaningful estimates of the rates at which such fires emit particles into the atmosphere on a worldwide basis. It is known, however, that smoke from forest fires can travel great distances. Smoke from such fires in Western Canada in 1950 was observed over the British Isles and caused the sun and moon to appear blue. The Royal Observatory in Edinburgh made spectroscopic studies of the solar radiation during this time and estimated an effective particle diameter for light scattering of about 1 μm.

There is considerable evidence that particles produced by forest fires not only contribute to the particulate loading of the atmosphere on a worldwide basis, but may have marked effects on the weather on a local and regional basis. For example, measurements made downwind from a simulated forest fire indicated that the fire increased the concentration of cloud condensation nuclei active at a supersaturation of about 1% by a factor of approximately 2.5. Similar results were observed for lower supersaturations [24].

Examination of the records for 60 years of rainfall in Hawaii during three months of the cane harvesting season indicated that rainfall decreased at inland stations with increased cane production. Cane fields are burned preliminary to harvesting. Apparently no such reduction occurred at a "control" station located upwind of smoke from the cane fires. This reduction suggests, though does not prove, that the smoke particles act as condensation nuclei and that by greatly increasing the concentration of such nuclei, the sizes of cloud droplets are reduced, thereby hindering the coalescence process of rain formation. Of course, other factors may have caused the observed climatic changes [25]. A study of sugar cane fires demonstrated that smoke is a prolific source of cloud nuclei and that it does greatly increase the number concentration of droplets formed well downwind from the fires [26].

The suggestion has often been made that meteoritic dust may influence the behavior of the earth's atmosphere. For example, the ablation from large meteors or meteorites may introduce substances such as sodium or sodium vapor into the ionosphere. Bowen [27] has suggested that extraterrestrial material may serve as freezing nuclei, triggering the formation of rain by the same mechanism that apparently makes cloud seeding effective. The meteoritic dust may be of at least two kinds. Submicron-size extraterrestrial particles from interplanetary space can reach the earth's surface essentially unchanged. The partial or complete melting of larger objects may cast off droplets or vapors. The solidified droplets and condensed vapors may eventually reach the troposphere and serve as condensation or freezing nuclei. Bowen [27] also suggested that rainfall should increase on a worldwide basis following major meteor showers. However, attempts to determine the validity of this second hypothesis have at best been inconclusive.

A number of attempts have been made to estimate the amount of extraterrestrial material which reaches the earth's surface. This cannot be precisely estimated from studies of meteors and meteorites alone if for no other reason than that the size of a small meteor passing through the atmosphere is very difficult to estimate.

One of the more trustworthy attempts to estimate the annual weight of meteoritic dust reaching the earth's surface was undertaken by Pettersson [28,29], who collected and studied samples of material in ocean sediments and in the earth's atmosphere. The materials in the air were collected by filtering large volumes of air and analyzing the collected particles for a number of substances, but especially for nickel. Sampling was undertaken on Mauna Loa and Haleakala in the Hawaiian Islands, where the atmosphere is relatively free of particles originating on the continent. Since nickel is rare in terrestrial dust, it was assumed that any nickel in the collected particles was of meteoritic origin. The mountains on which the sampling stations were located are composed of basalt, which contains a few hundred parts per million nickel, but presumably contamination from this source was avoided.

The ocean sediment samples were treated to obtain ferromagnetic spherules that presumably were formed from the molten surface of meteorites as they passed through the earth's atmosphere or by condensation of vapor that was emitted by such meteorites. On the basis of observations of the behavior of dust from the Krakatoa eruption, Pettersson estimated that particles of meteoritic origin would require approximately two years to pass through the atmosphere to the earth's surface. He concluded that about 1.4 $\times$ 10^7 tons of such particles are added to the earth each year. His value for the residence time (two years) agrees well with much more recent estimates of the residence time of stratospheric particles calculated from measurements of the rate of deposition of radioactive fallout. Glasstone [30] states that the very small particles reach the earth ultimately at an estimated rate of 10^7 kg day^{-1}. This is about 3.4 $\times$ 10^6 tons yr^{-1}, which is in reasonable agreement with the value estimated by Pettersson. It is noteworthy that although these seem like large amounts of material, they would increase the earth's mass by a factor of only about 4×10^{-6} in one billion years if Pettersson's estimate is correct. Other estimates of extraterrestrial particles reaching the earth include those of Öpik [31], who suggested that the total annual influx of interplanetary dust over the entire earth is about 2.5×10^5 tons, while Fiocco and Colombo [32] estimated approximately 2×10^7 tons, again in fair agreement with Pettersson's value.

Calculations have been made by Rosinski and Snow [33] of the size distribution of particles produced by the condensation of vapors from meteors ablating upon entering the earth's atmosphere. They suggested that the diameters of the particles should be approximately proportional to the size of the meteor and for the most part be less than 100 Å. They calculated median volume diameters ranging from 4.5 to 80 Å at 1 min following evaporation. The average concentrations of the condensed particles formed from meteor showers were calculated to be higher than the concentrations from the steady influx of sporadic meteors.

Rosinski [34] has studied the relationship between meteoric activity and extraterrestrial particle concentrations in the troposphere using magnetic spherules as an indicator of the flux of such particles. Spherules were collected by high-volume air filtration at 0.5°S, 19.5°, 40°, 47.5°, and 65°N latitude from August to October 1967. The simultaneous occurrence of concentration peaks at all of the stations precluded a terrestrial origin for the collected spherules. Sizes and densities of spherules were used to estimate falling times and thus to identify the parent meteor streams. Chemical analyses using electron microprobe techniques suggested that meteor streams vary in the amounts of stony and iron fragments they introduce into the atmosphere or may consist wholly of one or the other.

As every hay fever sufferer knows, biological materials constitute part of the particle loading of the atmosphere. These materials are especially important, of course, because of their biological action. For the most part, the biological particles, which may range in size from viruses to large birds, have diameters greater than 1 μm. This is true even for viruses, the individual crystals of which may be very small, but which are usually present in the atmosphere either as clumps of such particles or agglomerated with other particles.

Many investigators have observed that particles of biological origin often travel great distances. For example, spores of fungi have been found above the Caribbean Sea at least 600 miles from the nearest source, and pollen has been observed 1500 miles from its probable origin. Conversely, marine bacteria have been observed 80 miles inland. They probably were associated with sea salt particles formed by the breaking of bubbles as described earlier in this chapter. Particles of biological origin have also been found at very high altitudes. A classical experiment was that conducted by the manned balloon Explorer II; spores of a number of molds were caught in a spore trap released at 72,500 ft and set to close at 36,000 ft [35].

Of great practical importance is the extent to which man pollutes his atmosphere, both locally and worldwide, with the particles he emits into the atmosphere. It is useful, though somewhat artificial, to consider air pollution as consisting of two general types. One is the widespread type found in the atmospheres of many large cities. The other type is single-source air pollution, usually industrial, which refers to single-source or nearly single-source pollution emitted from a single factory or similar operation. Obviously, the division is arbitrary since factory effluents may contribute to city smog.

It is also convenient to define two types of smog. One of these is the so-called photochemical smog in which the unpleasant constituents are to a very large extent produced by photochemical reactions involving sunlight acting on air contaminated with organic vapors, especially unburned gasoline, and oxides of nitrogen. The other type might be called coal-burning smog since it consists largely of a combination of coal smoke and fog (from which the term "smog" was derived) or just coal smoke. The term "smog" was originally applied to the dirty atmospheres of cities such as London and Pittsburgh, and was quite appropriate. Now the term is also commonly applied to photochemical smog, although with less justification. Smog in most cities today is a combination of the two types, although one may predominate. Also, as our use of fossil fuels for energy has changed from coal-burning to petroleum-burning, smog of the photochemical variety has increased considerably. Of course, man-produced particles from sources other than the two just mentioned are always mixed with the smog and become part of it. Even the wearing of rubber tires on automobiles contributes somewhat to our smog problems.

The composition of the particles in photochemical smog has been studied extensively. In a light smog, minerals and other inorganic substances may constitute as much as 50% of the particulate material. Several studies have been made of the organic fraction of smog, usually after collecting the particles on a filter. The filter may be extracted with some organic solvents such as benzene or a paraffinic hydrocarbon and the resulting solution examined by spectroscopy. Infrared spectroscopy is particularly appropriate for this purpose; particles collected from Los Angeles smog produced an infrared spectrum which in many respects resembled the spectrum of the reaction products of ozone with unsaturated hydrocarbons similar to those found in automobile exhaust gases.

The author and his co-workers in 1950 found that a very large part of the material collected from intense Los Angeles smog consisted of droplets. Some of these were dark brown, gummy, water-insoluble, organic material. This material occasionally liberated iodine from acidulous aqueous solutions of potassium iodide and at times formed a blue color when diphenylbenzidine reagent was added to it. These are tests for oxidizing substances such as hydrogen peroxide and organic peroxides, and of course, when applied to the gas phase, ozone. This material was also often strongly acidic.

Many of the droplets seemed to consist largely of impure water. On standing after collection on a microscope slide, they very slowly evaporated and a film formed over the surface of the droplets which attained the appearance of partially collapsed tents. Presumably this film was in part responsible for the slow rate of evaporation even at relative humidities as low as 30%.

Crystalline material was also found on such slides and had either been in the air in the form observed or crystallized from the droplets of water. Much of the former material was insoluble in water, and probably consisted of the minerals of natural origin mentioned above. Ammonium sulfate and needles of calcium sulfate were commonly observed. Occasionally, hexagonal figures were found on the slides and these were tentatively identified as fluorides using a micrurgic technique [36,37]. These crystals had a refractive index of slightly greater than 1.43, apparently were inorganic, and from the refractive index and crystalline form, were possibly schairerite or pachnolite, probably the former. Whitby and his co-workers [38] studied Los Angeles smog in the summer of 1969 and they also observed the presence of large amounts of droplets. Whitby estimated that between 15 and 50% by number of the particles were droplets which were volatile and slowly evaporated on collection. He also observed that the mass medium diameter was between 0.2 and 1 μm, that particles were formed at night as well as during the day, and that the particles when formed were extremely small nuclei, probably less than 0.02 μm in diameter.

Smog of the coal-burning type tends to contain larger particles, especially flakes of soot, than does photochemical smog; when the smog is truly smoke and fog, the size distribution is of course that of the droplets which for the most part are very much larger than most particles found either in smoke or in photochemical smog. The type of particles in smog of the coal-burning type also varies more from city to city and from time to time than do those in the photochemical smog. In most coal-burning cities, various dusts, smokes, and fumes that have little or nothing to do with coal are of course emitted as mentioned above. Many of these are industrial, but backyard incineration and burning at city dumps also may contribute. Most of the studies of particles in the air of highly industrialized cities have been concerned with total particle loading, that is, with the mass concentrations.

Extensive studies of the concentrations of atmospheric pollutants in urban and nonurban environments have been made by the National Air Sampling Network of the United States Public Health Service. The network, established in 1953, consists of a very large number of urban and rural stations in all 50 states, Puerto Rico, and the District of Colombia. The urban stations are located in central business districts whereas the nonurban stations are located as remotely as possible, such as on ocean and lake shores, in deserts, forests, and mountain and farmland areas. The sampling network therefore covers a very large assortment of smoggy and nonsmoggy conditions.

TABLE 2. Suspended Particulate Matter (μg m^{-3}) at Urban Stations in Various Regions of the United States, 1957–1958

Region	Number of Samples	Min.	Max.	Arith. avg.	Geo. mean	Std. geo. dev.
New England	595	20	326	100	86	1.739
Mid-Atlantic	714	23	607	146	125	1.772
Mideast	516	27	745	123	103	1.698
Southeast	578	15	640	125	104	1.689
Midwest	967	11	978	158	139	1.629
Great plains	503	22	722	136	120	1.622
Gulf South	516	14	630	118	100	1.687
Rocky Mountain	247	15	466	99	84	1.809
Pacific Coast	704	11	639	136	109	2.026
Grand total	5340	11	978	131	111	1.772

Particles are collected by filtering 70,000–80,000 ft^3 of air through glass fiber filters for 24-hr time periods using the so-called high-volume samplers to draw the air through the filters. All the samples are analyzed for the weight concentration of particles, for the benzene-soluble fraction, and for gross radioactivity. Some samples are also analyzed for nitrate, sulfate, and for a large number of metals [39–42]. Some of the results for the concentrations of suspended particulate material at urban stations are shown in Table 2.

It is especially informative to compare data from the Pacific coast with those of the Midwest and mid-Atlantic since all three are noted for smog and of course that on the Pacific coast is mainly photochemical. Mass concentrations of suspended particulate material are lower for the Pacific coast than for other areas and this conforms with the fact that the number concentrations of large particles, that is, those that make an overwhelming contribution to the mass concentrations, are relatively much lower in photochemical smog than that of the coal-burning communities. More detailed information is given in Table 3, which shows that in the four western cities which have primarily photochemical smog, the percentages of benzene-soluble organic matter and nitrate are higher and those of sulfate are lower than for the eastern cities.

Smog of the coal-burning variety contains a particularly high concentration of carcinogenic compounds; polynuclear hydrocarbons in polluted atmospheres are especially effective carcinogens and a tracer for such carcinogens used by many laboratories is 3,4-benzpyrene. It melts at 180°C and boils at about 500°C at atmospheric pressure; therefore, it exists as particles, almost always associated with particles of other materials. Because it is formed when numerous organic substances are burned, it is found in most or all city smogs. Sawicki et al. [43] studied the concentration of 3,4-benzpyrene and found that the concentrations were tens and hundreds

TABLE 3. Particulate Concentrations for Selected Cities for 1958*

Station location	Suspended particulate, A $\mu g\, m^{-3}$	Benzene-soluble organic matter $\mu g\, m^{-3}$	% of A	Sulfate $\mu g\, m^{-3}$	% of A	Nitrate $\mu g\, m^{-3}$	% of A
Los Angeles	213	30.4	14.2	16.0	7.5	9.4	4.4
San Francisco	80	10.6	13.3	6.2	7.7	2.6	3.3
San Diego	93	12.2	13.1	7.7	8.3	4.2	4.5
Denver	110	11.0	10.0	6.1	5.5	2.3	2.1
New York	164	14.3	8.7	23.0	14.0	2.2	1.3
Pittsburgh	167	13.0	7.8	15.1	9.0	2.6	1.6
Cincinnati	143	13.7	9.6	12.2	8.5	2.6	1.8
Louisville	228	18.0	7.9	20.6	9.1	4.9	2.1

*Values are arithmetic means.

of micrograms per thousand cubic meters of air. Concentrations in cities where the smog was predominantly photochemical were much lower than in other cities. As is to be expected, the concentrations in nonurban regions were very much lower than those in urban regions.

We know much less about the particle size distributions of particles in smog of the coal-burning variety than of those in photochemical smog. What is known suggests that unless fog is present, the distributions of the two types of smog are very similar except at the large end of the scale. Many types of equations are useful for describing the size distributions of particles, but a type which is often very useful for particles in contaminated atmospheres has the form

$$dN/dr = ar^b \tag{1}$$

where dN is the number of particles per unit volume in the increment of size range dr; a and b are constants. Steffens and Rubin [44] obtained a value of -4.5 for b for particles in Los Angeles smog. It is often convenient to plot $dN/d(\log r)$ against r for particles in contaminated atmospheres. When this is done, a slope of about -3 on a logarithmic scale has often been found over much of the size range. Since

$$dN/d(\log r) \equiv r\, dN/dr$$

this slope corresponds to a value of about -4 for b, which agrees fairly well with the result obtained by Steffens and Rubin during their early studies of photochemical smog. Figure 4 is a typical plot of this type. A wide range of sizes and concentrations can be represented by such a graph and it can be used to estimate the concentrations of the particles in any size range. Thus, if we define

$$n(r) = dN/d(\log r) \quad cm^{-3} \tag{2}$$

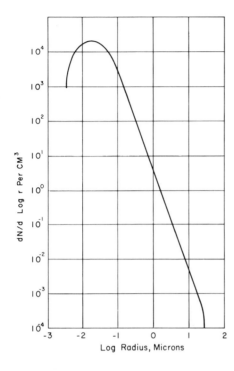

Fig. 4. Particle size distribution for continental air (U. S. Air Force, "Handbook of Geophysics," New York, Macmillan, 1960).

then the number of particles ΔN per cubic centimeter between the limits of the interval $\Delta(\log r)$ may be obtained from such a plot by

$$\Delta N = n(r)\,\Delta(\log r) \tag{3}$$

The corresponding log radius-surface (s) and log radius-volume (v) distributions are then obtained from the equations

$$s(r) = dS/d(\log r) = 4\pi r^2 \, dN/d(\log r) \tag{4}$$

and

$$v(r) = dV/d(\log r) = (4/3)\pi r^3 \, dN/d(\log r) \tag{5}$$

Defining ρ as the density of the particles, we have

$$m(r) = \rho v(r) \tag{6}$$

The curve shown in Fig. 4 is typical for all continental surface air. Possible reasons for this shape are discussed later; hower, the curves obtained are not always this smooth, particularly when the data are obtained near strong sources of pollution which contain particles having a very different size distribution from that usually observed in smog.

Chemical reactions play a very important role in the manner in which cities contribute to the particle loading of our atmosphere. To a certain extent, chemical reactions are important in smog of the coal-burning type since sulfur dioxide, for example, is continually being oxidized up to sulfur trioxide which then is hydrated to form sulfuric acid droplets; but chemical reactions are especially important in photochemical smog, as the name implies, and only a small fraction of the particles that are found in such smog would be present in the absence of such chemical reactions. The most important triggering reaction in photochemical smog seems to be the photo-chemical decomposition of nitrogen dioxide to form nitric oxide and atomic oxygen. This results from the absorption of sunlight by the nitrogen dioxide which is always present in such smog. Atomic oxygen can react with molecular oxygen to form ozone, but ozone also reacts with nitric oxide to form nitrogen dioxide and oxygen. If these were the only reactions occurring, the amount of ozone produced would be very much smaller than the amount often detected in such smog. However, the reaction of atomic oxygen with various hydrocarbons, such as olefins emitted in the exhaust of automobiles, forms free radicals. These in turn undergo a very large sequence of reactions, some of which involve the oxidation of nitric oxide which is liberated by various combustion processes to nitrogen dioxide. Other reactions may produce ozone and many of them produce very unpleasant organic compounds such as the aldehydes, formaldehyde, and acrolein, or the nitrogen-containing compound peroxyacetyl nitrate. Formaldehyde, acrolein, and peroxyacetyl nitrate are probably the primary compounds responsible for the eye irritation that is produced by smog, but no single compound seems to be largely responsible. These reactions are summarized as follows:

$$NO_2 + h\nu \rightarrow NO + O \tag{7}$$

$$O + O_2 + M \rightarrow O_3 + M \tag{8}$$

$$O_3 + NO \rightarrow NO_2 + O_2 \tag{9}$$

$$O + \text{Olefins} \rightarrow R + R'O \quad \text{or} \quad R\overset{O}{\diagup\!\diagdown}R' \tag{10}$$

$$O_3 + \text{Olefins} \rightarrow \text{Products} \tag{11}$$

$$R + O_2 \rightarrow RO_2$$

$$RO_2 + O_2 \rightarrow RO + O_3 \tag{12}$$

Reactions (10) and (11) lead to a long series of reactions producing various organic acids, aldehydes, ketones, and nitrogen-containing compounds. Reaction (12) may be the one mainly responsible for ozone formation. This sequence of reactions may be the most important one, but other reactions that are triggered by sunlight are undoubtedly also very important. For instance, it has been found that to some extent aldehydes can take the place of nitrogen dioxide in such reactions.

Judging from the ratio of organic particle mass to organic vapor mass in photochemical smog, such as that in Los Angeles, only about 5 % of the organic vapors, which are mainly hydrocarbons, are converted to particles. The results of several investigations indicate that aerosol is formed when many, if not most, 6-carbon and larger straight-chain, branch-chain, and cyclic olefins as well as a number of aromatic hydrocarbons mixed with nitrogen dioxide in air in the parts per million concentration range are irradiated. When automobile exhaust–air mixtures are irradiated, the concentration of particles in the mixtures increases. However, it is important to note that automobiles can cause some smog even in the absence of irradiation. The exhaust gases themselves contain high concentrations of particles and they also contain various aldehydes, including formaldehyde, that are obnoxious for various reasons.

When sulfur dioxide is added to a mixture of air, oxides of nitrogen, and hydrocarbons and the mixture is irradiated, the sulfur dioxide is very rapidly oxidized to sulfur trioxide, which in turn reacts with any water vapor to form sulfuric acid droplets. This reaction is much more rapid than would be produced merely by the irradiation of sulfur dioxide mixed in air in the absence of the other trace constituents and it may result from the reaction of the peroxy radical with the sulfur dioxide. Ammonia, often present in polluted atmospheres, will react with sulfuric acid droplets to form ammonium sulfate. The latter is often an important constituent of smog and of the aerosol in relatively uncontaminated atmospheres.

Automobile exhaust gases contain particles of high lead concentrations due to the presence of tetraethyllead in the gasoline. Gasoline also contains organic bromine compounds which are designed to serve as scavengers for the lead. Theoretically, the lead in the gasoline combines with the bromine to form nonvolatile lead bromide which will condense out on the walls of the exhaust system and ultimately be swept into the atmosphere as large particles that fall harmlessly to the ground. However, recent studies of the nature of the lead particles in smog and in the exhaust gases indicate that a large percentage of the particles actually escape into the atmosphere in very fine form. Such particles are drawn far down into the respiratory system of man and may cause long-term damage.

For many years, it was believed that lead from automobile exhaust gases was not a problem. A recent concern has developed over the possibility of lead poisoning from this source. Studies by scientists at the Ethyl Corporation have shown the presence of compounds in exhaust gases such as $PbCl \cdot Br$ and $PbO \cdot PbCl \cdot Br \cdot H_2O$ and smaller amounts of lead ammonium halide complexes. Extremely large particles, those of diameter greater than about 0.5 mm, contain high concentrations of iron oxides and lead principally as a sulfate and $PbO \cdot PbSO_4$.

Robinson and his co-workers [45] made bulk sample analyses of aerosol particles collected from smog and showed that only about 10% of the total lead in the air is soluble in water at 40°C. Thus the more soluble lead compounds such as $PbCl_2$, $PbBr_2$, and $PbSO_4$ probably constitute only a small part of the atmospheric lead aerosol. A typical size distribution for this aerosol had a mass medium diameter of 0.2 μm, with 25% of the mass being accounted for by particles smaller than 0.1 μm and another 25% by particles greater than 0.5 μm in diameter.

Concentrations of particles having diameters larger than about 0.2 μm in dense photochemical smog were found to range from about 3000 to 10,000 particles cm^{-3}.

Single-source pollution is, of course, another very important source of particles in the atmosphere. The pollutants from any industrial activity may be in the form of solids, gases, or even noise. When the pollutants affect only workers in a plant, the problem is generally classified as one of industrial hygiene; if the pollutants escape the plant, they are termed air pollutants. The operations of many industries make them especially subject to air pollution problems; there is no point in trying to discuss or list all such industries, but a few will be mentioned. A number of industrial processes result in the emission of fluorine and various gaseous and solid fluorine compounds. Such operations have had many problems resulting from the air pollution which they produce. The fluorides do considerable damage to plants and grazing animals. Industrial operations producing fluorides include the production of ceramics, certain fertilizers, and metals. The production of aluminum is a well-known example, and the open-hearth method of steel-making is another.

The public utility industry has also had many problems with air pollution. These arise almost entirely from combustion products and from the handling and storage of fuels. The utility companies that produce electricity from steam have had serious pollution problems. The steam is produced in boilers by burning fuel and the fuels generally are chosen on the basis of availability and cost and are practically always coal, natural gas, or fuel oil. Some plants are now being powered by nuclear energy. Control of the particles is mainly by the use of mechanical collectors such as cyclones, electrostatic precipitators, or both. The disposal of the collected fly ash is also a serious problem since the utility industry may produce as much as a million tons per year.

Agriculture and the control of pests may also be considered producers of single-source pollution. The effects of the use of insecticides such as DDT on our environment are receiving increasing attention. Since most insecticides have an appreciable vapor pressure, it is not known how much of such emissions when diluted with extremely large volumes of air are in particulate form and how much are in the gas phase.

Chemical reactions that may produce and alter airborne particles may occur in the open, ambient atmosphere as well as in smog. Many of these reactions are similar to those observed in smog and described above. As in smog, atomic oxygen is produced by the photochemical decomposition by sunlight of nitrogen dioxide and of ozone. The atomic oxygen will react rapidly with any sulfur dioxide which may be present to form sulfur trioxide which is hydrated to form sulfuric acid droplets. With the possible exception of some types of smog, concentrations of atomic oxygen in the troposphere are too low for this reaction to be of particular importance. Nonetheless, the reaction may be of importance in the lower stratosphere as discussed later.

A more important reaction is the photochemical oxidation of sulfur dioxide to sulfuric acid. Sulfur dioxide absorbs solar radiation in the troposphere quite strongly in the wavelength range 2900–3300 Å. The energy associated with the individual photons is too low to cause the sulfur dioxide to dissociate but does produce electronically excited SO_2. The latter reacts much more rapidly with molecular oxygen of the air than it does when in the ground or unexcited state. The rate at which such reactions occur in the atmosphere can be estimated from the photochemical yield Φ (the ratio of molecules of product formed per photon absorbed), the absorption coefficients of this reactant, the intensity of sunlight and the concentrations of the reactants. Integration must be performed over the wavelength range in which absorption occurs. Unfortunately, laboratory studies of this reaction have yielded a very wide range of values for Φ. The author and co-workers have recently obtained results that indicate that Φ is about 1.7×10^{-2} at 3130 Å, but over the wavelength range from 2800 to 4200 Å, Φ is about 2×10^{-3}. This difference may in part explain the previous discrepancy. The rate of oxidation by this mechanism in the atmosphere must be quite low, approximately one sulfur dioxide molecule per thousand oxidized per hour based on the latter value for Φ.

At least two other means of oxidation of sulfur dioxide by molecular oxygen are of possible importance. Urone and his co-workers [46] have found that even in the absence of sunlight, sulfur dioxide in air is oxidized very rapidly in the presence of powdered oxides of various metals such as aluminum, calcium, and iron. Airborne particles may have a similar effect. When sulfur dioxide dissolves in fog or cloud droplets, it forms sulfurous acid and this is oxidized by molecular oxygen much more rapidly than is sulfur dioxide by the gas-phase photochemical reaction mentioned above [47]. When this liquid-phase oxidation occurs in droplets that were nucleated by sea salt particles, the sulfuric acid produced may cause gaseous hydrogen chloride to be liberated when the droplets dry. As a result, part of the sodium chloride is converted to sodium sulfate.

Nitrates may also be formed in sea salt particles. High nitrate contents have been found in particles exceeding 1 μm in diameter collected over coastal areas of the northeastern part of the United States. The particles seem to result from the chemical reaction of salt particles with nitrogen dioxide in the air, and the concentrations of nitrate in the particles are very high when the concentrations of nitrogen dioxide are especially high due to air pollution. This reaction was studied by Robbins et al. [48] in the laboratory using mixtures of reactant vapors and aerosols in air in a 10-m^3 chamber. Samples were collected periodically by filtration from the air in the chamber and analyzed. The nitrogen dioxide reacted rapidly with the sodium chloride crystals to form nitrate but no nitrite. The first step in the reaction appears to be the hydrolysis of NO_2 to form nitric acid vapor:

$$3NO_2 + H_2O \rightarrow 2HNO_3 + NO \tag{13}$$

The HNO_3 then interacts with the sodium chloride to produce sodium nitrate and some desorption of the resulting hydrogen chloride.

Organic vapors in the ambient atmosphere undoubtedly undergo photochemical reactions similar to those that occur in smog. There are many sources for such vapors, including the smog itself, but an important source seems to be many types of vegetation. Examples are coniferous trees, sagebrush, and creosote bush. Such organic vapors absorb solar radiation very weakly, if at all, and traces of oxides of nitrogen in the ambient atmosphere may serve as the primary photochemical reactant just as they do in photochemical smog. Went [49] believes that summer heat haze results from such reactions and points out that a more or less dense haze exists all year over the jungles of South America and the highlands of southeastern Mexico. In fact, in the summer, a haze covers most of the United States. Such reactions may partially explain the fact that the atmosphere absorbs more sunlight in summer than in winter and that the night sky is so much clearer in winter.

Nuclear explosions have been a source of radioactive and nonradioactive particles in the atmosphere ever since the first nuclear explosion in 1945. Their nature has been studied with unusual thoroughness because of their possible physiological effects. Both the physical and chemical features of the particles depend to a large extent on conditions existing at the time of the explosion. For example, particles that are produced by a nuclear explosion on the surface of coral, such as on a coral atoll, consist largely of white calcium compounds from the coral plus traces of bomb debris containing radioactive products. Under these conditions, much of the total mass and radioactivity are associated with large particles that fall out close to the source. On the other hand, particles that are produced by a high-altitude explosion consist almost entirely of material condensed from the vaporized bomb and are so small that they can remain suspended in the atmosphere for very long periods of time.

Calculations have been made of the approximate size range produced by the condensation of vaporized bomb debris from a high-altitude nuclear explosion. The calculation is based on an assumed diffusion coefficient of 0.1 cm^2 sec^{-1} at temperatures of 1000–3000°K, on the assumption that the driving force for the diffusion is the initial concentration of the condensing material in the vapor phase, and a rate of gas cooling of the order of 1000°K sec^{-1} when the weapon yield is small and 10°K sec^{-1} for a very large weapon yield. The results suggest that the bomb debris forms particles in the range of

0.03–0.3 μm diameter. Of course, this assumes self-nucleation, and condensation of the radioactive vapor on particles already existing in the atmosphere is probably important. Furthermore, these various small particles will agglomerate with each other and with other particles in the air.

If the explosion takes place sufficiently close to the earth that soil is sucked into the "fireball" or if it is on or below the earth's surface, very large as well as small radioactive particles are produced by a number of processes and much of the radioactive fission products will fall to earth very close to the explosion center. Thus acute radiation damage from fission products is much more likely to occur as a result of a low- rather than a high-altitude nuclear explosion. Conversely, a high-altitude explosion will contribute more "worldwide fallout" than a low-altitude burst.

A number of studies have been made of the nature of particles produced by surface and underground nuclear explosions. The study of thin sections of such particles has been particularly informative. Such sections can be examined by chemical microscopy and by preparing autoradiographs. A combination of the two techniques demonstrates the composition of the portions of the particles that are more strongly radioactive. Many of the particles are spheres, but others have a multitude of shapes, including pear-shaped, dumbbell shapes, and agglomerates of particles of various forms. Explosions on or in silicate soils produced particles that were composed largely of transparent glass which ranged from colorless to almost black. Some particles contained numerous bubbles. The black material was probably magnetite (Fe_3O_4) and in general was the most radioactive. Tower shots on coral atolls produced particles that were black, spheroidal, magnetic, and at times cracked and veined with calcium salts [50].

2.2. Composition of Particles Collected from the Atmosphere

So far we have considered the sources of particles and the composition of the particles from these sources; however, if the particles are dispersed and become part of the particle loading in the ambient atmosphere, they are changed in a number of ways. Some are sufficiently large that they settle from the atmosphere or are impacted on various surfaces such as trees. Others are modified by various chemical reactions as described above, and almost all of them undergo modification as a result of agglomeration with other particles. In this section, we consider the nature of the particles as they exist in the ambient atmosphere, away from major sources of the particles. Also, as implied above, there is considerable variation in the nature of the aerosol particles over various portions of the earth. This variation is especially apparent with respect to oceans and continents, but there are also great differences on a regional basis over both of these types of the earth's surface. For example, the entire European continent is much more affected by pollution from the standpoint of the particle loading of the atmosphere than is the North American continent, and composition of the air over South America is undoubtedly greatly influenced by particles produced from the jungles. Furthermore, particles in the atmospheres above polar regions are different from those in the air over midlatitudes, in part at least because of the great influence of the continents on the latter.

Recently, a study was undertaken at the National Center for Atmospheric Research of the chemical composition of aerosol particles in the upper tropical troposphere [51]. Particles were collected in the vicinity of the Philippine Islands by means of filters carried on a U. S. Air Force aircraft C-135 flying out of Japan. The filters were those furnished the Air Weather Service by the Institute of Paper Chemistry. They consisted of submicron-diameter cellulose fibers impregnated with the oil diethylbutoxyphthalate. The filters and their calibration have been described by Friend [52]. The material collected on the filters was analyzed by wet-chemical and neutron-activation methods. The former method involved an aqueous extraction of one-half of the filter using Soxhlet extractors without the thimbles. The resulting aqueous solutions were analyzed for sulfate, ammonium ions, and total fixed nitrogen as described by Lazrus et al. [53]. Nitrate ion was determined specifically by the method of Brewer and Riley [54]. Sodium, magnesium, potassium, bromine, chlorine, and silicon were determined by nondestructive neutron activation techniques on the other one-half of the filter. The concentration of sulfate was very much higher than the concentration of either sodium or chlorine (Table 4), which of course was present in the atmosphere as the chloride ion, showing that sea salt was not a major constituent of the aerosol. This finding was also indicated by the marked variation in the ratios of sodium to chlorine from one sample to another and in general were not that for sodium chloride. The ratios of chlorine to bromine varied from about 21 to 95 and in all samples were considerably lower than the comparable ratio for sea salt in the oceans, which is about 300. The concentrations of silicon were for the most part low relative to those for sulfate, sodium, or chlorine. Since the most likely sources for silicon are various silicates from the minerals of the continents, this finding indicates that such minerals constituted a minor part of the particulate loading of the high oceanic troposphere.

The concentrations of nitrates shown in Table 4 are surprisingly high. Possibly the filters absorbed HNO_3 vapor in addition to collecting particles containing NO_3^-.

The ratios of chlorine to bromine are much lower than those found by Duce et al. [55] for aerosol particles close to the sea surface near Hawaii. They reported values of about 1000 for the particles over water and about 100 for those over land. They emphasized that the atmospheric chemistry of the halides is very complex but that the difference may result from bromine compounds in automobile exhaust gases. The lower ratios we found may be due to worldwide air pollution from automobiles.

TABLE 4. Analyses of Particles Collected on IPC Filters in the Upper Tropical Troposphere Near the Philippine Islands*

Latitude, N	Longitude, W	Altitude, ft × 10^{-3}	Concentrations, $\mu g\ m^{-3}$ ambient								Sampling time, min
			$SO_4^=$	Si	Na	Cl	(NO_3^-)†	NH_4^+	Mn	Br	
35°04'	139°35'	26	0.041	—	—	—	0.025(0.020)	0.0064	—	—	120
26°21'	120°40'	26	0.051	0.011	0.055	0.085	0.026(0.019)	0	0.00034	0.0014	84
19°32'	120°40'	25	0.10	0.016	0.016	0.052	0.054(0.042)	0.0013	0.00033	0.0018	36
16°30'	117°00'	25	0.15	0.022	0.00072	0.029	0.050(0.022)	0.013	0.00048	0.00096	72
13°00'	124°00'	25	0.073	0.009	0.010	0.013	0.046(0.039)	0.0033	0.00025	0.00063	68
21°00'	127°54'	25→39	—	0.014	0.019	0.018	—	—	0.00032	0.0006	105
11°00'	122°00'	39	0.23	0.0067	0	0.057	0(0)	0	0.00041	0.0006	20
11°00'	119°00'	39	0.16	0	0.019	0.060	0.0074(0.011)	0	0.00026	0.00073	44
16°52'	117°15'	39	0.068	0.0033	0.016	0.052	0.025(0.026)	0	0.00008	0.00056	47

*All samplings were made on November 19, 1969. Longitudes and latitudes are for start of sampling. No potassium ion or nitrite ion was detected in any of the samples.
†The NO_3^- concentration was calculated from the total combined nitrogen less that combined as NH_4^+ and also (the value in parentheses) by the method specific for NO_3^-.

Winchester, Duce, and their co-workers made a number of other studies of the ratios among halides in the atmospheres of Hawaii, northern Alaska, and Massachusetts [56]. They have found, for example, that both iodine and bromine appear to be associated with particles of smaller particle sizes and longer residence times than particles that are rich in chlorine. In many samples, the ratio of iodine to bromine is about 0.1–0.2.

A large part, perhaps as much as half, of the chlorine in the atmosphere over the oceans is in gaseous form, probably either existing as hydrogen chloride or as Cl_2. The possibility that chlorine is liberated from sea salt particles by their reaction with ozone has often been suggested; however, this reaction is much too slow to be of any importance in the atmosphere [57]. The process is much more likely to be one of two mentioned earlier, i.e., the oxidation of sulfur dioxide dissolved in droplets which are aqueous solutions of sodium chloride followed by the liberation of hydrogen chloride when the droplets dry out, or the reaction of nitric acid vapor with droplets containing sodium chloride or with solid sodium chloride particles. Junge [1] has suggested that the latter process often occurs close to the coast. He found the highest nitrate content in "giant" particles in coastal areas of northeastern United States, which he points out also contain most of the chloride. He concluded that the production of aerosol nitrates depended on the presence of both sea spray particles and nitrogen dioxide, the latter having especially high concentrations in certain coastal areas because of pollution. In coastal areas near Boston, he found ratios of nitrate to chloride in giant particles of about 0.5, but no nitrite was found.

An important unanswered question concerning the chemistry of the atmosphere is whether the oceans serve as a source or as a sink for sulfur dioxide. Evidence supporting a source theory is given in unpublished results obtained by Pate and his co-workers on the coast of Panama and Barbados. They found that during an onshore breeze off the ocean, the concentration of sulfur dioxide decreases as one goes inland from the shore. On the other hand, it has been suggested by Junge that any sulfur dioxide dissolved in sea water would very rapidly be converted by oxidation to sulfate. If the oceans are actually a source of sulfur dioxide, much of the sulfate in the particles over the oceans may result from the oxidation in air of this SO_2.

A few studies have been made of the nature of the particles in the air in the polar regions. Fenn et al. [58] found that about 40% of the mass of aerosol particles in the air above the Greenland ice cap consisted of sulfate particles. Cadle et al. [59] analyzed samples of particles collected by impaction from the antarctic atmosphere near the earth's surface in November and December 1966. The samples contained much higher concentrations of sulfur than similar samples collected in most parts of the world. This sulfur was largely in the form of sulfate, but some $S_2O_8^=$ may also have been present. The cations were largely ammonium and hydrogen ions. Most samples contained little sodium chloride. If the oceans are indeed sources of sulfur dioxide, the sulfate may have been produced by the oxidation of this sulfur dioxide in the atmosphere. Another possibility results from the belief that the antarctic is an area of strong atmospheric surface divergence from a center of subsidence in about midcontinent. This may tend to keep marine aerosols from entering the continent in any significant degree except during occasional storms, and would tend to bring down particles from the stratospheric "sulfate layer" which is described later in this chapter. The results obtained in the arctic might be similarly explained.

Numerous studies have been made of the chemical composition of particles collected over the continents. Recently, Blifford and Gillette of the National Center for Atmospheric Research collected particles by impaction at various altitudes and locations over the continents and the oceans [60]. The collected samples were analyzed by x-ray fluorescence techniques which give an elemental composition but not the nature of the compounds in which the elements occurred.

The elements determined were chlorine, sulfur, potassium, sodium, silicon, calcium, and titanium (Table 5). Although the sources for these elements undoubtedly varied considerably from place to place, they were all found in relative abundance at all of the places sampled. Such results again emphasize the great distances that particles can travel in the atmosphere and the large extent to which particulate material is thoroughly mixed into the atmosphere on a worldwide basis. As might be expected, the highest concentrations of particles were usually found in the first few meters above the surface of the ground or of the ocean.

TABLE 5. Average Concentrations and Standard Deviations of Seven Elemental Constituents of Aerosols ($0.01 \leq r \leq 10\ \mu$m) Collected at Varying Altitudes[a]

Alt., km	Scottsbluff, Nebraska	North Pacific Ocean	Death Valley, California	Chicago, Illinois	Orinoco Valley, South America	Line Islands Pacific Ocean
Chlorine						
0.015	—	1.52 ± 0.73 (4)	0.44 ± 0.28 (5)	—	—	—
0.915	—	0.24 ± 0.14 (5)	0.40 ± 0.26 (6)	0.79 (1)	0.44 ± 0.55 (5)	—
1.5	0.64 ± 0.42 (2)	—	—	—	—	—
1.8	1.01 ± 1.04 (3)	0.21 ± 0.17 (6)	0.31 ± 0.15 (6)	—	—	—
2.3	0.98 (1)	—	—	—	—	—
3.0	0.28 ± 0.24 (4)	—	—	0.19 (1)	—	—
3.7	—	0.25 ± 0.09 (7)	0.25 ± 0.14 (5)	—	—	—
4.6	0.18 ± 0.10 (2)	—	0.13 (1)	—	—	—
6.1	0.26 ± 0.42 (7)	0.22 ± 0.19 (8)	0.23 ± 0.15 (7)	0.37 (1)	—	0.34 (1)
7.6	0.54 ± 0.73 (2)	0.28 ± 0.22 (5)	0.23 ± 0.09 (7)	—	—	0.63 (1)
9.1	0.19 ± 0.22 (5)	0.35 ± 0.23 (6)	0.15 ± 0.08 (6)	0.49 (1)	—	—
Sulfur						
0.015	—	0.24 ± 0.14 (3)	0.31 ± 0.15 (5)	—	—	—
0.915	—	0.09 ± 0.05 (4)	0.32 ± 0.23 (5)	0.20 (1)	0.28 ± 0.42 (5)	—
1.5	0.11 ± 0.05 (2)	—	0.23 (1)	—	—	—
1.8	0.24 ± 0.24 (5)	0.11 ± 0.08 (6)	0.24 ± 0.06 (5)	—	—	—
2.3	0.19 (1)	—	—	—	—	—
3.0	0.11 ± 0.12 (4)	—	—	—	—	—
3.7	—	0.04 ± 0.03 (6)	0.14 ± 0.10 (5)	—	—	—
4.6	0.09 (1)	—	0.03 (1)	—	—	—
6.1	0.06 ± 0.04 (4)	0.05 ± 0.02 (5)	0.07 ± 0·03 (6)	0.08 (1)	—	0.22 (1)
7.6	0.05 (1)	0.03 ± 0.02 (5)	0.12 ± 0.16 (6)	—	—	0.05 (1)
9.1	0.10 ± 0.15 (4)	0.06 ± 0.06 (4)	0.06 ± 0.05 (5)	—	—	—
Potassium						
0.015	—	0.34 ± 0.10 (4)	0.29 ± 0.14 (4)	—	—	—
0.915	—	0.05 ± 0.04 (4)	0.17 ± 0.06 (6)	0.58 (1)	0.26 ± 0.34 (5)	—
1.5	0.31 ± 0.15 (3)	—	1.49 (1)	—	—	—
1.8	0.17 ± 0.11 (7)	0.09 ± 0.04 (5)	0.13 ± 0.07 (5)	—	—	—
2.3	0.08 ± 0.06 (3)	—	—	—	—	—
3.0	0.11 ± 0.10 (7)	—	—	0.07 (1)	—	—
3.7	—	0.03 ± 0.01 (4)	0.10 ± 0.02 (6)	—	—	—
4.6	0.03 ± 0.02 (3)	—	0.60 (1)	—	—	—
6.1	0.17 ± 0.19 (6)	0.06 ± 0.04 (8)	0.07 ± 0.04 (7)	0.09 (1)	—	0.31 (1)
7.6	0.09 ± 0.11 (3)	0.04 ± 0.04 (6)	0.13 ± 0.12 (5)	—	—	0.16 (1)
9.1	0.05 ± 0.02 (8)	0.06 ± 0.02 (6)	0.12 ± 0.05 (4)	0.10 (1)	—	—
Sodium						
0.015	—	0.38 (1)	0.28 ± 0.33 (2)	—	—	—
0.915	—	—	0.15 ± 0.15 (2)	0.35 (1)	0.26 ± 0.46 (5)	—
1.5	1.27 (1)	—	—	—	—	—
1.8	0.64 ± 0.79 (2)	0.14 (1)	0.07 ± 0.08 (3)	—	—	—
2.3	1.07 (1)	—	—	—	—	—

TABLE 5—contd.

Alt., km	Scottsbluff, Nebraska	North Pacific Ocean	Death Valley, California	Chicago, Illinois	Orinoco Valley, South America	Line Islands, Pacific Ocean
3.0	0.29 ± 0.22 (2)	—	—	0.08 (1)	—	—
3.7	—	0.04 ± 0.02 (3)	0.03 ± 0.03 (2)	—	—	—
4.6	0.32 (1)	—	—	—	—	—
6.1	0.08 ± 0.09 (2)	0.01 ± 0.01 (3)	0.05 ± 0.05 (3)	0.11 (1)	—	0.16 (1)
7.6	0.66 (1)	0.04 ± 0.02 (2)	0.08 ± 0.06 (4)	—	—	0.01 (1)
9.1	0.34 ± 0.35 (2)	0.04 ± 0.06 (3)	0.07 ± 0.10 (2)	0.17 (1)	—	—
Silicon						
0.015	—	0.48 ± 0.39 (4)	0.54 ± 0.48 (5)	—	—	—
0.915	—	0.07 ± 0.03 (4)	0.27 ± 0.07 (6)	0.48 (1)	0.41 ± 0.61 (5)	—
1.5	0.74 ± 0.45 (3)	—	0.90 (1)	—	—	—
1.8	0.27 ± 0.29 (7)	0.11 ± 0.06 (6)	0.27 ± 0.11 (6)	—	—	—
2.3	0.35 ± 0.13 (3)	—	—	—	—	—
3.0	0.28 ± 0.37 (5)	—	0.76 (1)	—	—	—
3.7	—	0.06 ± 0.04 (5)	0.14 ± 0.09 (6)	—	—	—
4.6	0.03 ± 0.01 (3)	—	0.13 (1)	—	—	—
6.1	0.39 ± 0.58 (5)	0.06 ± 0.04 (7)	0.11 ± 0.08 (7)	0.04 (1)	—	0.19 (1)
7.6	0.14 ± 0.18 (2)	0.07 ± 0.06 (6)	0.22 ± 0.33 (8)	—	—	0.04 (1)
9.1	0.18 ± 0.32 (7)	0.08 ± 0.07 (6)	0.20 ± 0.17 (6)	0.12 (1)	—	—
Calcium						
0.015	—	0.13 ± 0.11 (3)	0.44 ± 0.35 (4)	—	—	—
0.915	—	0.03 ± 0.02 (4)	0.18 ± 0.06 (5)	2.22 (1)	0.32 ± 0.54 (5)	—
1.5	0.20 ± 0.12 (3)	—	0.58 (1)	—	—	—
1.8	0.28 ± 0.39 (8)	0.06 ± 0.03 (5)	0.20 ± 0.06 (5)	—	—	—
2.3	0.06 ± 0.04 (3)	—	—	—	—	—
3.0	0.04 ± 0.03 (5)	—	—	—	—	—
3.7	—	0.02 ± 0.01 (4)	0.10 ± 0.07 (6)	—	—	—
4.6	0.01 ± 0.01 (2)	—	0.01 (1)	—	—	—
6.1	0.13 ± 0.19 (5)	0.04 ± 0.03 (6)	0.08 ± 0.05 (8)	0.05 (1)	—	—
7.6	0.02 ± 0.02 (2)	0.06 ± 0.03 (6)	0.10 ± 0.11 (6)	—	—	0.15 (1)
9.1	0.03 ± 0.03 (5)	0.04 ± 0.04 (4)	0.30 ± 0.41 (4)	0.42 (1)	—	—
Titanium						
0.015	—	0.06 ± 0.06 (4)	0.12 ± 0.10 (5)	—	—	—
0.915	—	0.02 ± 0.01 (5)	0.03 ± 0.01 (6)	0.13 (1)	0.07 ± 0.21 (5)	—
1.5	0.02 ± 0.02 (3)	—	0.04 (1)	—	—	—
1.8	0.08 ± 0.15 (7)	0.04 ± 0.03 (5)	0.05 ± 0.04 (6)	—	—	—
2.3	0.02 (1)	—	—	—	—	—
3.0	0.01 ± 0.01 (6)	—	—	0.03 (1)	—	—
3.7	—	0.01 ± 0.01 (6)	0.02 ± 0.01 (6)	—	—	—
4.6	0.004 ± 0.003 (3)	—	0.005 (1)	—	—	—
6.1	0.03 ± 0.06 (5)	0.01 ± 0.01 (7)	0.01 ± 0.01 (7)	0.01 (1)	—	0.03 (1)
7.6	0.005 ± 0.006 (2)	0.03 ± 0.03 (4)	0.02 ± 0.01 (7)	—	—	0.04 (1)
9.1	0.02 ± 0.02 (5)	0.03 ± 0.02 (5)	0.10 ± 0.15 (6)	0.03 (1)	—	—

[a]Number of observations in parentheses; concentrations are $\mu g\ m^{-3}$ ambient.

Junge [61-63] used a cascade impactor to collect samples of particles from the troposphere over Europe and the United States. He separated two particle size ranges, 0.08–0.8 and 0.8–8 μm radius, representing approximately the large and giant particles, respectively. Analyses were made for the ions ammonium, sodium, magnesium, sulfate, chloride, nitrate, and nitrite, using micro techniques, mainly color reactions [1]. Samples collected in and near Frankfurt contained ammonium ion and sulfate. These ions were an important constituent of the large particles, and the concentration ratio was about that for $(NH_4)_2SO_4$. Only small amounts of ammonium ion were in the giant particles, on the other hand, and apparently much of the sulfate was bound by other cations. Chloride was present in both size ranges.

Junge pointed out that when fresh maritime air passed over central Europe, the chloride concentration increased for the giant particles and decreased for the large particles, suggesting that the giant sea-salt particles can penetrate far inland, and that chloride particles smaller than 0.8 μm are formed over land and accumulate in continental air masses. Similar results were obtained on the east coast of the United States about 50 miles south of Boston. The region is rural in contrast to the sampling sites in central Europe, although, as Junge points out, the air near Boston is still to some extent influenced by the big cities and industrial activities in that region. The sodium and chloride ions were found almost exclusively in the giant particles, even when the wind came directly from the ocean, only about 100 m away. Junge commented that the sea salt component of airborne particles can be clearly distinguished from the components of continental origin and that the influence of the latter can also be traced to the center of large oceans.

Atmospheric precipitation (rain, snow, and hail) washes large amounts of particulate gaseous material from the air and thus provides another means of studying various trace constituents of the atmosphere. Although the relationships between the concentrations of constituents in air and in precipitation are not known quantitatively, the concentration patterns of trace atmospheric constituents in precipitation do appear to reflect their patterns in air. From 1960 to 1966, a precipitation sampling network was operated throughout the United States. It consisted of 33 stations, each of which was collected in a polyethylene bucket was transferred after each open only during periods of precipitation, thereby excluding dry fallout from the sample. The moisture-sensing grid which activated the lid was thermostated so that snow and sleet as well as rain were collected. The precipitation which was collected in a polythylene bucket was transferred after each precipitation period to a polyethylene bottle which was sent to the National Center for Atmospheric Research or, during the early period of the operation of the network, to the U. S. Public Health Service for analysis. The polyethylene bottles contained a small amount of toluene to minimize changes resulting from biological activity in the samples. Analyses were made for total fixed inorganic nitrogen, sulfate, chloride, ammonia, and a number of metal

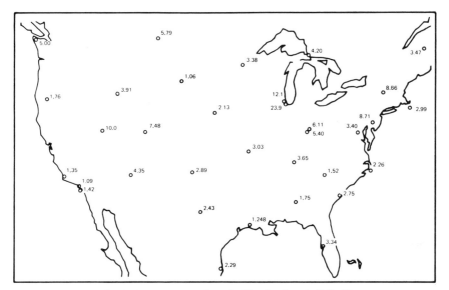

Fig. 5. Mean concentration of sulfate ion (ppm), 1963–66.

ions, including potassium, sodium, calcium, magnesium, zinc, iron, manganese, copper, lead, and cadmium [64,65].

Sulfate in precipitation water may result from entrainment of particulate sulfate or from absorption and solution of sulfur dioxide with subsequent oxidation to sulfate as explained above. If the two processes occur in the cloud itself, they are referred to as "rainout"; if they occur as the rain drops fall from the clouds to earth, they are referred to as "washout."

A map of the average monthly concentrations of sulfate in precipitation (Fig. 5) suggests the presence of regions of high concentration in arid areas of the West, the Midwest, and the Northeast. Lodge et al. [64] suggest that the high concentration in the arid areas originates from soil dust carried aloft and entrained in rain water. They point out that calcium carbonate and calcium sulfate occur in high concentrations in the top soil layers of arid zones and that the calcium carbonate may be converted to calcium sulfate by reaction with sulfuric acid droplets. It must be emphasized, however, that a weakness in comparisons based on analyses of rainfall is that the average concentrations tend to vary inversely with the quantity of precipitation. Furthermore, evaporation of rain drops as they fall, especially in arid atmospheres, further contributes to uncertainties in the interpretation of the analyses.

The giant sea-salt particles which seem to constitute such a large part of the sodium chloride content of the atmosphere over continents are probably incorporated into droplets by rainout with almost 100% efficiency. As Lodge et al. [64] point out, efficient rainout combined with efficient

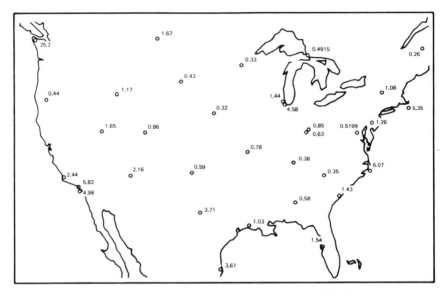

Fig. 6. Mean concentration of chloride ion (ppm), 1960–66.

transport of sea-salt particles by advection of maritime air inland is partially responsible for maintaining relatively high chloride levels in precipitation over the North American continent as shown in Fig. 6. Of course, anthropogenic and soil contributions make up part of the chloride content of the continental atmospheres. As for sulfates, the high chloride values in the arid regions may be caused by airborne soil which is washed out by the precipitation. The sodium chloride content of soil in the Southwest is quite high because of salt flats and deposition of sea salt transported from the Gulf of Mexico. Since the rainfall rate is low, the dissipation by leaching of such salts from the surface is minimized. Similar to observations in Europe, high concentrations of chloride in the United States were found in areas of heavy industrialization. The chloride probably is largely generated by combustion of fuels. This was confirmed by the seasonal variations, which indicated that high concentrations of chloride occurred during the winter months in urban areas. The chloride-to-sodium ratio by weight in sea water is about 1.8. This ratio was very high for industrialized areas, which further indicates anthropogenic sources of chloride. The highest value was about 2.9 in the vicinity of Chicago, whereas values as low as 0.83 were found in the western United States. The excess of sodium is believed to have been contributed by airborne soil particles. The values obtained at coastal stations were close to the sea water ratio of 1.8.

The nitrate and the ammonium ions in rain water were originally partially gaseous and partially particulate in nature. Nitrogen dioxide dissolves readily in water to form nitrous acid and nitric acid, and the nitrous acid may be oxidized to nitric acid. Concentrations of nitrogen dioxide are much higher in industrial areas than in nonurban areas; values as high as 80 μg m^{-3} have been observed in Los Angeles. The concentration of ammonia in the air over the continents varies from 4 to 20 μg m^{-3}, but may be as low as 2 μg m^{-3} over the oceans. Such ammonia dissolves in rain drops to form ammonium ion. The other source of ammonium ion in rain drops is, of course, ammonium salts in the atmosphere. The average concentrations of ammonia in rain water were highest in the industrialized areas of the East and concentrations, calculated as inorganic nitrogen, varied from about 0.01 ppm by weight near the coast of Oregon to over 5 ppm in the Chicago area.

Sodium and potassium in precipitation originate largely from the ocean and from the earth's surface. The ratio of sodium to potassium in sea water is 27.8. However, even at coastal stations, the highest ratio found in precipitation by Lodge et al. [64] did not reach this value, the highest value found being 13.2 on the Atlantic coast. Inland values were much lower, generally between 1 and 3.

The ratio of magnesium to calcium in sea water is 3.18. At Nantucket and at Cape Hatteras, the ratios in particles were less than one-half of this value. These data are consistent with an enrichment of calcium in these maritime areas by material originating inland. The ratio of magnesium to calcium decreased inland, which gives further evidence for a continental source for the calcium. This ratio was relatively constant for inland stations, which suggests that the bulk of the magnesium found inland, as well as the calcium, is of mineral origin.

TABLE 6. Particulate Concentrations for Nine Nonurban Stations, 1958*

Station location	Suspended particulate, A μg m^{-3}	Benzene-soluble organic matter μg m^{-3}	% of A	Sulfate μg m^{-3}	% of A	Nitrate μg m^{-3}	% of A
Acadia National Park, Me.	27	2.5	9.3	5.6	21	0.9	3.3
Baldwin Co., Ala.	27	2.7	10	3.5	12.9	0.9	3.3
Bryce Canyon Park	83	2.2	2.7	1.9	2.3	0.4	0.48
Butte Co., Idaho	23	1.4	6.1	2.2	9.6	0.6	2.6
Cook Co., Minn.	44	2.4	5.5	4.0	9.1	0.5	1.1
Florida Keys	36	2.0	5.6	4.7	13.1	1.2	3.3
Huron Co., Mich.	44	1.6	3.6	6.5	14.8	1.5	3.4
Shannon Co., Mo.	37	2.0	5.4	6.5	17.5	1.8	4.8
Wark Co., N. D.	28	1.9	6.8	3.2	11.4	0.9	3.2

*Values are arithmetic means.

The concentrations of lead, zinc, copper, iron, manganese, and nickel in rain water were determined by atomic absorption. Values for each station averaged over approximately six months during 1966 and 1967 indicated that human activity was the primary source of these materials in atmospheric precipitation. A strong positive correlation was observed between the concentration of lead in precipitation and the amount of gasoline consumed in the area in which the sample was collected.

The monitoring network of the U. S. Public Health Service determined particle concentrations for nonurban stations as well as for urban ones. Some of the results are shown in Table 6. Note that as much as 10% of the total weight of particles consisted of benzene-soluble organic material. Part of this material undoubtedly originated from combustion, either natural or man-made, although some of it may result from photochemical reactions involving organic vapors that are given off by trees and other plants. A third source for this organic material is the oceans, i.e., some of it may have been associated with sea salt particles as described earlier in this chapter. It is noteworthy that most organic plant material, consisting largely of cellulose, is not soluble in benzene.

An investigation of the organic carbon and nitrogen associated with particles in the air over Sweden was undertaken by Neumann et al. [66]. Organic nitrogen was defined as nitrogen which distilled as ammonia in the Kjeldahl procedure from which had been subtracted the free ammonia as determined by an alkaline distillation. They defined total nitrogen as free ammonia nitrogen plus organic nitrogen.

One test series consisted of samples of rain and snow collected at nine stations in Sweden over a period of four months. Concentrations of organic carbon were constant or nearly constant over the sampling region and averaged about 160 mg carbon m^{-2} month^{-1}.

Analysis of a second series of rain and snow samples indicated that the ratio of organic carbon to total nitrogen was nearly constant regardless of the sampling location. Between 25 and 100% of the total ammonia was free ammonia, possibly indicating a varying degree of decomposition of the organic substances in the samples. These results suggested that there is a very extended source for organic carbon in the atmosphere and that this source is the layer of organic material on the sea surface. The authors also speculated that the ocean is a source of gaseous ammonia, nitrogen-containing organic material first being introduced into the air on particles consisting largely of sea salt followed by decomposition of the organic substances liberating gaseous ammonia.

Goetz [67] developed a method for determining the presence of reactive hydrocarbons in the ambient air that might be expected to undergo photo-oxidation to form particles. His method consists in converting such hydrocarbons into aerosol particles through partial photochemical oxidation by exposing a steady stream of air briefly to ultraviolet radiation. The particles are then collected with a moving slide impactor in order to determine the stability of the particles during exposure to prolonged irradiation, higher temperature, etc., which cause volatilization of the organic compounds into gaseous products. The application of this technique has demonstrated the existence of organic particle precursors both over continents and over the oceans. At the same time, particles that were shown by their instability to be organic were collected from the unirradiated gases.

2.3. Particle Concentrations and Size Distributions

Until recently, most of the knowledge concerning the concentrations and size distributions of particles in various parts of the atmosphere has been the result of work undertaken by Junge and his co-workers [1]. Figures 4 and 7 are plots of $dN/d(\log r)$ per cubic centimeter against the radius r in microns on a log-log scale, where N is the number of particles. Figure 4 is for continental air; Fig. 7 is for maritime air. These curves are, of course, averages for a large number of measurements and have been smoothed. Actual particle size distribution at a given time and place may differ some-what from these curves. Such curves yield not only the particle size distribution, but also the concentrations within various size ranges as described earlier.

During recent years, the size distribution and the concentration of atmospheric aerosol particles have been the subject of many investigations,

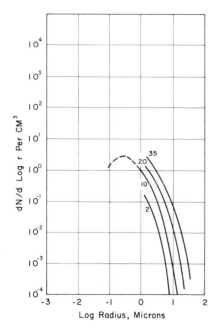

Fig. 7. Particle size distribution in maritime air. The numbers on the curves refer to air speed in miles per hour. (U. S. Air Force, "Handbook of Geophysics," New York, Macmillan, 1960).

including those of Pasceri and Friedlander [68], Clark and Whitby [69], and Blifford and Ringer [70]. These last investigators made a series of 22 aerosol particle collections in 1966 at altitudes to about 10 km in the vicinity of Scottsbluff, Nebraska. Additional samples were taken at other locations in the central United States when low tropopause conditions permitted collecting particles from the lower stratosphere. A few additional measurements were made over Limon, Colorado. Samples were collected with impactors from an aircraft. The collecting site at Scottsbluff, Nebraska was selected because of its convenience to the National Center for Atmospheric Research laboratory at Boulder, Colorado; its geographical location, which is approximately midway between the paths of cyclonic systems from the Arctic and the Gulf; and the altitude capability and range of aircraft. It is in the Great Plains of the United States and remote from most sources of industrial pollution.

Some of the results obtained are shown in Fig. 8 plotted, for ease of comparison, in the same manner as Figs. 4 and 7. Variations in both concentrations and typical shapes of the distribution curves as a function of altitude were observed.

To investigate hourly variations in the total particle concentration, three samples were taken near Limon, Colorado for successive 60-min periods at altitudes of 4.9, 6.7, and 8.5 km using two aircraft. The results suggested that variations by a factor of at least four may not be uncommon. The results do not, of course, differentiate between temporal and spatial variations (Table 7).

Several studies have been made of the variations in concentrations and size distributions of Aitken particles. For example, concentrations of Aitken particles in mountain regions have ranged between 0 and 155,000 particles cm^{-3} [71].

Twomey and Severynse [72] have investigated the rate of coagulation of such particles by collecting air samples in large metallized plastic bags and passing the samples through the particle size measuring device called the diffusion battery. The results indicated that most natural airborne particles are initially about 0.01 μm in radius and rapidly increase in size by agglomeration. The mode in the size distribution of the particles was about 0.08 μm radius. The investigators pointed out that this is about the size that airborne particles have the longest existence since they do not settle rapidly and are too small to be washed out effectively by rain, but are large enough to remain relatively unaffected by diffusion. When their data were plotted as in Figs. 4, 7, and 8, the slope for particles larger than 0.1 μm in radius was -2 to -2.5 rather than -3.

Little is known about the concentrations and size distributions of particles in polar regions. The results obtained by Cadle et al. [59] for the antarctic atmosphere using a cascade impactor showed that almost all of the particles were collected on the third and fourth stages of the impactors; most of the particles collected must have had diameters between about 0.2 and

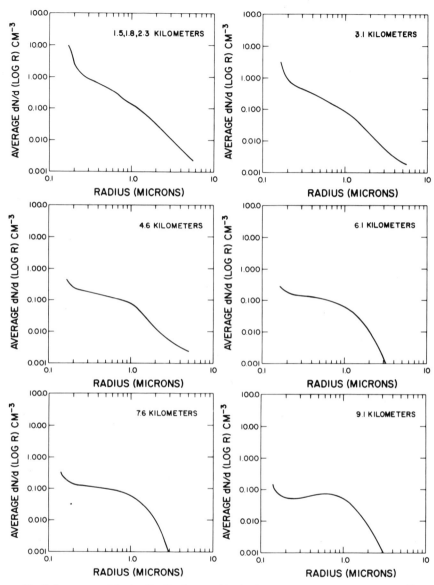

Fig. 8. Average particle size distributions at altitudes from 1.5 to 9.1 km over Nebraska [70].

2 μm. The former figure is about the lower limit for high collection efficiency by the impactor.

This conclusion was substantiated by electron micrographs of the collected particles. The number concentrations of the particles varied from about 0.1 cm^{-3} (upper Taylor Valley, Ross Ice Shelf) to 1 cm^{-3} (5000-ft level

TABLE 7. Short-Term Variations in Total Particle
Concentration for Particles with Radii in the Range 0.13–
5.5 μm on 6 September 1967 over Limon, Colorado

Time, GMT	Altitude, km	Particle concentration, cm⁻³
2340	4.9	0.14
2440	4.9	0.056
2540	4.9	0.047
1848	6.7	0.087
1948	6.7	0.11
2048	6.7	0.027
1906	8.5	0.26
2006	8.5	0.065
2106	8.5	0.11

on Mt. Discovery). At McMurdo Station, the air contained much larger concentrations of particles, largely dark-colored, that almost certainly consisted mainly of the powdered lava from the ground in that region.

Fenn et al. [58], whose studies of particles in the atmosphere above the Greenland icecap were mentioned earlier, found that the concentrations of the large particles collected with the impactor over Greenland varied from less than 1 to about 5 particles cm⁻³. All the particles were collected on the fourth stage of the impactor, so they could not have been larger than about 1.5 μm in diameter. Concentrations of Aitken particles were also measured and were much more variable, ranging from almost 0 to about 1500 cm⁻³.

More recent investigations of the nature of the tropospheric aerosols by Blifford [73] involved flights made over a period of 14 months at a location 250 km west of Santa Barbara, California and at Death Valley, California. He also collected samples in the central Pacific Ocean near Palmyra and at Barbados in the Atlantic. The individual size distributions in all of these places were quite variable, but on the average, showed a consistent trend toward uniformity with increasing altitude similar to the findings at Scottsbluff. At the lower altitudes over the oceans, the typical sea salt distributions were found similar to that observed by Junge. However, at altitudes above a few kilometers, the size distribution over the ocean became similar to the distribution over the midcontinental United States. Over the Pacific, a pronounced minimum was observed in the particle concentrations at about an altitude of 3 km which was attributed to a strong inversion that appeared to restrict the upward transport of the maritime aerosols. Blifford observed high particle concentrations at Barbados at altitudes up to about 6 km and suggested that the high concentrations resulted from large amounts of Sahara dust at that location. The systematic change in size with increasing altitude was explained on the basis that the upper tropopause contains aged particles whose distribution has been modified by chemical and physical processes. Coagulation and chemical growth, for example, will tend to reduce the number of very small particles, whereas sedimentation and rainout remove the larger ones. Blifford also suggested that repeated cycles of condensation and evaporation without rainout may enhance the middle range of the particle sizes considered.

A number of attempts have been made to describe the size distribution of particles in the atmosphere with theoretical models. The results have only been partially successful. One problem probably lies in the simplifying assumptions and another is that the size distributions seldom look precisely like those plotted in Figs. 7 and 8.

This subject has recently been reviewed by Hidy [74]. Earlier theories relied only on Brownian coagulation and sedimentation to control the evolution of the size distribution [75], but these two mechanisms are insufficient to account for tropospheric aerosol behavior. Processes such as turbulent coagulation, washout by rain, or rainout by cloud droplets must also be considered.

Friedlander [76] has postulated that a steady state may occur in the aerosols in the atmosphere. According to his concept, similarities in shape of size distributions result from a balancing between particles entering the lower end of the distribution that are undergoing Brownian coagulation, and the removal of particles from the other end of the distribution by sedimentation. A similarity hypothesis, somewhat different from the steady-state suggestion, was later suggested for an aerosol cloud undergoing coagulation [77]. Junge [78], however, has argued that neither of these approaches can take place for aerosols, at least in the lower troposphere. Instead, the apparent regularity of the particle size distributions for tropospheric aerosols is merely the result of the statistics in mixing particles together from many sources. The contributions from each source may have nearly the size distribution observed for the atmospheric aerosol and when mixed together, they produce the observed atmospheric particle size distribution. Hidy [74] points out that if the tropospheric aerosol spectrum begins in a self-preserving form, it may retain this shape by a quasistationary mechanism in the atmosphere. He states that under such circumstances, a simple model may be deduced from the variation of Aitken nuclei with altitude and that this model agrees satisfactorily with the very limited data available in the literature.

2.4. Mechanisms of Removal and Residence Times

A number of mechanisms are responsible for the removal of particles from the atmosphere and most or all of these have already been mentioned. Sedimentation is a very important mechanism for the removal of very large particles, but small particles settle so slowly that other mechanisms are much more important. When particles are smaller than 1 or 2 μm in radius, sedimentation, in the troposphere at least, can in general be ignored.

Atmospheric convection on various scales is another important mechanism for particle removal and is one that is often overlooked. Convection is especially important in connection with other processes of particle removal since it may bring the particles into some region of the atmosphere where other mechanisms are operating. For example, convectional processes can bring airborne particles into rainstorms where the particles may be removed by rainout or washout. The effectiveness of removal by precipitation varies considerably with particle size. Small particles diffuse more rapidly than larger ones and agglomerate quite rapidly with cloud droplets which may eventually be incorporated in rain. On the other hand, larger particles are more rapidly removed by falling rain drops than are smaller ones. Coalescence of particles among themselves, of course, plays an important role in the mechanism of removal by converting large numbers of small particles into much smaller numbers of relatively large ones. Coalescence of particles can occur in several ways. Perhaps the most important is Brownian diffusion, but others include the scavenging of smaller particles by larger ones during their fall and collisions between particles during atmospheric turbulence.

Impaction of particles on obstacles near the earth's surface is another removal mechanism, but knowledge of its effectiveness is only theoretical. Wind velocities and rates of vertical mixing play an important role in this mechanism. Once particles are brought near a surface by atmospheric mixing, diffusional processes may control the rate of deposition. This diffusion may be due to Brownian motion, but other mechanisms such as thermophoresis or diffusiophoresis may also be important. Thermophoresis results from the force that exists on a particle in a thermal gradient and is directed toward the region of lower temperature. Diffusiophoresis results from the force exerted on a particle by diffusing molecules of a vapor close to a point at which the vapor is condensing. Obviously these diffusion mechanisms are only important when the particles are brought very close to a surface.

Numerous methods have been used to estimate so-called residence times for particles in the troposphere. The use of the term residence time, however, is somewhat misleading since most of the methods are designed to determine the time required to decrease the concentrations of particles to $1/e$ of the initial value. Thus, if one considers a given amount of material introduced at a given moment, much of this material will still remain suspended in the atmosphere much beyond the so-called residence time. Some of the earliest estimates were based on measurements of the ratios of short- to long-lived decay products of radon. These yielded residence times of one to six days, but the results are not very reliable for reasons discussed by Junge [1]. More recently, a number of investigations of residence times of particles have been made using atmospheric properties such as the ratio in rain water of Be^7/P^{32}, the behavior of fission products from Nevada nuclear tests, and the ratio of

RaF/RaD in rain water. Such techniques yielded values of 17–40 days. Martell and Poet [79] have attempted to obtain improved information on the residence times of aerosols at various levels of the troposphere and lower stratosphere by measuring Pb^{210}, Po^{210}, Bi^{210}, and Sr^{90} in large air filter samples. Their results suggest residence times of one to two weeks, at least for the midcontinental United States, where most of their measurements were made.

3. THE LOWER STRATOSPHERE

The stratosphere lies immediately above the troposphere and is the region of nearly uniform or rising temperature, extending upward to about 50 km. The layer of demarcation between the troposphere and the stratosphere is known as the tropopause and is usually a region occupying a narrow altitude range. The tropopause is generally defined in terms of rate of temperature change. The temperature of the stratosphere is largely controlled by radiation, whereas that of the troposphere is governed by mixing and convection. Tropopause air is in part heated by contact with the surface of the earth, which is warmed by radiation and in part by the condensation of water vapor to form clouds. The stratosphere is generally much drier than the troposphere and has few of the rainstorms so characteristic of the troposphere, although large thunderstorms may penetrate the troposphere into the lower stratosphere. For the most part, the stratosphere is less turbulent than the troposphere, although severe turbulence is occasionally observed by high-flying aircraft.

3.1. The Sulfate Layer

Our knowledge of stratospheric particles has recently been reviewed by Rosen [80]. He has pointed out that twilight effects considered to be caused by the presence of dust in the stratosphere have been noted in the literature for more than 80 years. On the basis of a study of the purple light in the twilight, Bigg [81] concluded that a scattering layer existed in the lower stratosphere at about 15–20 km altitude. Similar studies have been made by Volz and Goody [82] and Volz [83].

Another indirect method which has been very useful for studying particles in the stratosphere has been with laser radar ("lidar"). The first of these studies was published by Fiocco and Grams in 1964 [84]. Many similar studies followed; for example, Schuster [85] recently studied tropospheric and stratospheric aerosol layers by lidar and found marked stratification shown by variations of the optical back-scattering to a height of 80 km. Short-lived layers with thicknesses and spacings of 30 m were routinely found throughout the range of observation. He observed a well-defined layer of temporal stability of several hours at the tropopause. He also observed that the 20-km layer actually is subject to variations in altitude, thickness, and apparently concentration. Quasistable layers were found to exist between 25 and 40 km, including one at 36 km which has been postulated to be responsible for the blue and white bands observed in photographs by astronauts.

Several studies have been made of particle concentrations and size distributions in the stratosphere as a function of altitude. Rosen [80,86,87], who made several such studies, used a photoelectric particle counter flown on a balloon. This counter was essentially a dark-field microscope equipped with a light source and a photomultiplier tube as a detector. The sample was of such a size that no more than one particle was likely to be in focus at a given moment. Thus the pulse height was a function of particle size and refractive index. The output from the photomultiplier was fed into a two-channel pulse height discriminator so that one channel counted all particles of diameter greater than $0.5\,\mu m$ and the other all particles of diameters greater than $0.75\,\mu m$. In some flights, the discriminator level was set to detect particles of about $0.2\,\mu m$ diameter and larger. The results indicated that concentrations of particles are considerably higher in the equatorial stratosphere than in the high-latitude stratosphere. The number concentrations of particles increased with increasing altitude above the tropopause to a maximum concentration at about 20 km near Panama and at about 16 km at midlatitudes. Nonetheless, these vertical profiles were highly structured indicating the presence of many layers rather than just a single layer of particles.

Junge et al. [88] undertook a field study of stratospheric aerosols up to 30 km altitude and Junge [89] investigated the vertical profiles of Aitken nuclei in the upper troposphere and stratosphere. Unlike that for larger particles, the concentration of the Aitken nuclei dropped slowly and irregularly with increasing altitude throughout the troposphere and stratosphere. A few kilometers above the tropopause, the concentration was a few particles per cubic centimeter and remained at this concentration to the peak elevation reached by the balloon (about 30 km). Again, considerable vertical structure was observed.

Junge et al. [88] also determined vertical profiles for particles larger than $0.1\,\mu m$ radius. They used a balloon-borne air sampler which collected solid particles from the air with jet impactors. Their results, like those obtained by Rosen using counters, indicated that a peak in the particle concentration existed in the stratosphere between 16 and 22 km. Thus the profiles for particles exceeding $0.1\,\mu m$ radius were very different from those for the Aitken nuclei.

Junge and his co-workers [88,90] were the first to study the chemical composition of the particles in the size range $0.1-1.0\,\mu m$ radius that are found at about 20-km altitude. An electron microprobe analyzer was used to determine the relative concentrations of various elements. A number of elements were detected among the stratospheric particles, but of these, only sulfur was present to an appreciable extent. The particles were hygroscopic, which suggests that they were soluble salts, probably sulfate. Nickel was not detected, indicating that little of the particulate material was extraterrestrial. Their suggestion that the sulfur was in the form of sulfate led to the term "sulfur" or "sulfate layer" which is commonly given to this layer of particles in the stratosphere. Friend [91] also studied the chemical composition, size distribution, and concentration of stratospheric aerosol particles. He collected particles with an impactor and determined the composition from electron diffraction patterns of single particles or a few small particles. The results suggested that most of the particles consisted of ammonium sulfate and ammonium persulfate. Friend stated that the sulfate particles were to a large degree flat rosettes and often were accompanied by haloes of very small particles.

Studies at the National Center for Atmospheric Research have revealed that rosettes accompanied by haloes of small particles were originally sulfuric acid droplets which may, of course, have contained dissolved impurities.

TABLE 8. Analyses of Particles Collected on IPC Filters in the Tropical Stratosphere above Central America*

At start of sampling		Alti-	Concentrations, $\mu g\ m^{-3}$ ambient							Radio activity ^{95}Nb Dpm m^{-3}
Latitude, N	Longi- tude, W	tude, ft × 10^{-3}	$SO_4^=$	Si	Na	Cl	NO_3^-	Mn	Br	ambient
31°40'	99°30'	55	0.11	0.052	0.020	0	0.058	0	0.0025	3.09
30°00'	96°05'	55	0.12	0	0	0	0.042	0	0.0014	2.39
28°00'	93°20'	55	0.13	0.022	0.025	0	0	0	0.0026	6.09
25°30'	90°30'	55	0.075	0	0.037	0.038	0	0	0.0020	0.69
23°00'	88°00'	56	0.10	0.058	0.004	0	0.16	0	0.0019	1.49
20°30'	85°20'	56	0.26	0	0.015	0.034	0	0	0.0030	1.73
19°20'	83°30'	57	0.092	0	0	0	0	0	0.0020	1.19
18°30'	80°00'	58	0.30	0.008	0.015	0	3.4	0	0.0068	8.77
11°00'	79°00'	58	0.11	0.015	0.049	0.017	0.20	0.00047	0.0030	4.84
8°00'	80°00'	58	0.20	0.0057	0	0	0.17	0	0.0018	3.18
4°00'	81°20'	58	0.17	0.040	0.053	0.053	0.058	0.00047	0.0015	1.57
2°00'	82°30'	58	0.14	0.025	0	0	0.049	0	0.0009	1.54
1°30'	83°30'	58	0.32	0.010	0	0.002	0.058	0.00002	0.0021	2.19
3°00' (S)	85°00'	62	0.13	0	0	0	0.021	0	0.0006	1.05
1°00' (S)	82°00'	62	0.29	0.0062	0.002	0.018	0.015	0.00004	0.0015	1.65

*All samplings were 30 min long. Longitudes and latitudes are for start of sampling. No nitrite, potassium, or ammonium ion was detected in any of the samples, and the NO_3^- concentration was calculated by the method specific for NO_3^-. The upper set of samples was obtained on 7/10/69 and the lower set on 7/23/69.

Thus Friend's results suggest that in addition to ammonium sulfate and ammonium persulfate, sulfuric acid droplets were present.

Cadle *et al.* [51] made a number of analyses of stratospheric air by methods similar to those that they used for analyzing particles in the high tropical troposphere. The Air Weather Service of the U. S. Air Force operated RB-57 F aircraft equipped to carry 12 42-cm-diameter filters and expose them sequentially in flight as described earlier. Routine flights by these planes out of Kirtland Air Force Base, flying at altitudes of about 18 km, permitted Cadle and his co-workers to collect on filters relatively large amounts of stratospheric particles and analyze them much more completely than previously.

Twenty samples were collected in the tropical stratosphere in the vicinity of Central America during trips to and from Panama and in the vicinity of that country. A number of samples were also collected at mid-latitudes over the United States. The filters used were mostly "IPC" filters [52], but a few samples were obtained for comparison with polystyrene fiber filters. These latter filters consist of submicron-diameter fibers and were prepared in a clean room in these laboratories by the method of Cadle and Thuman [92]. They have the advantage over the IPC filters of being exceptionally free of elements occurring in atmospheric aerosols. All the filters were handled in "clean" or "semiclean" rooms by workers wearing disposable plastic gloves. The analytical methods were the same as those used for the analyses of samples from the high troposphere, except that when polystyrene filters were used, which are not readily wet by water, one-half of each filter was dissolved in methylene chloride and the methylene chloride solution repeatedly extracted with deionized water.

Representative sets of results obtained for the tropics are shown in Table 8. The sulfate concentrations varied markedly during each flight and were usually much higher than those for any other constituent except nitrate, which at times was higher in concentration than the sulfate. Concentration ratios for pairs of constituents differed markedly from those for sea water. The large variations in concentration found for sulfate were also observed for the other constituents. No potassium, nitrite, or ammonium ions were detected on any of the filters.

The analyses of particles collected on IPC and polystyrene filters from the midlatitude stratosphere are shown in Table 9. The results obtained for the two types of filters were very similar except that little or no nitrate was found on the polystyrene filters and the values for silicon were much higher for the polystyrene than for the IPC filters. Ammonium ion was absent from the tropical stratospheric samples but was in the samples collected by both IPC and polystyrene filters in the midlatitude stratosphere. The chlorine concentrations were high for control IPC filters but extremely low for polystyrene filters. Therefore the Cl/Br ratios obtained with polystyrene filters are considered to be especially reliable. It is noteworthy that the

TABLE 9. Analyses of Particles Collected on IPC and Polystyrene Filters in the Midlatitude Stratosphere*

Filter Type	Latitude, N	Longitude, W	Altitude, ft × 10^{-3}	Concentrations, $\mu g\ m^{-3}$ ambient								Radioactivity (^{95}Nb) Dpm m^{-3} ambient
				$SO_4^{=}$	Si	Na	Cl	(NO_3^{-})†	NH_4	Mn	Br	
PS	34°30'	102°55'	55	0.32	0.18	0.004	0.042	0(0.0012)	0.0034	0.0036	0.0021	28.4
PS	37°20'	102°30'	58	—	0.19	0.003	0.023	—	—	0.0021	0.0019	37.5
IPC	40°20'	102°15'	58	0.21	0.037	0.054	0.071	0.31	0.026	0.0010	0.0028	40.9
IPC	43°25'	101°50'	59	0.37	0.035	0.030	0.052	0.41	0.0089	0.0009	0.0021	29.2
PS	46°30'	101°31'	59	0.24	0.19	0.003	0.041	0(0)	0.017	0.0025	0.0026	42.6
PS	47°48'	101°45'	60	0.22	0.17	0.002	0.030	0(0)	0.012	0.0012	0.0020	43.2
IPC	44°25'	101°45'	61	0.35	0.084	0.050	0.088	0.36	0.0067	0.0009	0.0030	49.3
IPC	41°00'	102°10'	60	0.32	0.031	0.001	0.046	0.35	0.0040	0.0004	0.0020	32.3
PS	37°30'	102°40'	60	0.36	0.17	0.003	0.051	0(0.0036)	0.043	0.0049	0.0024	59.7

*All samplings were 30 min long, made on December 4, 1969. Longitudes and latitudes are for start of sampling. No potassium ion or nitrite ion was detected in any of the samples.
†The NO_3^{-} concentration was calculated from the total combined nitrogen less that combined as NH_4^{+} (the value in parentheses) and also by the method specific for NO_3^{-}.

Cl/Br ratios obtained with the latter filters varied from 12 to 19 compared with 300 for sea water.

The relatively high concentrations of nitrate found in most of the samples collected with the IPC filters were surprising, especially in view of the low concentrations of ions other than H^{+} with which it may be associated. The results suggest that the nitrate exists largely as HNO_3 vapor which was absorbed by the filters. This apparent finding of high nitrate concentration is consistent with the recent spectroscopic discovery of HNO_3 vapor and of NO_2 in the stratosphere [93,94]. The much higher amounts of silicon, which presumably were in the form of silicates, on the polystyrene than on the IPC filters may be due to a higher collection efficiency of the polystyrene filters for such particles.

The sulfate concentrations are about 30-fold higher than those obtained by Junge and his co-workers, but in general agree with results of total particle concentrations obtained in more recent investigations, e.g., Shedlovsky and Paisley [95]. They also agree with concentrations calculated from Rosen's results. This apparent increase during the last decade has been attributed to an increase in volcanic activity, and especially to the eruption of Agung in Bali in 1963. At least part of this increase came from the Agung eruption in 1963, as discussed earlier in this chapter. However, the Junge collections of stratospheric aerosol particles were by means of impactors, which are notoriously inefficient for particle collections, especially when the particles are as small as those which exist in the stratosphere. Little correlation existed between the sulfate concentrations and those of other constituents, which suggests that the source of the sulfate differs from that of the other constituents. This, together with the relatively low concentrations of metallic cations, agrees with the theory that the sulfate is produced largely by the oxidation of sulfur dioxide in the stratosphere. The variations in the sulfate concentrations indicate the presence of clouds or tilted strata of sulfate particles.

The ratios Cl/Br are much lower than those for sea water (about 300). The low ratios may be caused by complex physical and chemical reactions in the atmosphere, but they might be a result of worldwide air pollution from automobiles, the bromine being produced from the lead-scavenging agents in gasoline.

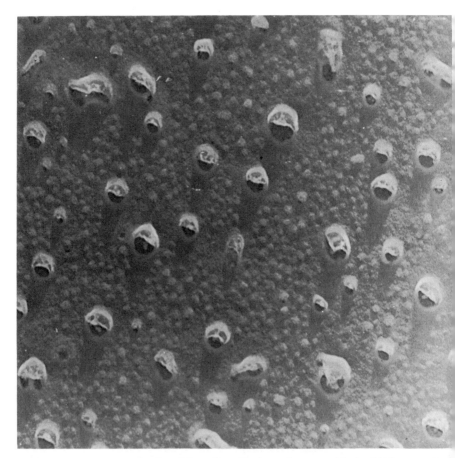

Fig. 9. Electron micrograph of particles collected by impaction at an altitude of 18 km over the mid-United States. Distance across the micrograph is 2 μm.

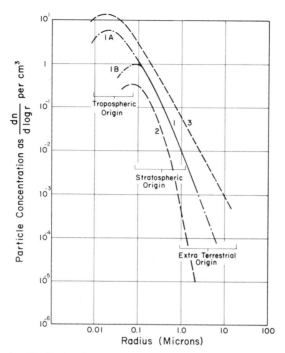

Fig. 10. Average size distribution for stratospheric aerosols. Curve 1A is for the lower strato-sphere and 1B for altitudes above 20 km. Curves 2 and 3 are estimated confidence limits [88]. (*Courtesy of the American Meteorological Society.*)

An impactor similar to that used by Junge but with some improvements to prevent contamination was flown during recent stratospheric flights. Electron microscope grids were mounted on the impactor surface. Figure 9 is an electron micrograph of particles collected in the stratosphere over the middle United States. It appears to represent two sizes of droplets; probably if fewer droplets had been collected, the appearance would be that of a large particle surrounded by much smaller ones such as those observed by Friend. This photograph suggests that the particles consisted largely of very highly acidic droplets, which is in good agreement with the analyses obtained.

Junge *et al.* [88] presented smoothed-out curves for average size distribu-tions for stratospheric aerosols (Fig. 10). During the studies of tropospheric aerosols over Scottsbluff, Nebraska, Blifford observed that some of the higher-altitude samples were taken when standard maps of tropopause height showed that the altitude of collection was as much as several thousand feet above the tropopause. The presence of stratospheric air was confirmed by high concentrations of ^{137}Cs and ^7Be. Some of the results obtained are shown in Fig. 11, which offers a comparison of the size distributions of tropospheric and stratospheric aerosols. Blifford concluded that the aerosol

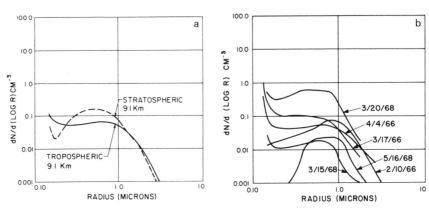

Fig. 11. (a) Average $dN/d(\log r)$ [cm^{-3}] versus particle radius for collections made in stratospheric air and in the troposphere at 9.1 km; (b) $dN/d(\log r)$ [cm^{-3}] versus particle radius for individual collections made in stratospheric air at 9.1 km [70].

distribution functions for stratospheric air seem to follow the general pattern of decreasing relative concentration of both small and large particles with altitude.

A number of suggestions have been made with regard to the sources of the stratospheric particles, especially those in the sulfate layer. The Aitken particles were probably largely transported upward from the troposphere, since their concentration constantly decreases with increasing altitude. The preponderance of evidence indicates that the sulfate was produced in the stratosphere by chemical reactions. Probably sulfur dioxide entering the stratosphere from the troposphere or injected into the stratosphere by violent volcanic eruptions is oxidized to sulfur trioxide which in turn is hydrated to sulfuric acid droplets. Ozone does not react with sulfur dioxide, at least when the sulfur dioxide is in the ground electronic state, and the photooxidation of sulfur dioxide is probably too slow to account for the formation of the sulfuric acid. However, sulfur dioxide does react with atomic oxygen in a three-body reaction. Cadle and Powers [96] have shown that this reaction is sufficiently rapid to account for the formation of sulfate in the stratosphere. Since sulfates also occur in volcanic eruption clouds [97], some sulfate particles may be injected by volcanoes directly into the stratosphere. Radioactivity in the stratosphere seems to be associated with these sulfate particles, probably because of agglomeration of extremely small radioactive particles with the larger sulfate ones. The source of the nitrate, if there really is much nitrate present, is unknown. The silicon and most of the metals presumably originate from the earth's surface. Data obtained by Shedlovsky and Paisley [95] indicate that the stratospheric aerosols contain no more than 10% of meteoritic material, while Junge suggests that only the giant particles in the stratosphere could be of meteoritic origin.

3.2. Mechanisms of Removal and Residence Times

The mechanisms of removal of stratospheric aerosol particles are for the most part sedimentation, turbulent mixing, and exchange of air between the troposphere and the stratosphere by various processes. Some large thunderstorms penetrate into the stratosphere and these can also remove particles by rainout and washout.

The mean residence time of particles in the stratosphere increases considerably with altitude between the tropopause and the top of the stratosphere. Estimates are about one month at the tropopause, about one to two years at 20 km altitude, and about 4–20 years at 50 km altitude. These estimates of stratospheric residence times are based on observations of nuclear bomb debris, natural atmospheric radioactivity, and the results of two high-altitude radioactive tracer experiments. These were a ^{102}Rh tracer produced in a high-altitude thermonuclear explosion in August 1958 and a ^{109}Cd tracer produced in a high-altitude thermonuclear explosion in July 1962. It is noteworthy that all of these radioactive substances had become attached to a considerable extent to particles in the 0.1–1-μm radius size range. It takes about two years for a particle of 0.3 μm radius and a density of about 2.5 to settle from the 20-km altitude region to 10 km. Thus the residence times measured actually relate to the residence times for particles and the residence times for the stratospheric air may be considerably longer than the estimates given.

REFERENCES

1. Junge, C. E., "Air Chemistry and Radioactivity," Academic Press, New York, 1963.
2. Aitken, J., "Collection of Scientific Papers," Cambridge Univ. Press, London, 1923.
3. Randall, R. E., McGill Univ. Dept. of Geography, Climatological Bull. No. 3, pp. 23–35, January 1968.
4. Bierly, W. B., *J. Environ. Sci.* **8**, 15 (1965).
5. Cadle, R. D., "Particles in the Atmosphere and Space," Reinhold, New York, 1966.
6. Mason, B. J., *Nature* **174**, 470 (1954).
7. Day, J. A., in "Proc. Int. Conf. on Cloud Physics, Tokyo and Sapporo," May 1965, pp. 34–36.
8. MacIntyre, F., Ion Fractionation in Drops from Breaking Bubbles, Ph.D. thesis, MIT, September 1965.
9. Eriksson, E., *Tellus* **11**, 375 (1959).
10. Garrett, W. D., *J. Geophys. Res.* **73**, 5145 (1968).
11. Meinel, M. P., and Meinel, A. B., *Science* **142**, 582 (1963).
12. Newell, R. E., *Nature* **227**, 697 (1970).
13. McClaine, L. A., Allen, R. V., McConnell, Jr., R. K., and Surprenant, N. F., *J. Geophys. Res.* **73**, 5235 (1968).
14. Cadle, R. D., Wartburg, A. F., Frank, E. R., and Lodge, Jr., J. P., *Nature* **213**, 581 (1967).
15. Cadle, R. D., and Frank, E. R., *J. Geophys. Res.* **73**, 4780 (1968).
16. Cadle, R. D., Lazrus, A. L., and Shedlovsky, J. P., *J. Geophys. Res.* **74**, 3372 (1969).

17. Naughton, J. J., Heald, E. F., and Barnes, Jr., I. L., *J. Geophys. Res.* **68**, 539 (1963); **68**, 545 (1963).
18. Rittmann, A., "Volcanoes and Their Activity," Wiley, New York, 1962, pp. 38, 153–157; Sapper, K., "Vulkankunde," Engelhorn, Stuttgart, 1927.
19. MacDonald, G. A., Geological Survey Bull. 1021-B, U.S. Gov. Printing Office, Washington, D.C., 1965, pp. 97–100.
20. Shepherd, E. S., *Am. J. Sci.* **235A**, 311 (1938).
21. Robinson, E. and Robbins, R. C., Report No. PR-6755, Stanford Research Institute, Menlo Park, Calif., 1968, pp. 38–45.
22. Mason, B. J., "The Physics of Clouds," Clarendon Press, Oxford, 1957.
23. Neuberger, H., *Mech. Eng.* **70** (3), 221 (1948).
24. Hobbs, P. H., and Rodke, L. F., *Science* **163**, 279 (1969).
25. Warner, J., *J. Appl. Meteor.* **7**, 247 (1968).
26. Warner, J., and Twomey, J., *J. Atmos. Sci.* **24**, 704 (1967).
27. Bowen, E. G., *Nature* **177**, 1121 (1956).
28. Pettersson, H., *Nature* **181**, 330 (1958).
29. Pettersson, H., *Scientific American*, **1960** (February), p. 123.
30. Glasstone, S., "Sourcebook on the Space Sciences," Van Nostrand, New York, 1965.
31. Öpik, E. J., *Irish Astron. J.* **4**, 84 (1957).
32. Fiocco, G., and Colombo, G., *J. Geophys. Res.* **69**, 1795 (1964).
33. Rosinski, J., and Snow, R. H., *J. Meteorol.* **18**, 736 (1961).
34. Rosinski, J., *J. Atmos. Terres. Phys.* **32**, 805 (1970).
35. Rogers, L. A., and Meier, F. C., *Nat. Geog. Soc. Contrib. Papers, Stratosphere Ser.* **2**, 146 (1936).
36. Cadle, R. D., Rubin, S., Glassbrook, C. I., and Magill, P. L., *Arch. Ind. Hyg. Occ. Med.* **2**, 698 (1950).
37. Cadle, R. D., "Micrurgic Identification of Chloride and Sulfate," Monograph No. 3, American Geophysical Union, 1959, pp. 18–21.
38. Whitby, K., Paper presented at the Particulate Workshop conducted at the National Center for Atmospheric Research, Boulder, Colo., August 1970.
39. U. S. Public Health Service Publication No. 637, GPO, "Air Pollution Measurements of the National Air Sampling Network," Superintendent of Documents, Washington, D. C., 1958.
40. Zimmer, C. E., Tabor, E. C., and Stern, A. C., *J. Air Poll. Control Assoc.* **9**, 136 (1959).
41. Am. Ind. Hyg. Assoc., "Air Pollution Manual, Part I—Evaluation," Detroit, 1960.
42. U. S. Department of Health, Education, and Welfare, Public Health Service, Consumer Protection and Environmental Health Service, National Air Pollution Control Administration Publication No. AP-49, "Air Quality Criteria for Particulate Matter," Washington, D. C., 1969.
43. Sawicki, E., Elbert, W. C., Hauser, T. R., Fox, F. T., and Stanley, T. W., *Am. Ind. Hyg. Assoc. J.* **21**, 443 (1960).
44. Steffens, C., and Rubin, S., in "Proc. First Nat. Air Pollution Symp.," Stanford Research Institute, Menlo Park, Calif., 1949.
45. Robinson, E., and Ludwig, F. L., Final report, Project PA-4788, Stanford Research Institute, Menlo Park, Calif., 1964; Robinson, E., Ludwig, F. L., DeVries, J. E., and Hopkins, T. E., Final report, PA-4211, Stanford Research Institute, Menlo Park, Calif., 1963.
46. Urone, P., Lutsep, H., Noyes, C. M., and Parcher, J. F., *Environ. Sci. and Technol.* **2**, 611 (1968).
47. Johnstone, H. F., and Coughanowr, D. R., *Ind. Eng. Chem.* **50**, 1169 (1958).
48. Robbins, R. C., Cadle, R. D., and Eckhardt, D. L., *J. Meteorol.* **16**, 53 (1959).
49. Went, F. W., *Scientific American* **1955** (May), 63.
50. Adams, C. E., Farlow, N. H., and Schell, W. R., *Geochim. et Cosmochim. Acta* **18**, 42 (1960).

51. Cadle, R. D., Lazrus, A. L., Pollock, W. H., and Shedlovsky, J. P., in "Proc. meeting on Tropical Meteorology," American Meteorological Society, Honolulu, June 1970.
52. Friend, J. P., High Altitude Sampling Program VI. HASP Purpose and Methods. DASA Report 1300, 1961.
53. Lazrus, A., Lorange, E., and Lodge, Jr., J. P., "Trace Inorganics in Water," Adv. in Chem. Ser. No. 73, Am. Chem. Soc., Washington, D.C., 1968, pp. 164–171.
54. Brewer, P. G., and Riley, J. P., *Deep-Sea Research* 12, 765 (1965).
55. Duce, R. A., Winchester, J. W., and Van Nahl, T. W., *J. Geophys. Res.* 70, 1775 (1965).
56. Winchester, J. W., and Duce, R. A., *Tellus* 18, 287 (1966).
57. Cadle, R. D., and Robbins, R. C., *Disc. Faraday Soc.* 30, 155 (1960).
58. Fenn, R. W., Gerber, H. E., and Wasshausen, D., *J. Atmos. Sci.* 20, 466 (1963).
59. Cadle, R. D., Fischer, W. H., Frank, E. R., and Lodge, Jr., J. P., *J. Atmos. Sci.* 25, 100 (1968).
60. Blifford, I. H., and Gillette, D. A., Personal communication.
61. Junge, C., *Tellus* 5, 1 (1953).
62. Junge, C., *J. Meteorol.* 11, 323 (1954).
63. Junge, C., *Tellus* 8, 127 (1956).
64. Lodge, Jr., J. P., Pate, J. B., Basbergill, W., Swanson, G. S., Hill, K. C., Lorange, E., and Lazrus, A. L., Final report on the National Precipitation Network, National Center for Atmospheric Research, Boulder, Colo., August 1968.
65. Lazrus, A. L., Lorange, E., and Lodge, Jr., J. P., *Environ. Sci. and Technol.* 4, 55 (1970).
66. Neumann, G. H., Fonselius, S., and Wahlman, L., *Int. J. Air Poll.* 2, 132 (1959).
67. Goetz, A., *Staub.* 29, 357 (1969).
68. Pasceri, R. E., and Friedlander, S. K., *J. Atmos. Sci.* 22, 577 (1965).
69. Clark, W. E., and Whitby, K. T., *J. Atmos. Sci.* 24, 677 (1967).
70. Blifford, Jr., I. H., and Ringer, L. D., *J. Atmos. Sci.* 26, 716 (1969).
71. Landsberg, H., "Atmospheric Condensation Nuclei," *Ergebn. Kosm. Phys.* 3, 207 (1938) (Leipzig Akademische Verlagsgesellschaft).
72. Twomey, S., and Severynse, G. T., *J. Atmos. Sci.* 20, 392 (1963); 21, 558 (1964).
73. Blifford, Jr., I. H., *J. Geophys. Res.* 75, 3099 (1970).
74. Hidy, G. M., The Dynamics of Aerosols in the Lower Troposphere, presented at the Third Conference on Environmental Toxicology, New York, June 1970.
75. Friedlander, S. K., *J. Meteorology* 17, 479 (1960).
76. Friedlander, S. K., The Similarity Theory of the Particle Size Distribution of the Atmospheric Aerosols, in "Aerosols, Physical Chemistry and Applications," K. Spurny, ed., Czechoslovak Acad. Sci., Prague, 1964, pp. 115–130.
77. Swift, D., and Friedlander, S. K., *J. Colloid Sci.* 19, 621 (1964).
78. Junge, C., *J. Atmos. Sci.* 26, 603 (1969).
79. Martell, E. A., and Poet, S. E., Personal communication.
80. Rosen, J. M., *Space Sci. Rev.* 9, 58 (1969).
81. Bigg, E. K., *Tellus* 16, 76 (1956).
82. Volz, F. E., and Goody, R. M., *J. Atmos. Sci.* 19, 385 (1962).
83. Volz, F. E., *Science* 144, 1121 (1964).
84. Fiocco, G., and Grams, G., *J. Atmos. Sci.* 21, 323 (1964).
85. Schuster, B. G., *J. Geophys. Res.* 75, 3123 (1970).
86. Rosen, J. M., *J. Geophys. Res.* 69, 4673 (1964).
87. Rosen, J. M., *J. Geophys. Res.* 73, 479 (1968).
88. Junge, C. E., Chagnon, C. W., and Manson, J. E., *J. Meteorol.* 18, 81 (1961).
89. Junge, C. E., *J. Meteorol.* 18, 501 (1961).
90. Junge, C. E., and Manson, J. E., *J. Geophys. Res.* 66, 2163 (1961).
91. Friend, J. P., *Tellus* 18, 465 (1966).
92. Cadle, R. D., and Thuman, W. C., *Ind. Eng. Chem.* 52, 315 (1960).
93. Murcray, D. G., Kyle, T. G., Murcray, F. H., and Williams, W. J., *Nature* 218, 78 (1968).

94. Goldman, A., Murcray, D. G., Murcray, F. H., Williams, W. J., and Bonomo, F. S., *Nature* **225**, 443 (1970).
95. Shedlovsky, J. P., and Paisley, S., *Tellus* **18**, 499 (1966).
96. Cadle, R. D., and Powers, J. W., *Tellus* **18**, 176 (1966).
97. Cadle, R. D., Lazrus, A. L., and Shedlovsky, J. P., *J. Geophys. Res.* **74**, 3372 (1969).

Chapter 3

REMOVAL PROCESSES OF GASEOUS AND PARTICULATE POLLUTANTS

G. M. HIDY

North American Rockwell Science Center, Thousand Oaks, California
and
California Institute of Technology, Pasadena, California

1. INTRODUCTION

In an age of increasing pressures placed on the earth's resources, there is great concern for the atmosphere's ability to cleanse itself of anthropogenically generated material. Indeed, the earth's atmosphere has been overloaded locally with pollution, particularly over cities, to create rather unhealthy and aesthetically unpleasant living conditions. On larger scales, there is conflicting evidence of man's intrusion into the natural composition of the atmosphere. Perhaps the best evidence for anthropogenically stimulated imbalances between production and removal of trace constituents are the data on the continuing increase of carbon dioxide in the air. A variety of different observations has indicated that the atmospheric CO_2 concentration has increased at a rate of about 0.2% per year during the period between 1958 and 1969 [1]. Other evidence such as that reported for global decreases in visibility is less well documented, and independent evidence of emission surveys does not confirm a serious increase in global haze development by anthropogenic sources [1,6].

The disturbance of the average composition of the atmosphere as we know it depends on upsetting a delicate balance between production of trace materials from a variety of sources and the removal of such substances by diverse mechanisms. If the dynamical processes in the atmosphere cannot rid the air rapidly enough of pollutants, they will increase in concentration to higher and higher levels. Since man's activities will continue to accelerate the emission of pollutants, it is now urgent that we develop a much better ability to assess and forecast the consequences of increased concentration levels of such materials. Thus, it has become important not only to identify and quantify natural and anthropogenic emissions, but also to formulate quantitative models for estimating the lifetime of gaseous and particulate emissions from the atmosphere.

Sources of Trace Constituents. The trace gases and aerosols that are found in the earth's atmosphere come from many diffuse natural and anthropogenic sources. However, virtually all of such materials in significant amounts presently emanate from the planet's surface. They may enter the air after production from (a) biological activity, (b) the winds raising spray from the sea or dust from the continents, (c) volcanoes, (d) combustion, (e) mining, smelting, and other industrial activity, and (f) explosions, for example, nuclear bomb tests.

Some important nonradioactive trace constituents are listed with their source classifications in Table 1. Here, the significance of substances identified with the carbon, sulfur, and nitrogen chemistry cycles is apparent. From the summary of sources in the table, the natural sources of many gases and particles far exceed the anthropogenic sources on a global basis. However, the anthropogenic emissions of many important pollutants are much larger than the natural ones near urban complexes. This emphasizes the fact that the major impact of air pollution still remains a local one where centers of population have evolved. It is of interest, too, that trace substances may be inorganic or organic in nature. The inorganic materials and many of the low-molecular-weight organic compounds have been identified and studied, but only about 5% of the high-molecular-weight organic materials have been characterized.

Although radioactivity in the atmosphere is currently emphasized less than the materials in Table 1, it is nevertheless of considerable consequence as a potential health hazard. A list of some natural and anthropogenic radionuclides found in the atmosphere is shown in Table 2. Three classes of radioactive material are given, covering natural substances from the soil and ocean, cosmic ray spallation products, and nuclear bomb debris of man-made origin. The substances detected come from a wide range of natural and man-made sources. An interesting feature about some of them is their uniqueness as tracers of such sources or of atmospheric processes. Radioactive bomb debris especially has been used to further understanding of certain stratospheric processes.

Removal of Trace Constituents. The removal of material from the air can take place by many different mechanisms. These include absorption or deposition on the earth's surface, cleansing by rain cloud processes, etc. In a sense, one also can classify chemical reactions causing a transition of a material to other constituents as a removal mechanism. The rate of removal

TABLE 1. Summary of Sources, Estimated Annual Emissions, Concentrations and Major Reactions of Nonradioactive Atmospheric Trace Cases

Contaminant	Major pollution sources	Natural sources	Estimated emissions, tons		Atmospheric background concentrations	Calculated atmospheric residence time	Removal reactions and sinks	Solubility in water*	Remarks
			Pollution	Natural					
SO_2	Combustion of coal and oil	Volcanoes	146×10^6	None	0.2 ppb	4 days	Oxidation to sulfate by ozone, or after absorption, by solid and liquid aerosols	11.3	Photochemical oxidation with NO_2 and H may be the process needed to give rapid transformation of $SO_2 \rightarrow SO_4$
H_2S	Chemical processes, sewage treatment	Volcanoes, biological action in swamp areas	3×10^6	100×10^6	0.2 ppb	2 days	Oxidation to SO_2	0.385	Only one set of background concentrations available
CO	Auto exhaust and other combustion	Forest fires, terpene reactions, oceans (?)	274×10^6	75×10^6	0.1 ppm	<3 yr	Ocean and soil speculated; large sink necessary	0.00284	Ocean contributions to natural source probably low
NO/NO_2	Combustion	Bacterial action in soil (?)	53×10^6	NO: 430×10^6 NO_2: 658×10^6	NO: 0.2–2 ppb NO_2: 0.5–4 ppb	5 days	Oxidation to nitrate after sorption by solid and liquid aerosols, hydrocarbon photo-chemical reactions	0.00618 (NO) (†)	Very little work done on natural processes
NH_3	Waste treatment	Biological decay	4×10^6	1160×10^6	6–20 ppb	7 days	Reaction with SO_2 to form $(NH_4)_2SO_4$, oxidation to nitrate	62.9	No quantitative rate data on oxidation of NH_3 to NO_3, which seems to be a dominant process in atmosphere
N_2O	None	Biological action in soil	None	590×10^6	0.25 ppm	4 yr	Photodissociation in stratosphere, biological action in soil	0.121	No information on proposed absorption of N_2O by vegetation
Hydrocarbons	Combustion exhaust, chemical processes	Biological processes	88×10^6	480×10^6	CH_4: 1.5 ppm non-CH_4: <1 ppb	16 yr (CH_4)	Photochemical reaction with NO/NO_2, O_3; large sink necessary for CH_4	6.45×10^{-4} (CH_4)	"Reactive" hydrocarbon emissions from pollution $= 27 \times 10^6$ tons
CO_2	Combustion	Biological decay, release from oceans	1.4×10^{10}	10^{12}	320 ppm	2–4 yr	Biological absorption and photosynthesis, absorption in oceans	0.164	Atmospheric concentrations increasing by 0.7 ppm per year

*In g/100 g H_2O at 20°C.

†NO_2 or dimer reacts in $H_2O + 4NO_3$.

TABLE 2. Summary of Some Radioactive Trace Contaminants in the Atmosphere and their Sources [3]

Element	Symbol	Source	Radioactive half-life	Estimated tropospheric residence time*
Uranium	^{238}U	Earth's crust	4.5×10^9 yr	5–30 days
Radium	^{226}Ra	Earth's crust	1.58×10^3 yr	5–30 days
Radon	^{222}Rn	Earth's crust	3.83 days	~ 3.8 days
Radium A	^{218}Po	Earth's crust	3.05 min	~ 3 min
Thorium	^{232}Th	Earth's crust	1.65×10^{10} yr	5–30 days
Thoron	^{220}Tn	Earth's crust	54.5 sec	~ 50 sec
Beryllium	^{10}Be	Cosmic-ray-produced	2.7×10^6 yr	5–30 days
Carbon	^{14}C	Cosmic-ray-produced	5.7×10^3 yr	—
Sulfur	^{35}S	Cosmic-ray-produced	87 days	—
Beryllium	^7Be	Cosmic-ray-produced	53 days	5–30 days
Phosphorous	^{32}P	Cosmic-ray-produced	14.3 days	—
Cesium	^{137}Cs	Bomb debris	28.8 yr	5–30 days
Strontium	^{90}Sr	Bomb debris	27.7 yr	5–30 days
Zirconium	^{95}Zr	Bomb debris	65 days	5–30 days
Strontium	^{89}Sr	Bomb debris	51 days	5–30 days
Barium	^{140}Ba	Bomb debris	12.8 days	~ 13 days

*Based on assumption of rapid attachment to aerosols.

of gases then will depend on the chemical reactivity of the molecular species as well as its solubility in a condensed phase. The importance of rain clouds is well illustrated in this case, since water-soluble gases such as ammonia are removed by rainfall much more quickly than more insoluble gases such as carbon monoxide. Mechanisms for removal and sinks for pollutants are tabulated in Table 1 along with typical background concentrations in the atmosphere. In general, the less reactive trace species have higher background levels and much longer lifetimes. It is of interest that evidence accumulated to date suggests that some nonreactive, insoluble gases like methane and nitrous oxide remain in the atmosphere for years. More reactive species like sulfur dioxide and nitrogen dioxide remain for only a few days.

The removal of radioactive material from the atmosphere takes place by mechanisms similar to those of nonradioactive species. Of importance, of course, is the attachment process where "primary" particles such as small ions are removed by agglomeration with larger aerosol particles. Rain cloud processes also are believed to be of great significance in removal of radioactive materials from the troposphere.

The chemistry of radionuclides in the atmosphere is complicated by their ability to undergo decay processes. Therefore, like other chemically reactive species, such radioactive materials may be "removed" from the air in the sense that they will be transformed chemically into "secondary" or product species.

Lifetimes of Trace Materials. Our discussion so far has indicated briefly what a complicated subject the lifetimes of trace gases and aerosols are in the atmosphere. The actual lifetime τ is some function of many time scales. For example, at least three times are of interest in air pollution. The first, of course, is the overall lifetime τ. The second is a characteristic chemical half-life of a chemically reactive material, τ_c. The third is a time scale for removal of a pollutant in an urban atmosphere, τ_m. The second scale will depend only on the chemical properties of the material of interest, in addition to the chemistry of its environment. The last scale, on the other hand, will be a function of both microscale processes, including chemistry, and larger-scale meteorological variables like the wind field and the inversion base. The parameter τ_m should be most useful in evaluating the exposure of individuals to pollutants. Though conceptually simple as such a measure, τ_m is difficult to obtain quantitatively. Indeed, any time scale averaged over days or seasons may have little meaning as an indicator for health hazards. Statistically, the fluctuations in τ_m for any day or urban location are likely to be larger than the mean value itself. Therefore, one has to make rather prudent use of such time scales in the assessment of air pollution hazards.

The overall lifetime in the atmosphere τ probably remains the single most important time scale in the geochemistry of trace constituents. The definition of this time scale is the ratio of the concentration in air to the removal rate. When the removal rate is linearly proportional to the concentration of the trace material, τ is ordinarily given as that time required for the material to be reduced to $1/e$ of its original concentration. This implies, of course, that a residual concentration of any material may exist for very long times in the atmosphere. The meaning of τ is complicated by the fact that observations near a source may indicate very high concentrations; ambient air measurements in a city may show pollutants in the ppm range, but such materials may be observed in ppb concentrations far from their sources. Thus a potentially significant fraction of emissions may be present for many months despite the fact that its original concentration has decayed rapidly to a "background" level in a few days.

The lifetime τ for various trace constituents is deduced in different ways from a variety of evidence from the ambient levels of concentration, magnitude of sources, and chemical reactivity. The estimated values for the materials summarized in Tables 1 and 2 are also tabulated for comparison. In the case of nonreactive, relatively insoluble gases, the lifetimes are generally considerably longer in the atmosphere than those of more reactive and soluble species. The importance of rain processes in removal is illustrated qualitatively by noting that the more soluble the gas, the shorter its lifetime in the atmosphere. The lifetimes of gases such as SO_2, NO, and NO_2 are diminished by the relatively high chemical reactivity in the atmosphere, as will be discussed later in Section 2.

The lifetime of small particles in the atmosphere, *aerosol particles*, is comparatively short in contrast to those of many gases. Removal of aerosols depends on additional factors that do not influence gases; among these are sedimentation because of the relatively large mass of particles, and agglomeration through collisions by turbulent air motion, or differential settling. These mechanisms will be outlined in detail in Section 3. The lifetimes of aerosols in the lower troposphere have been found to vary between a few hours to over a month.

Objectives of Review. Our introductory comments have stressed that the removal of trace constituents from the earth's lower atmosphere involves many factors from microscale activity of physics and chemistry to macroscale activity of meteorological processes. In this chapter, these interacting mechanisms will be discussed further and several theoretical ideas as well as observational evidence will be reviewed. Since pollutants are of main interest in this book, our attention will be centered on gases and aerosols that originate at least partly with man's activity. Because the behavior of radioactive materials has been treated at length elsewhere [3] and is essentially similar to that of nonradioactive material, we shall not consider these trace contaminants further. In the succeeding sections, our discussions will begin with consideration of relatively insoluble gases to avoid initially the complexity of rain cloud processes. Later, the more soluble constituents will be treated. Finally, the last sections will deal with the atmospheric removal of aerial suspensions both in a dry atmosphere and in the presence of rain clouds.

2. THE PHYSICAL CHEMISTRY OF REMOVAL OF TRACE GASES

The loss of material from the atmosphere may be defined in several ways. In this discussion, loss or removal will be referred to changes occurring in a control volume of air traveling with the time-averaged wind field. In other words, consideration will be given to the losses in an "identified" volume element as it evolves in space and in time. With such a definition, one can write a material balance for the rate of change in concentration of any arbitrary gaseous species, based on transport theory. That is,

$$dn_\alpha/dt \quad = \quad \partial n_\alpha/\partial t \quad + \quad \mathbf{q} \cdot \nabla n_\alpha \quad = \quad \nabla \cdot D_T \nabla n_\alpha + \quad S_\alpha \quad + \quad L_\alpha$$

net rate of change of species α	net rate of change relative to fixed position in space	convective or advective change	net loss by diffusion	net gain from sources inside the volume element	net loss by removal processes inside the volume element

Here, n_α is the time-averaged concentration* of gaseous species α ($= $ A, B, ...); q is the time-averaged vector wind field; t is time; and D_T is the turbulent eddy diffusivity. Consistent with previous investigations, the molecular diffusivity of materials is disregarded in the atmosphere since it is ordinarily much smaller than the eddy diffusivity in free air away from boundaries. The last two terms on the right account for losses or gains by removal or sources in the volume element. Losses or gains at the earth's surface are considered in specification of the lower boundary conditions for this differential equation. The discussion below will deal mainly with these boundary conditions as well as the source and removal terms S_α and L_α. Turbulent transport and convection in the lower atmosphere are beyond the scope of this chapter. For a description of the fluid dynamics of the atmospheric surface layers, the reader is referred to Hidy [80], Munn [81], and Lumley and Panofsky [61].

2.1. Water-Insoluble Constituents

The water-insoluble contaminants include a variety of materials, ranging from hydrocarbons to one of the oxides of nitrogen, nitrous oxide. Because of their low solubility, the hydrological cycle plays a relatively insignificant role in their removal from the troposphere. The main removal mechanisms that have been identified are: (1) homogeneous chemical reactions in the atmosphere; (2) attachment, absorption, and chemical reaction on aerosols; (3) absorption and chemical reaction on objects at the earth's surface; (4) escape and subsequent removal by chemical reactions into the stratosphere.

Unless absorption on suspended material or at the earth's surface is irreversible and is accompanied by fallout, the trace material may not actually be removed from a volume element permanently. Thus the role of either homogeneous or heterogeneous chemical reactivity plays a significant role in mechanisms for removal of the insoluble trace gases.

Removal at the earth's surface may involve absorption and reaction in the ocean, in soil, or on exposed surfaces of rock. Some gases such as CO, N_2O, and CH_4 also are believed to be removed as a result of their link with the biochemistry of vegetation. In most cases, insoluble trace constituents will be removed from air by more than one of the listed mechanisms.

Homogeneous Chemical Reactions. Perhaps the best-known example of removal of trace pollutants from the lower atmosphere is the photochemical oxidation of hydrocarbons in the presence of nitrogen oxides. These reactions have been studied extensively in recent years because of their implication to photochemical smog (see for example, Leighton [16]; Niki et al. [82]).

The olefins emitted into the atmosphere by combustion and the terpenes from vegetation undergo similar rapid chemical transition in the presence of

---

*In turbulent media, the molecular transport equations still apply at any instant to the behavior of any species. However, it is normally easier to discuss the properties of a turbulent fluid by time-averaged equations whose derivation is discussed, for example, in detail by Bird et al. [4].

nitrogen oxides and sunlight. These materials may be degraded by successive oxidation steps to such materials as aldehydes, ketones, and carbon dioxide, or may undergo a series of free-radical polymerization reactions possibly ending in the condensed phase as aerosol particles. The limited data available [1,5-7] would suggest that 1–10% by weight of the reactive hydrocarbons emitted in urban air are converted to aerosols, while the remainder is probably oxidized eventually to carbon dioxide.

Paraffinic hydrocarbons have been found to be much less susceptible to photochemical removal than unsaturated materials. By far the most prevalent paraffinic gas in the atmosphere is methane, coming mostly from the biosphere. The nonreactive hydrocarbons emitted in cities as air pollution can be locally high in concentration, but are less significant in terms of total tonnage present in the atmosphere than methane.

Since the nonreactive hydrocarbons are oxidized only very slowly in the lower atmosphere, they must be removed by other mechanisms to be consistent with their estimated residence times. In particular, there is reason to believe that bacteriological processes and vegetation are of significance in their removal [8].

Attachment, Absorption, and Chemical Reaction on Aerosols. The incorporation of trace gases into aerosols in the atmosphere can be a significant removal process under some circumstances.

If the rate of absorption is rapid at the surface of a suspended particle, the rate of removal of material will be controlled by the thermal motion of that species to the particle surface. The theoretical model for the transfer rate takes a different form, depending on the size of the aerosol particle. Consider the case of component A diffusing through B to the velocity particle. The particle is assumed to move with the mass-average velocity surrounding gas. In the extreme where the particle radius is the order of or less than the mean free path of the gas, the rate of transfer to a sphere is given by [2]

$$\Phi_A = 2\pi R^2 \alpha_{CA} \frac{[n_A^+(2kT^+/\pi m_A)^{1/2} - n_A^-(2kT^-/\pi m_A)^{1/2}]}{1 + (\alpha_{CA}\xi R/\lambda_B)} \tag{1}$$

where λ_B is the mean free path of the suspending gas, α_{CA} is the condensation coefficient, n_A is the number density of diffusing species A, T is the absolute temperature, k is Boltzmann's constant, R is the radius of the absorbing particle, ξ is a parameter depending on the molecular masses and diameters of A and the suspending gas (values between 0.02 and 3.2), and the plus and minus signs denote properties for molecules directed away from or toward the particle surface.

For conditions where the particle radius is much greater than the mean free path of the suspending medium, the rate of transfer can be written

$$\Phi_A = 4\pi D_A R(n_A - n_{A\infty})(1 + C\lambda/R) \tag{2}$$

where C is the slip coefficient, which is approximately unity for most gases, and D_A is the diffusivity of the vapor A in B. Here, the transfer rate reduces to the classical diffusion rate for a continuum when the *Knudsen number* λ/R approaches zero.

The difference between the two extremes of behavior is evident from the form of Eqs. (1) and (2). For example, Eq. (1) gives Φ_A proportional to R^2, $\lambda/R \to \infty$, while Eq. (2) predicts a proportionality to R, $\lambda/R \to 0$.

If aerosols travel at speeds differing from that of the surrounding medium, the rate of mass transfer to the particle surfaces may be enhanced by convection. When these velocity differences are small, the differences in mass transfer rate are negligible provided that the Mach number of the particle is small (implying the particle Reynolds number is less than unity, too). In general, this will be the case for *atmospheric aerosols*, which are mainly smaller than 20 μm radius. Hence, the transfer rates given by Eqs. (1) and (2) need not be corrected for velocity differences between air and the atmospheric particles for most practical calculations.

An important consequence of the diffusional mechanism for attachment of trace vapors or gases is their distribution on aerosols. This is of particular interest for the absorption of toxic materials since the smaller aerosols, $\lesssim 0.5$ μm, penetrate far into the respiratory system. Brock [9] has estimated from Eq. (1) the expected spectrum of attachment for typical aerosol size distributions illustrated in Chapter 2. His calculated distributions of attached gases without coagulation of the aerosol particles are indicated for maritime and urban aerosols in Fig. 1. Here, Δ represents the ratio of the amount of A condensed on particles of size $R \lesssim 0.1$ μm to the amount of A condensed on particles with radius $R = 1.0$ μm.

On the basis of this comparison, it is expected that the maritime aerosol would contain the largest amounts of condensed A in the range 0.2–2 μm. The situation for the urban aerosol is quite different because of the much greater population of small particles less than 0.1 μm compared with typical maritime aerosols, combined with the increased transfer rate below 1 μm radius. These differences are dramatically reflected in the observation that the attached species A should be concentrated almost exclusively below 0.5 μm. There is some limited support for this conclusion from experiments such as Soilleux's [10], who found that 90% of the radon daughters were attached to particles in the range 0.025–0.1 μm in samples of continental aerosol.

The effect of coagulation of aerosols absorbing trace gases will complicate the simple picture described by Brock. Qualitatively, it is expected that coagulation will broaden the attachment spectrum at a faster rate than attachment alone.

In many cases, however, coagulation in the atmosphere takes place at a comparatively slow rate, as will be seen later in Section 2.3. Therefore, the simple arguments for attachment spectra solely on the basis of absorption and condensation should have more than limited applicability to the behavior of atmospheric aerosols, especially at higher altitudes away from strong sources.

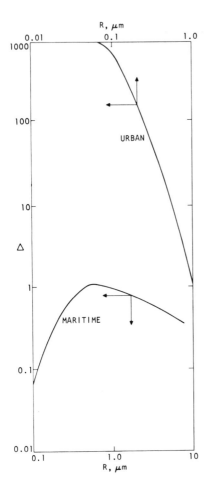

Fig. 1. Relative attachment spectra for maritime and urban aerosols [9].

Catalytic Activity of Particles. If the absorption rate is slow at a particle surface compared with the rate of diffusion of gas into a particle, the removal rate estimated by the gas-phase diffusion model has to be corrected. Let us consider as an illustration a large, porous particle in which a first-order reaction in the particle is removing component A from the air. Using diffusion theory, a correction factor may be derived for such a situation. The concentration distribution of the reacting component A inside the particle is uniform radially, but the steady-state distribution in the particle is given by the equation

$$\frac{1}{r^2} \cdot D_p \frac{d}{dr}\left(r^2 \frac{dn_A}{dr}\right) - k_A n_A = 0 \tag{3}$$

with boundary conditions $n_A = n_{A0}$, $r = R$, and $dn_A/dr = 0$, $r = 0$. Here, r is the radial direction from the center of the particle, D_p is the pore diffusivity

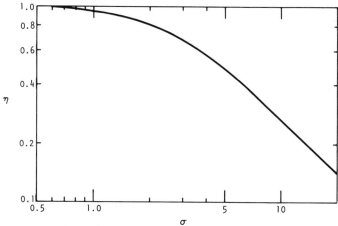

Fig. 2. Reaction rate parameter η as a function of reaction-to-diffusion ratio σ [69] in molecules per cm³ per second, per mole per cm³.

of species A, and k_A is the reaction rate constant at the particle surface in molecules per cm³ per second, per mole per cm³. The rate of removal L is given by

$$L = -k'_A n_A A_s R\eta/3a \qquad (4)$$

where A_s is the total pore cross-sectional area on the external surface of the particle and a is the hydraulic radius of the pores. The correction factor η is given by Thiele [69] as

$$\eta = \frac{3}{\sigma}\left(\frac{1}{\tanh \sigma} - \frac{1}{\sigma}\right) \qquad (5)$$

where $\sigma = R(k'_A/D_p a)^{1/2}$, and k'_A is the reaction rate constant k_A per cm² of particle pore surface. The functional form of Eq. (5) is shown in Fig. 2. From this result, it can be seen that the reaction is diffusion-controlled and occurs throughout the particle for small values of σ, whereas it is outer-surface-dominated for large σ, where the reaction rate is much larger than the pore diffusivity. Thus, for effective removal of gaseous pollutants in the atmosphere by absorption in aerosols, the porosity is quite important. One can readily see from Fig. 2, for example, that the presence of tarry or oily organic material in particles in the atmosphere, such as those emitted in an urban area, may clog pores in a particle and inhibit the ability of otherwise porous aerosols to remove reactive pollutants from the air.

There is little information on the physical properties of atmospheric aerosols that bears on their catalytic activity or their porosity. Some information may be obtained for particles by comparison between their surface areas as measured by the well-known Brunauer, Emmett, and Teller (BET) adsorption method and their superficial surface area as estimated from the aerosol size distribution. This has been done in one case recently by Corn et al. [11]. They found that aerosol particles collected from air over Pittsburgh,

Pennsylvania had a mean surface area longer than the superficial area. The surface area was seasonally dependent, ranging from 1.98 to $3.05 \, \text{m}^2 \, \text{g}^{-1}$. This cannot be considered highly porous material compared with adsorbents such as activated charcoal or silica gel, but nevertheless may be a significant factor in gaseous removal.

Removal at the Earth's Surface. Objects at the Earth's surface itself present an enormous area of contact with the lower atmosphere. The trace constituents contained in air have been found to be immediately connected with biological and chemical processes taking place at the planet's surface. For example, the ocean acts as an enormous sink for gases that can be dissolved and subsequently bound in sea water or removed by biological or chemical activity. Gases like CO are known to be oxidized to CO_2 by certain bacteria *Bacillus oligocarbophilus* [12]. Robinson and Robbins [7] also believe interaction with vegetation to be involved in the removal of gases like CO, CH_4, and N_2O.

Buildings and stone outcroppings may also absorb and oxidize catalytically trace gases. They may be of particular significance in removal of such gases as SO_2. An important but sobering consequence of the uptake of gas in stone is the degradation of surfaces by SO_2 of buildings and art objects in cities like Venice.

The uptake rate of gases on vegetation or other objects at the earth's surface is often expressed in terms of an empirical parameter called the deposition velocity q_g. The parameter is defined from the relation

$$J = n_{A0}q_g \qquad (6)$$

$$\frac{\text{rate of deposition}}{\text{unit area}} = (\text{trace concentration at the surface})(\text{deposition velocity})$$

The functional form of the deposition velocity may be derived ideally from correlations of mass transfer through turbulent boundary layers. The empirical relations stem from a variety of experimental results and will differ somewhat depending on the roughness of the absorbing surface [13,14]. Some typical values for the deposition velocity of gases are shown in Fig. 3.

The results of a number of experiments suggest that molecular diffusivity still enters into the mass transport mechanism in a turbulent gas flow whether the boundary is rough or smooth.

The transfer of gases or aerosol particles from a turbulent gas to an underlying boundary depends on the properties of the flow near the surface as well as the nature of the surface itself. In a simplified picture, transfer takes place through a turbulent boundary layer into a viscous or laminar sublayer adjacent to the boundary. The boundary layers are identified with a velocity gradient and with a gradient in concentration of diffusing trace species. In the turbulent layer, these gradients will be essentially similar in thickness because of the dominance of the eddying motion of the turbulence. In contrast,

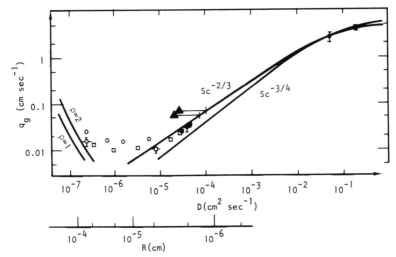

Fig. 3. Deposition velocity as a function of particle diffusivity for flow over smooth boundaries. Open circles, polystyrene latex spheres; darkened squares, sodium chloride aerosols; darkened triangles, purified water aerosols; darkened circles, gas values (water vapor and carbon dioxide); lines on far left are calculated sedimentation velocities for particle densities of one and two [70].

the viscous sublayer, defined by gradients of properties, and the nature of the transfer mechanism will depend primarily on molecular properties of the fluid. Its thickness then will vary with the molecular diffusivity of the species being transferred. Since molecular transfer is quite slow compared with that in the turbulent layer, the mechanism of transport in the sublayer is the rate-limiting mechanism in the process, and deposition of gases or aerosols on surfaces then will vary with the molecular (or Brownian) diffusivity as indicated in Fig. 3.

When the surface is roughened sufficiently, there is considerable evidence that the viscous sublayer associated with momentum transfer develops a "turbulence-like" character. Under such conditions, the drag on the surface markedly increases, resulting in increased momentum transfer. In the case of mass transfer in the sublayer, roughness does not produce an increase in transport rate corresponding to the increase in drag. It is believed that the viscous character of the diffusion sublayer is retained in the presence of surface roughness, so that within a few microns of the surface, molecular processes remain significant in the determination of the deposition rate. It is anticipated that the relation

$$\mathrm{St}_d = q_g/U_\infty \propto (x^+)^{-1/9}\mathrm{Sc}^{-2/3}(c_f/2)^{-1/2} \tag{7}$$

will be a satisfactory rough estimate for q_g on the basis of extrapolation from known correlations. Here, $\mathrm{Sc} = \nu/D_\alpha$, with ν the kinematic viscosity of the gas and D_α the molecular diffusivity of species α; $x^+ = \int_0^x [(u^*)^2 \xi / v] \, d\xi$, $u^* = U_\infty \sqrt{c_f}$; and $c_f = F/\rho_G u_\infty^2$, with F the surface stress, U_∞ the free stream velocity, and ρ_G the gas density. The constant of proportionality in Eq. (7) has been found from a variety of data to be dependent on the nature of the surface.

Transport to the Stratosphere. In addition to a potential net loss to the lower boundary of the troposphere, gases can escape by diffusion into the stratosphere if there is a sink (removal by chemical reaction) at high altitude.

It is assumed that the stratosphere is a perfect absorber for trace constituents. For a turbulent atmosphere, the maximum rate of transport then should depend on the turbulent diffusivity of the troposphere, taken crudely as a constant, 10^4–10^5 cm^2 sec^{-1}, the sedimentation velocity, and the rate of chemical reaction in the stratosphere.

Applications to Carbon Monoxide, Methane, and Nitrous Oxides. With the considerations outlined briefly in this section, let us carry through the current arguments for the removal of three illustrative insoluble gases, CO, CH_4, and N_2O. These gases are comparatively inert and are long-lived trace constituents in the troposphere. At the outset, our evaluation will disregard any absorption or attachment on aerosols because no known particle suspension in the atmosphere removes these gases by adsorption and chemical reaction.

At the present rate of global emission of CO from natural and combustion sources at the earth's surface, one would expect to see a gradual increase in ambient CO levels. However, the observations available fail to indicate any such changes. Therefore, there must be major sinks for CO in the troposphere or at its boundaries.

Although our knowledge about ambient concentrations of CO is incomplete, evidence reviewed by Robinson and Robbins [8] and Seiler and Junge [15] suggest that the mean ambient level is about 0.1 ppm. For an emission rate of $\sim 2.6 \times 10^{11}$ kg yr^{-1} as estimated by Robinson and Robbins [8], the annual removal rate to retain the ambient concentration of 0.1 ppm by volume is about 6×10^{11} kg since the total mass of the atmosphere is 5.3×10^{18} kg. This implies that the lifetime of CO is roughly three years.

There are gas-phase reactions that could contribute to CO removal by oxidation. The reaction with molecular oxygen

$$2CO + O_2 \rightarrow 2CO_2$$

is possible, but evidence indicates that it is unimportant in the atmosphere. More likely candidates are reactions with atomic oxygen, but Leighton [16] has concluded that these, too, are not significant. Ozone will also oxidize CO to CO_2, but Leighton [16] has also indicated that this reaction is very slow at atmospheric temperatures and concentrations. Oxidation by NO_2 in the reaction

$$NO_2 + CO \rightarrow CO_2 + NO$$

has been ruled out by Robinson and Robbins [8] based on consideration of high activation energies. Workers have then reached the conclusion that no tropospheric gas-phase reactions can be significant sinks for removal of CO.

The possibility that CO is scavenged by exchange with the oceans has been argued against on grounds that the solubility of CO is too low to account for more than 2% of the annual production of this gas [8]. Furthermore, Seiler and Junge [15] have reviewed evidence that the oceans may actually be an important source of CO in some regions. There is no apparent chemical reaction in the sea that would remove CO as another constituent.

In the absence of identifiable sinks for CO, Robinson and Robbins [8] have invoked the bacterial action of soil as a possible sink, as mentioned earlier. If we assume that the soil is a perfect absorber and q_g is 1 cm sec^{-1} from Fig. 3, then for the earth's land surface of 5.12×10^8 km^2 about 10^{11} kg ideally could be removed in a year by the soil. This is about the same as the emission rate into the atmosphere. Therefore, it must be concluded that surface mechanisms are potentially significant for carbon monoxide removal.

Robinson and Robbins [8] have rationalized the lack of an identifiable sink for CO in terms of the biosphere. They feel that there is some circumstantial evidence to link CO with the respiration of plants in an analogous way as for CO_2. On the other hand, Seiler and Junge [15] have noted that their limited measurements of global CO concentrations in the upper troposphere are roughly uniform but decrease sharply with penetration into the stratosphere. On this basis, these workers have argued that the oxidation of CO in the stratosphere, possibly by free-radical reactions involving OH, is the significant sink for this gas.

As a third alternative, Brock and Hidy speculate that a tropospheric reaction to oxidize CO to CO_2 may take place in thunderstorms. There are as many as 50,000 thunderstorms going on globally during one day. It appears that CO may be excited to sufficiently high energies in electrical microdischarges between colliding hydrometeors to allow oxidation to take place. There is some sketchy evidence based on laboratory kinetic data to support such a hypothesis, but further work is required before accepting such a model.

Before passing on to the question of methane, it is of interest that CO behavior can be used to estimate roughly the lifetime τ_m of a polluted air mass over an urban area. As an illustrative example, consider the Los Angeles metropolitan area, a basin surrounded on three sides by mountains. The meteorological conditions are dominated much of the year by a marine layer with a sea breeze blowing from the Pacific Ocean eastward keeping pollutants moving toward the enclosing mountains. The inversion over the city is a strong one, with a base at 300 m on the average. The area of the Los Angeles County is 10,000 km^2 and the mean ambient concentration is ~ 10 ppm in the City. The Los Angeles Air Pollution Control District gives the daily emission of CO as 8.79×10^6 kg. This allows calculation of the average lifetime τ_m of an air mass over the metropolitan Los Angeles area of about one day.

The methane cycle in the troposphere is not unlike that of CO in some respects. Both gases are quite inert chemically in the atmosphere. An average concentration for methane of 1.5 ppm gives about 4.4×10^{12} kg of the gas in the atmosphere. Robinson and Robbins [8] have reviewed the estimates of emissions and estimated that CH_4 is produced mainly from decay of organic material at an annual rate of about 14.5×10^{11} kg. This, in combination with the ambient concentration level, suggests that the mean residence time of CH_4 is the order of three years, an estimate much lower than Koyama's [18] value of 16 years. Bainbridge and Heidt [19] have estimated from limited data for the vertical distribution of CH_4 that only about 10 % of this gas is transferred into the stratosphere. Without any other identifiable sinks in the atmosphere, Robinson and Robbins [8] have speculated that methane is oxidized rapidly on the surfaces of vegetation, in analogy to a known reaction for ammonia. However, the details for such a removal process have not been established.

Another gas that falls into the category of inert trace constituents in the atmosphere is nitrous oxide, though it has virtually no known anthropogenic sources. The ambient levels measured in different places are reasonably consistent and suggest an ambient level of 0.25–0.30 ppm. Schütz et al. [20] have presented an extensive review of N_2O sources and sinks and have verified that this gas is removed partially by photodissociation in the stratosphere. With this mechanism alone, these workers have estimated the atmospheric lifetime of N_2O to be about 70 yr. There is evidence that the oceans may be involved in the N_2O production–elimination cycle, and it appears that there is rather variable destruction of N_2O in the soil [21]. The uncertainties in the postulated mechanisms have led Bates and Hays [21] to believe that N_2O, like CO and CH_4, is absorbed in plant tissue during the photosynthesis process. If the rate of loss to the biosphere of N_2O is scaled to CO_2 production by vegetation, Robinson and Robbins [8] conclude that the lifetime of N_2O may be as low as ~ 1 yr. On the other hand, Schütz et al. [20] have concluded that the lifetime of N_2O should be less than 10 yr. They feel that the sink added to the photodissociation process is most likely a land one, particularly the microbiological action in soil.

Based on our discussions here, we may only conclude that the major removal mechanisms for the inert, insoluble atmospheric gases CO, CH_4, and N_2O remain largely a mystery. There is circumstantial evidence, however, that the biosphere is intimately involved with atmospheric processes in the balance of these trace gases.

2.2. Water-Soluble Constituents

Many trace gases are known to be removed effectively from the atmosphere by processes involving rain clouds. It is commonly observed that certain trace constituents are usually reduced significantly after a rainfall

TABLE 3. Average Decrease of Aerosol and Gas Concentrations in
Ground Air Due to Rain for Various Constituents in Frankfurt/Main,
June 1956–May 1957 [22]

	Air concentration, $\mu g\ m^{-3}$						
	Aerosols			Gases			
	NH_4^+	NO_3^-	$SO_4^=$	NH_3	NO_2	SO_2	Cl_2
Before rains	6.7	6.0	16.7	21.6	21.6	238	14.3
After rains	4.7	1.6	9.7	11.0	9.1	212	5.3
Per cent decrease	30	73	42	49	24	35	63

exceeding very scattered shower activity. To illustrate this, the average
decrease in gas concentration over Frankfurt/Main, Germany is tabulated
in Table 3. The apparent decrease in pollutants ranges annually from 24%
to 73% over this city, which has an average rainfall comparable with some
cities in the eastern United States.

The data cited for removal of trace materials by rainfall such as those
in Table 3 are somewhat ambiguous. Such results represent ground-level
concentrations and not conditions at cloud base. Furthermore, the degree
of apparent removal is clouded by the fact that storm systems are advected
as they develop. Thus, a measurement of a decrease at a fixed station at the
surface may reflect only the passage of the new, stormy air mass over the
urban area where sampling took place. Despite these reservations, the
measurement of the chemical composition of rain water in combination
with considerations of differing residence times for soluble and insoluble
gases (Table 1) provides ample evidence that rain clouds play an important
part in removal of the more soluble trace gases.

The development of precipitation can remove trace constituents in two
regimes. The first, termed *rainout*, involves the various processes taking
place inside the cloud. The second, *washout*, refers to the removal of trace
materials below cloud base by falling hydrometeors. In the case of gases,
both rainout and washout involve effectively the same mechanisms of absorption, with and without chemical reaction, while aerosol removal may differ
somewhat between these two classifications.

Rainout of gases can take place in three different ways: (a) by simple
solution in cloud water according to a solubility law such as Henry's law;
examples include N_2O and CH_4; (b) by dissolving with reversible hydration
and dissociation; examples include CO_2 and NH_3; (c) by solution and
subsequent irreversible conversion or reaction with other dissolved substances in cloud water; examples are SO_2 and NO_2.

Washout of gases will take place if the gas concentrations under cloud
level are higher than those within the cloud and/or if the solubilizing reaction
within the cloud is only partially complete.

In the first category listed, the dissolving of the gas generally will follow a law in which the absorption coefficient of the gas in water, H, is independent of the partial pressure of that component (i.e., Henry's law). The equilibrium ratio of dissolved to undissolved gas in cloudy air may be determined by the following considerations [3]. Consider the volume ratio of liquid water to cloudy air given by the ratio of the liquid water content of the cloud $\mathscr{L}$ (g cm^{-3}) to ρ_w, the density of water. Essentially, $\mathscr{L}/\rho_w$ is $\mathscr{L}$ and the ratio of dissolved to undissolved gas will then be $H\mathscr{L}$. In the atmosphere, $\mathscr{L} \approx 10^{-6}$, so that only gases with absorption coefficients greater than 10^4 cm^3 dissolved per cm^3 water will be removed to any appreciable extent by clouds. For inert gases in the first category, A ranges between 0.01 and 0.1 at normal temperatures, making rainout an insignificant mechanism for removal of such materials. Such a consideration is reinforced, of course, by the fact that removal is reversible here in the sense that the dissolved gas is released whenever the droplets evaporate.

In cases where gases follow category (b), absorption coefficients may be considerably larger than in category (a), and removal by precipitation will become important. Gases in category (b) form ions in solution and the solubility consequently is highly dependent on the pH of the water. Two well-known examples are the CO_2–H_2O system and the NH_3–H_2O system. In the former case, the CO_2 concentration is given by

$$[CO_2] = A_1[H_2CO_3] \qquad (8)$$

in the absence of other ions since only the undissociated part obeys Henry's law. But carbonic acid, H_2CO_3, dissociates by stages in water, according to the equilibrium constants

$$K_1 = \frac{[H^+][HCO_3^-]}{[H_2CO_3]}, \qquad K_2 = \frac{[H^+][CO_3^=]}{[HCO_3^-]}, \qquad K_3 = [H^+][OH^-] \quad (9)$$

Noting that

$$[H^+] = [OH^-] + [HCO_3^-] + 2[CO_3^=]$$

Junge [3] calculates the CO_2 concentration to be

$$[CO_2] = A_1([H^+]^3 - K_3[H^+])/K_1([H^+] + 2K_2) \qquad (10)$$

For rainwater, the pH often ranges from five to six, dominated by acids such as HCl, H_2SO_4, and HNO_3. Junge [3] uses the following data for air at 10°C:

$$[CO_2] = 1.34 \times 10^{-5} \text{ mole liter}^{-1} \text{ (300 ppm)}$$

$$A_1 = 1.2$$

$$K_1 = 3.4 \times 10^{-7} \text{ mole liter}^{-1}$$

$$K_2 = 3.2 \times 10^{-11} \text{ mole liter}^{-1}$$

$$K_3 = 3.6 \times 10^{-15} \text{ mole}^2 \text{ liter}^{-2}$$

to find at pH = 5.6 that the following equilibrium concentrations in rain water can be expected:

$$[H_2CO_3] = 0.71 \text{ mg liter}^{-1} \text{ (as } CO_2)$$

$$[HCO_3^-] = 0.14 \text{ mg liter}^{-1}$$

$$[CO_3^=] = 0.19 \times 10^{-5} \text{ mg liter}^{-1}$$

At equilibrium, the amount of H_2CO_3 in rain water then remains fixed by the absorption constant; however, the ion concentrations may vary considerably with the pH of the rain water.

The other important case in category (b) is ammonia in water. For this case, the concentration of ammonia in air at equilibrium with water at 10°C is

$$[NH_3] = A_2[NH_4OH], \qquad A_2 = 3.6 \times 10^{-4} \qquad (11)$$

and

$$K_4 = [NH_4^+][OH^-]/[NH_4OH], \qquad K_4 = 2.0 \times 10^{-5} \text{ mole liter}^{-1} \qquad (12)$$

Under atmospheric conditions, the pH of cloud water often ranges from 5 to 6. For NH_3 concentrations in air of 3 μg m^{-3}, approximately 97 % of the NH_3 then could be absorbed in cloud water in the absence of influence of CO_2. Even with carbon dioxide present, the absorption of NH_3 is calculated to be 87 % at equilibrium [3]. Thus, the removal of ammonia from the atmosphere by rain clouds should be rather efficient if equilibrium solution is achieved. This does not appear to be the case. Thus, the *rate* of transport of the gas to the surface and away from the droplet surface in the liquid also should be considered.

Mechanisms in category (c) are more complex and as yet are not fully understood. Perhaps the best studied so far is the SO_2–H_2O system, in which the SO_2 absorbed in water may be oxidized to $SO_4^=$ in the presence of metal ions or NH_4^+. The case of SO_2 wet removal is of sufficient interest in comparison to dry processes that later we shall discuss this trace gas further as an illustration of ideas introduced in the first two sections of this review.

Diffusion of Gases and Removal Rates. Thermodynamic considerations of equilibria are important in discussing rain water removal in that they yield an estimate of the maximum amount of material than can be disposed of by precipitation processes.

The rate of absorption of a gas into a falling droplet may be controlled by the rate of diffusion of the gas through to the droplet surface, or it may depend on the rate of diffusion of the soluble material away from the droplet surface into its interior.

In general, the time required to achieve a nearly equilibrium solution in a raindrop will be less than estimated from molecular diffusion alone, because the droplets move relative to the air. Under these circumstances, the rate of transfer of gas to the droplet surface is enhanced by convection. To estimate the rate of convection transport, the mass transfer is introduced,

$$4\pi R^2 m_A \mathscr{K}_g (n_{A\infty} - n_{A0}) = \int_S m_A D_A \left(\frac{\partial n_A}{\partial r} \right)_{r=R} dS, \tag{13}$$

where S = surface area.

Here, the subscript ∞ refers to conditions in the bulk gas far from the droplet and 0 refers to the surface condition.

For droplets of small radius whose Reynolds number $\mathrm{Re} \, (= q_s R/v)$ is small, Acrivos and Taylor [23] have reported the correlation

$$\mathrm{Sh} = 2\mathscr{K}_g R/D_A = 2 + \mathrm{Pe} + \mathrm{Pe}^2 \ln \mathrm{Pe} + \cdots \tag{14}$$

Here, $\mathrm{Pe} = q_s R/D_A < 1$ and q_s is the sedimentation velocity of a drop of radius R. For larger droplets, the well-known Frössling [24] relation for the Sherwood number predicts

$$\mathrm{Sh} = 2(1 + 0.33 \, \mathrm{Sc}^{1/3} \mathrm{Re}^{1/2}), \qquad \mathrm{Re} \gtrsim 100 \tag{15}$$

Whenever the droplet travels at a speed differing from that of the surrounding air, external disturbances on large drops can create circulation inside the drops. This motion aids in mixing dissolved gas away from the surface, decreasing the time calculated from consideration of gas-phase transport alone.

The concentration of absorbed species in a drop will depend on the rate of transfer across the gas–liquid boundary as well as the conversion rate to product species. As an example, suppose a gas is being removed in rain that is being converted by an irreversible first-order chemical reaction to a nonvolatile species. Assuming that the equilibrium liquid concentration of this absorption species A follows Henry's law, an estimate of the droplet concentration of A can be made from diffusion theory. Let us assume that the drop is spherical and of constant size. The liquid inside is stagnant and the drop is initially solute-free. Equation (3) can be applied with replacement of D_p by the diffusivity of A in a pure liquid ($\mathscr{D}$). Two classes of surface conditions are considered. The first refers to the case of a very soluble gas where gas-phase resistance to transfer is important. Then the flux of A at the boundary must be constant on both sides of the boundary and Eq. (13) is appropriate. On the other hand, if the trace constituent is sparingly soluble, gas-phase resistance will be a minor factor and the transport in the drop will be dominated by liquid-phase processes. Under such circumstances, the condition

$$r = R, \qquad n_{A\infty} = n_{A0}$$

is applicable. In the case of a very soluble gas, the Henry's law or "partition" coefficient A would exceed 10, but in the latter sparingly soluble case, the coefficient $A \lesssim 10$. The problem of diffusion of species A reacting in a drop has been solved for the first-order reactions by Dankwerts [25]. In the case of a highly soluble gas, the dimensionless time required to achieve the equilibrium concentration is

$$6\mathcal{K}_g t/2AR \approx 1 \tag{16}$$

Using Eq. (14), we can estimate for typical gas diffusivities that the equilibrium concentration should be achieved in less than 10^{-4} sec for highly soluble gases absorbing in 10 μm droplets.

For sparingly soluble gases, the dimensionless time to achieve steady-state concentration is

$$\mathcal{D}_A t/R^2 \approx 1 \tag{17}$$

The solute diffusivity in liquid is about 10^{-5} cm^2 sec^{-1}, so the steady-state concentration is achieved in about 10^{-1} sec for 10 μm radius droplets in contrast to the highly soluble case. Where the time scale $k_A t \to 0$, the equilibrium concentration of solute is realized at a steady state, but the liquid concentration is enhanced beyond that of Henry's law for $k_A t > 0$. Thus, it is anticipated that the liquid diffusion limit is not controlling irreversible absorption in this model (see also, Postma [26]).

Recently, Postma [26] has applied Dankwerts's theory to absorption of SO$_2$ and iodine and has found that for gases that ionize in solution, disproportionation may cause enhancement of the apparent partition coefficient. The extent of ionization then must be known before the role of precipitation can be estimated.

On empirical grounds, the concentration of a soluble gas C_A in cloud water in mg liter^{-1} has been given by Junge [3] as

$$C_A = \sigma_1 \rho_A/\mathcal{L} \tag{18}$$

where ρ_A is in g m^{-3} and σ_1 is the rainout efficiency. Junge [3] and Beilke and Georgii [27] suggested that σ_1 might be a linear function of liquid water content. Their results are based only on very light rainfall. More recently, Summers [28] has reported values of σ_1 for severe storms near Alberta, Canada with much larger liquid water content than earlier studies. These results are shown in Fig. 4. Summers finds that a reasonable approximate relation for rainout efficiency is

$$\sigma_1 = 10^{-1}\mathcal{L} \tag{19}$$

Calculation of Removal Rates by Precipitation. The methods to date for estimating removal rates by precipitation rely on a simple model employing empirical rainout and washout coefficients. The model is considered to be

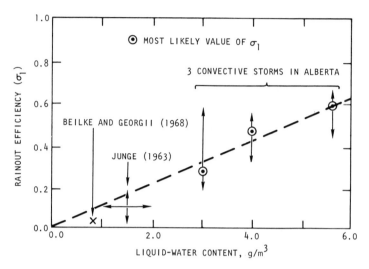

Fig. 4. Rainout efficiency and liquid water content in clouds [28].

first order in the rate of decrease in contaminant concentration n_A, or

$$dn_A/dt = -\omega_1 n_A - \omega_2 n_A \tag{20}$$

where ω_1 is rainout rate coefficient and ω_2 is the washout rate coefficient. Integrating the equation for conditions where the contaminant concentration is initially constant n_{A0} with height below clouds, we find [29],

$$n_A(t) = \alpha n_{A0} + \beta n_{A0} e^{-\omega_1 t} + (h'/h)\gamma n_{A0}\, e^{-\omega_2 t} \tag{21}$$

The coefficients α, β, and γ are constants and h' is the virtual height of the cloud base and depends on the decay rate of the concentration profile with altitude. In the case of washout in the subcloud layer, we have

$$\gamma = (h/I)\omega_2$$

where h is the height of cloud base and I is the intensity of precipitation.

Andersson [29] has generalized Eq. (21) to include the case where the contaminant levels decay exponentially with height so that the time-averaged concentration during a rainfall (the concentration measured in a sample of some duration) is given by

$$\bar{n}_A = n_{A0}\left[\alpha + \frac{\beta}{\omega_1 t}(1 - e^{-\omega_1 t}) + \frac{h'}{h}\frac{\gamma}{\omega_2 t}(1 - e^{-\omega_2 t})\right] \tag{22}$$

A plot of the second and third terms in the brackets of Eq. (22) gives an indication of the relative importances of rainout and washout based on this simple model. For a typical case where $\beta = 10\gamma$ and $\omega_1 = 10\omega_2 = 10^{-4} \sec^{-1}$, the relative changes in trace constituent concentration are indicated in Fig. 5 for two different values of the ratio h'/h. When this ratio is unity, corresponding

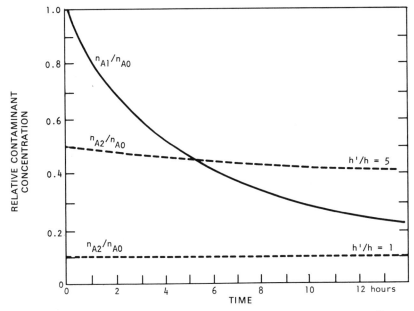

Fig. 5. Contaminant concentration in relative units for rainfall of different durations [29]. Here, n_{A1}/n_{A0} is the in cloud removal, given by the second term in brackets of Eq. (22). n_{A2}/n_{A0} is the below-cloud removal, given by the third term in the brackets of Eq. (22).

to the case where n_{A0} is constant with height, rainout is more important for removal than washout. However, if the ratio $h'/h = 5$, or the ground level concentration is 15 times larger than the ambient subcloud level concentration n_{A0}, washout may become more important, particularly during a persistent period of precipitation. The latter case is believed to correspond roughly to conditions over a city.

The marked change in rainout shown in Fig. 5 with duration of precipitation has been observed in rain sampling data. Earlier studies have been reviewed by Junge [3]. Georgii and Wötzel [30] have reported newer data confirming such a decrease in trace element concentration with rate of rainfall (or duration if the storm is considered an advected air mass).

The arguments using such a model for precipitation removal as Eq. (20) are highly simplified in several respects, including exclusion of time dependence in the coefficients and disregard of the sensitivity of rainout and washout to droplet size. There is ample evidence in the literature that differences in trace element concentration exist with droplet size. Recent data reported by Georgii and Wötzel [30], for example, indicate that the highest concentrations of the ions $SO_4^=$, Cl^-, NH_4^+, Ca^{++}, NO_3^-, K^+, and Na^+ are found in the smallest droplets originating from clouds with tops below the freezing point. Qualitatively, this is consistent with the attachment pattern based on considerations of diffusion rates in air in combination with the presence of a larger number of smaller droplets in clouds. Georgii and Wötzel also noted that evaporation of drops below the cloud base may be significant in shaping the trace constituent spectrum in precipitation. The relationship between nucleation and hygroscopic particles also may influence the distribution of constituents by raindrop size.

2.3. Chemical Reactivity and Sulfur Dioxide

In many circumstances, the fate of water-soluble trace gases in the atmosphere is complicated by their chemical reactivity in dry air. Perhaps the case studied most extensively is that of sulfur dioxide. Although precipitation unquestionably accounts for the removal of substantial amounts of SO_2, direct oxidation of this gas in the troposphere also is an important factor in its disappearance. Robinson and Robbins [8], for example, estimate that 80% of the atmospheric sulfur deposited on the earth comes from rain, while the remainder is from dry deposition. There is considerable evidence from the study of natural aerosols that both dry and wet processes involving aerosols play a role in sulfur removal.

Some discussion is included in Chapter 4 on the removal of SO_2 by oxidation and aerosol formation. However, this process is of sufficient importance that it should be considered here also.

Oxidation without Liquid Water. In the absence of water droplets, the oxidation of sulfur dioxide in the atmosphere may take place by homogeneous gas-phase reactions or by reactions on the surface of aerosol particles. Several years ago, Leighton [16] reviewed the possible reactions in the atmosphere, finding that the literature up to that time was difficult to interpret. The studies reported pictured a complicated kinetic scheme with differences of more than an order of magnitude in oxidation rates depending on reactants, reactant concentration, types of irradiation, and types of reaction vessels.

The bond dissociation energy of $SO_2 \rightarrow SO + O - 135$ kcal by photochemical or thermal excitation is much too large to occur under atmospheric conditions. Therefore, Leighton proposed that the oxidation mechanism would follow a chain mechanism involving a peroxide SO_4:

$$SO_2 + hv \rightarrow SO_2^*$$
$$SO_2^* \rightarrow SO_2^{**} + M \rightarrow SO_2 + M \qquad (M = N_2, O_2, \text{ or } SO_2)$$
$$+$$
$$O_2 \rightarrow SO_4$$
$$SO_4 + SO_2 \rightarrow 2SO_3$$
$$SO_4 + O_2 \rightarrow SO_3 + O_3$$
$$SO_3 + H_2O \rightarrow H_2SO_4$$

This reaction sequence has yet to be verified, and the species SO_4 has not been detected.

Recently, Urone and Schroeder [31] have reviewed the available rate studies for SO_2 oxidation and have concluded that the reaction rates in pure air, with irradiation extrapolated to atmospheric conditions, are at least an order of magnitude too low to account for some observed removal rates. Observations of the rate of consumption of SO_2 are limited, but the reported results suggest a rate of $0.1–10\%$ min^{-1} depending on SO_2 concentration and other factors.

The early work of Schuck and Doyle [32] showed that the consumption rate of SO_2 by photooxidation could be enhanced considerably in the presence of NO_2 and hydrocarbons. Free-radical photochemical mechanisms are evidently involved, as indicated in the work of Ogata et al. [33].

The importance of water vapor in the consumption rate of SO_2 was elucidated in relation to its nucleation with SO_3 [34]. Recent studies such as those of Quon et al. [35] and Katz [36] show a significant increase in SO_2 oxidation beyond relative humidities of 50%. The former study suggested that SO_2 oxidation in moist air without hydrocarbons yielded an aerosol end-product. Yet, Urone and Schroeder [31] refer to work of Harkins and Nicksic [37] that indicates that little or no hydrocarbon was found in aerosol particles involving SO_2 oxidation in the presence of propylene and ethylene. This is also suggested in recent evidence of infrared spectra of aerosols in smog described by Stephens [38].

Photochemical oxidation with nitrogen oxides, water vapor, and hydrocarbons in polluted atmospheres appears to offer one avenue for increased oxidation rates compatible with atmospheric reactions. Another possibility is surface-catalyzed reactions on solid particles containing metal salts. Little is known about this case, but the capability for such reactions will be limited by adsorption on the surface area of atmospheric aerosol particles. For mass loadings found in the atmosphere of $100–200\ \mu g$ particles m^{-3}, the total surface area will be of the order of $10^{-3}–10^{-4}\ m^2\ m^{-3}$. Thus, one would expect that the aerosols in the atmosphere would have only a limited adsorption capacity for SO_2, with concentrations of $< 200\ \mu g\ m^{-3}$, especially with a variety of other adsorbable trace materials present.

Oxidation in Hydrometeors. A third possibility lies in the reaction of SO_2 in aerial suspension of aqueous particles. In the 1950's, Johnstone and Coughanowr [39] studied the oxidation of SO_2 in droplets containing metal salts. The results of this investigation indicated that 1 ppm of SO_2 could be oxidized at a rate of 1% min^{-1} for 0.2 g water m^{-3}, more than one hundred times the rate for photochemical oxidation in the gas phase. Such liquid water contents are not excessive in a cloudy atmosphere.

Parallel work of Junge and Ryan [40] combined with later studies of Van den Heuvel and Mason [41] and Scott and Hobbs [42] has suggested that ammonia is an important factor in increasing the rate of SO_2 oxidation in cloud and fog drops. Based on equilibrium considerations, there is ample ammonia available to help neutralize the sulfuric acid in droplets as it is formed.

Van den Heuvel and Mason measured the rate of formation of ammonium sulfate in water droplets exposed to air containing known concentrations of sulfur dioxide and ammonia. They determined that the mass of sulfate formed was proportional to the product of the surface area of the droplets and the time of exposure, giving possible evidence of a gas-diffusion-limited rate mechanism. When SO_2 alone was in the air, the sulfate formed was at least two orders of magnitude less than with gaseous ammonia present.

Considering the different equilibria potentially involved in the catalytic influence of aqueous ammonia on SO_2 oxidation, Scott and Hobbs [42] postulated that the rate-determining step for sulfate formation involves the the $SO_3^=$ ion. The rate of production of $SO_4^=$ then is given by the first-order rate expression

$$d[SO_4^=]/dt = k_A[SO_3^=] \tag{23}$$

where $[SO_3^=] = 0.84 \times 10^{-10} p_{SO_2}/[H^+]^2$, k_A is taken to be 0.1 min^{-1}, and p_{SO_2} is the partial pressure of SO_2. The constant on the right side comes from equilibrium constants in several possible reactions involving H^+, $SO_3^=$, HSO_3^-, SO_2, and $SO_2 \cdot H_2O$. The electroneutrality relation

$$[H^+] + [NH_4^+] = [OH^-] + [HSO_3^-] + 2[SO_3^=] + [HCO_3^-] + 2[SO_4^=]$$

yields a relation for hydrogen ions, sulfate ions, and SO_2 partial pressure that allows Eq. (23) to be integrated numerically. The theoretical curves for the production rate of sulfate are shown in Fig. 6. These results are

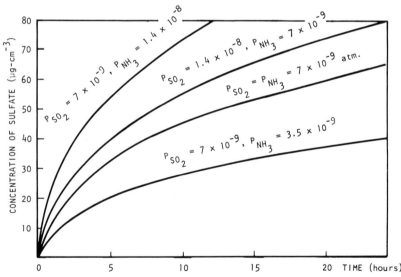

Fig. 6. Absorption rates in water droplets for the following respective sulfur dioxide and ammonia partial pressures [42]: (a) 7×10^{-9}, 1.4×10^{-8}, (b) 1.4×10^{-8}, 7×10^{-9}, (c) 7×10^{-9}, 7×10^{-9}, (d) 7×10^{-9}, 3.5×10^{-9} atm.

qualitatively similar to data of Junge and Ryan, and Van den Heuvel and Mason, but do not reflect a limiting value of sulfate or a direct proportionality between $[SO_4^=]$ and p_{SO_2}. The theory does provide insight, however, into the behavior of the SO_2–NH_3–liquid water system.

The calculations of Scott and Hobbs did not take account of the other potential rate-limiting step, the diffusion of SO_2 or NH_3 from the air to droplet surfaces. Experiments of Terraglio and Manganelli [43] determined the rate of absorption of SO_2 in aqueous solutions. Over the concentration range of 0.8–8.7 mg SO_2 m^{-3} (0.3–3 ppm), the absorption rate was a nonlinear function of SO_2 concentration. The pH of water droplets achieved a value of 4.0 or less, comparable with some observations in fog and cloud droplets over large industrial complexes, but the total solubility of SO_2 did not follow Henry's law, as expected from Postma's [26] conclusions. The recent calculations of Miller and de Pena [44], modifying the Scott–Hobbs [42] results for gas diffusion, are consistent with the experiments of Terraglio and Manganelli [43].

There is some evidence supporting the significance of clouds in the removal of SO_2 from the atmosphere. Georgii [45], for example, has reported distinct differences in the vertical distributions in arid areas of the western United States as contrasted to areas having more rainfall in Europe. In the latter case, the $SO_2/SO_4^=$ ratio decreases much faster than in the former case.

The rate of removal of SO_2 in the atmosphere in the absence of fog and clouds appears to be too rapid to be explained solely by "dry" oxidation processes, so that work is continuing on the study of the photochemical oxidation of SO_2 in the presence of other pollutants. The recent investigations of Katz [36] and Urone et al. [46] are of interest in that there is evidence that photochemical excitation of SO_2 during the day may be sufficiently long-lasting to allow for continued oxidation at night. Furthermore, Urone et al. [46] feel that the possible solid intermediate product nitrosylbisulfate, $NOHSO_4$, may be of significance in the relation between SO_2 and NO_2 in the atmosphere.

Regardless of the details of the complex mechanisms involving sulfur dioxide in the atmosphere, it appears that the main route for removal of this reactive gas is through a chain ending either in aerosols or by precipitation. A similar fate is expected for the nitrogen oxides NO and NO_2, leading to nitrate formation. However, much less is known about these materials and how they interact with water and aerosols in the atmosphere.

If the main source of sulfate in the atmosphere is associated not with soil or sea salt, but with SO_2, the sulfate should be found to be most rich in the small raindrops and in the smaller aerosol particles. The observations available tend to confirm this, as mentioned earlier, in the work on rain of Georgii and Wötzel [30] and is indicated for aerosols in Fig. 7. Here, an average distribution curve of $SO_4^=$ as a function of particle size is given. This curve is compared with that for chloride, a material that is believed to

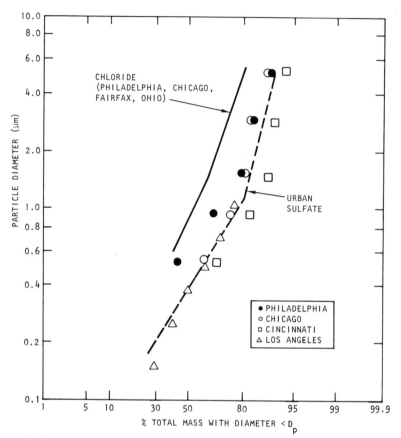

Fig. 7. Distribution of sulfate and chloride with respect to size in urban aerosol (data from Refs. 71–73).

be at least partially maritime in origin, associated more with larger aerosols by virtue of the spray and bubble-breaking generation process at sea.

The residence time for SO_2 in the atmosphere has been estimated by Robinson and Robbins [8] from emissions and from ambient concentration levels to be of the order of a few days. Assuming on the basis of evidence available that the rate of consumption of SO_2 is first order in this reactant, then a half-life based on a reaction rate of 0.1% min^{-1}, identified either with dry oxidation processes involving NO_x and hydrocarbons or ammonia-catalyzed reactions in clouds or rain, is compatible with the other geochemical calculations.

3. REMOVAL MECHANISMS FOR AEROSOLS

The discussion so far has indicated the significance of the presence of aerosols in the atmosphere as they relate to the removal of gaseous pollutants.

Of course, it is well known that the aerosols themselves also represent an important natural or anthropogenic pollutant. The removal of aerosol particles follows a pattern similar to gases in that the earth's surface is the great source and ultimate receptacle for such material. Precipitation is an important factor in returning material to the earth and its oceans, though in some circumstances it is not the overriding factor. The removal of aerosol particles is more complicated than that of gases in that it involves the continuous "transport" of small particles up the size spectrum to larger ones by collision processes.* As in the cases of gases, we can construct a theory for aerosol removal in the atmosphere by considering the concentrations of particles in many size ranges as individual "species" of trace contaminants. If one considers the highly diverse chemical composition as another identifiable classification in addition to particle size, the bookkeeping on aerosol species or components becomes almost hopelessly complicated.

To formulate a dynamical model for aerosols in the atmosphere, one can proceed by introducing the composition probability density function, an analogy to the concentration of a trace gas (e.g., Ref. 47). The state of such a system can be described mathematically by introducing a probability density function (pdf) for the chemical composition of the aerosol: Let dN be the number of particles per unit volume of gas containing molar quantities of each chemical species in the range between c_i and $c_i + dc_i$ with $i = 1, 2, \ldots,$ k, where k is the total number of chemical species. Then

$$dN = N_\infty f(c_1, c_2, \ldots, c_k, t)\, dc_1\, dc_2 \cdots dc_k, \qquad (24)$$

with N_∞ the total number of particles per unit volume of gas, defines the composition probability density function of the aerosol. Since the integral of dN over all values of N gives N_∞, the integral of f over all of the phase space represented by the variables $c_1, c_2, \ldots$ must be equal to unity:

$$\int \cdots \int f(c_1, c_2, \ldots, c_k)\, dc_1\, dc_2 \cdots dc_k = 1 \qquad (25)$$

It is clear that the composition phase space can be extended to include electrical charge.

The density of any species in the atmosphere (mass per unit volume of air) is given by the expression

$$\rho_j = M_j N_\infty \int c_j f\, dc_i \qquad (26)$$

where M_j is the molecular weight of species j.

---

*The residence time of an airborne particle can most simply be defined as the time for that particle to be deposited on the earth, so that "removal" from a given particle size class really is only a step in final deposition (see also p. 167).

In certain cases, it is more convenient to employ a size-composition pdf, $g(v, n_2, \ldots, n_k)$, defined as follows:

$$dN = N_\infty g(v, c_2, \ldots, c_k) \, dv \, dc_2 \cdots dc_k \tag{27}$$

The particle volume v is given by the expression

$$v = c_1 \bar{v}_1 + c_2 \bar{v}_2 + \cdots + c_k \bar{v}_k = \sum_i c_i \bar{v}_i \tag{28}$$

where $\bar{v}_i$ is the "effective" partial molar volume of species i. That is, $\bar{v}_i \delta c_i$ represents the change in the particle volume resulting from the addition of δc_i moles of species i. If the particles are solutions (solid or liquid) and not aggregates, $\bar{v}_i$ is the usual thermodynamic partial molar volume. The two pdf's are related by the expression

$$g(v, c_2, \ldots, c_k) = |\, \delta(c_1, c_2, \ldots, c_k)/\delta(v, c_2, \ldots, c_k)\,| \, f(c_1, c_2, \ldots, c_k)$$
$$= (1/\bar{v}_1) f(c_1, c_2, \ldots, c_k) \tag{29}$$

Here, the factor enclosed in vertical bars is the appropriate Jacobian for the coordinate transformation between the pdf's. The particle size distribution function of coagulation theory is obtained from g as follows:

$$n(v) = \int \cdots \int g(v, c_2, \ldots, c_k) \, dc_2 \cdots dc_k \tag{30}$$

3.1. Dynamical Processes and Aerosol Removal

Suppose a volume of gas is considered containing an aerial suspension specified by the pdf as specified by Eq. (24). This elemental volume is considered to travel with the turbulent wind field as an identifiable parcel. Then the change with time in species *in the identified* volume is given by*

$$d[N_\infty g(v, c_i, \mathbf{x}, t) \, dc_i \, dv]/dt$$

$$= \begin{pmatrix} \text{net gain by production from physical} \\ \text{and chemical processes (A)} \end{pmatrix} + \begin{pmatrix} \text{net gain or loss by collisions} \\ \text{between particles (B)} \end{pmatrix}$$

$$+ \begin{pmatrix} \text{net gain or loss by condensation} \\ \text{and absorption of trace gases (C)} \end{pmatrix} - \begin{pmatrix} \text{loss by deposition} \\ \text{or fallout (D)} \end{pmatrix} - \begin{pmatrix} \text{loss by diffusional} \\ \text{transport (E)} \end{pmatrix}$$

$$- \begin{pmatrix} \text{loss by scavenging} \\ \text{(collisions by fallout) (F)} \end{pmatrix} - \begin{pmatrix} \text{loss by raincloud} \\ \text{processes (G)} \end{pmatrix} - \cdots \tag{31}$$

---

*For the atmosphere, the additional independent variable, the space vector $\mathbf{x}$, has been added to the specification of g.

The processes represented by D–G actually remove aerosol material from the control volume. However, the terms A–C essentially either generate new aerosol material or transfer material to larger ranges of size *within* the control volume. We shall consider material as "lost" or removed from a given size and compositional class by either category of process.

It is worth recognizing at the outset that the size distribution of atmospheric aerosol particles is quite broad, covering effectively several orders of magnitude in size (e.g., Fig. 4, Chapter 2). Therefore, we expect that the controlling mechanisms for particle removal may differ markedly depending on the range of sizes under consideration. For example, the largest particles are likely to be lost by sedimentation, while the tiny Aitken nuclei with radii less than 0.05 μm will probably disappear mainly by coagulation or diffusional deposition on surfaces. It is likely that major differences also are reflected in different ranges of altitude. For example, removal mechanisms at the ground should be quite at variance with those at cloud level. The differences in removal rates or transport mechanisms will result in different profiles of concentration with altitude for certain size fractions. This can be seen readily by considering typical profiles for three successively larger classes of particles as indicated in Fig. 8. Although the data are very scanty, particles in the Aitken range retain lower concentrations at higher altitudes than the larger particles relative to their ground-level concentrations.

Let us consider first the dry processes leading to actual removal of particles from a volume element. Later, further attention will be devoted to

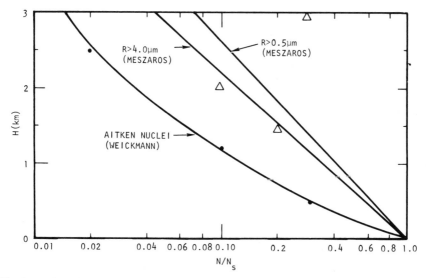

Fig. 8. Vertical distribution of Aitken nuclei compared with distributions of larger particles. For comparison, triangles are Blifford and Ringer's [74] results for $0.13 \lesssim R \lesssim 5.5 \mu$m. (Data from Refs. 75–77.)

processes contributing to the effective loss of particles by transfer to larger or smaller particle classes in the size-composition spectrum. For simplicity, the discussion below will cover only the simpler case of size distribution function for a chemically homogeneous aerosol. Thus the dynamics will be developed in terms of $n(v, \mathbf{x}, t)$ as defined in Eq. (30).* Equation (31) of course, can be rewritten conceptually for such a function.

Sedimentation. Perhaps the most common example of a removal process for an aerial suspension is fallout by gravity acting on a particle of larger mass than the surrounding medium. Since aerosols are normally defined rather loosely as aerial suspensions with particles less than about 20 μm radius, the calculation of the terminal settling velocity of such objects is relatively simple. If this is the maximum size range to be considered, the regime to be considered is that for particle Mach number less than unity [2]. This restriction implies that the particle Reynolds number is also less than unity for all Knudsen numbers of interest. In other words, the sedimentation velocity is given by

$$q_s = Bmg \quad \text{for} \quad \text{Re}_p \equiv Rq_s/v < 1 \tag{32}$$

where B is the particle mobility, m the particle mass, and g the gravitational acceleration. The particle mobility has been deduced empirically from experiments as

$$B = (6\pi\mu R)^{-1}\{1 + [1.257 + 0.400\exp(1.10\lambda/R)]\} \tag{33}$$

where λ is the mean free path of the suspending medium and μ is the viscosity of the medium. For the particle Knudsen number λ/R approaching zero, Eq. (32) reduces to the well-known formula for the sedimentation velocity of spheres based on the Stokes drag, $6\pi\mu Rq_s$.

The values of the sedimentation velocity from Eqs. (32) and (33) are plotted in Fig. 9. As shown, the velocity of gravitational fallout drastically decreases for particles smaller than a few microns radius.

To evaluate the loss from an elemental volume, the sedimentation velocity in itself is not of direct interest. Instead, the more important calculation is the loss of particles per unit volume per unit time. In these units, the loss rate L_s is given by

$$L_s = -q_s(\partial n/\partial y) = -\mathscr{S}R^2(\partial n/\partial y) \tag{34}$$

where $\mathscr{S} = (2g/9\mu)\rho \cong 1.3 \times 10^6$ for particles of unit density (ρ) in air at 20°C. Here, y is the height (vertical) coordinate.

---

*Ordinarily, the assumption is made that the particles are spherical in deriving a quantitative form of Eq. (3). The development of the theory of removal processes becomes much more complicated for irregular particles.

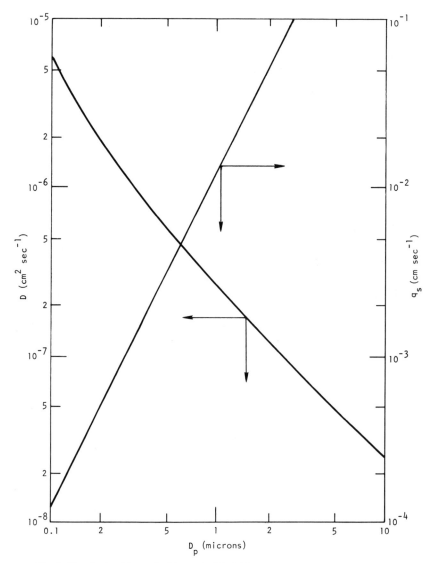

Fig. 9. Terminal settling velocities and diffusivities of spherical aerosol particles.

Diffusional Transport. In the same way that gas molecules are transported by diffusional processes, aerosol particles may migrate in and out of a control volume either by their own thermal agitation (Brownian motion) or turbulent eddying of air.

The Brownian diffusion coefficient for aerosol particles is given in terms of their mobility as

$$D = BkT \qquad (35)$$

where T is the absolute temperature. For air at 20° T, D is plotted as a function of particle size in Fig. 9. In general, these diffusivities are exceedingly small compared with the molecular diffusivities of gases. In the atmosphere, the turbulent eddy diffusivities D_T for mass transport may range from 10^2 to 10^5 cm^2 sec^{-1}, depending on the proximity to the ground and conditions of local hydrostatic stability of the air. Thus, turbulent diffusion will far exceed the effects of Brownian motion for atmospheric transport, except in very thin layers near surfaces like leaves or stones, where the air motion will be slow and laminar in nature.

In the absence of horizontal concentration gradients, the loss rate per unit volume for diffusional transport is given by [48]

$$L_D = -D_T(\partial^2 n/\partial y^2) \qquad (36)$$

Deposition on Obstacles and Surfaces. At the lower boundary of the atmosphere, particles may be removed permanently from the air by deposition on objects. Deposition may take place by one of three mechanisms: *sedimentation, diffusion,* or *impaction.* The first two already have been described briefly. The third involves the trapping of particles by virtue of the fact that they do not follow exactly the gas flow around an object. This is illustrated in Fig. 10 for two possible geometries in vegetation. Here, the inertial forces acting on particles with much larger masses than the surrounding gas molecules tend to prevent the particles from following the gas flow. This leads to a collision with the object at some location on its surface provided that the particle is near enough to a surface following a trajectory (A) upstream of the collecting body.

A fourth mechanism for deposition is sometimes separated from the other three. This is *interception.* Interception only refers to the fact that, geometrically, the particles are large in radius compared with gas molecules.

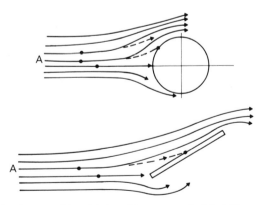

Fig. 10. Schematic diagram of inertial deposition on a cylindrical fiber and a flat surface (a leaflet).

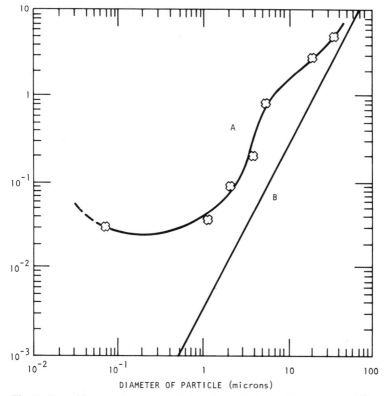

Fig. 11. Deposition velocity and terminal settling velocity for flow over grass [49].

Hence estimates of impaction rates assuming point particles will be too low by a factor of the distance normal to the collecting surface equal to the particle radius.

As in the case of gases, the removal of aerosols on surfaces may be expressed in terms of a deposition velocity q_g as defined by Eq. (6). Some values of particle deposition velocities over smooth, flat surfaces have been shown as a function of Schmidt number v/D in Fig. 3. For flat surfaces roughened with grass, Chamberlain [49] has reported some values of q_g which are compared with sedimentation velocities in Fig. 11. The values of q_g for aerosol particles are always greater than or equal to q_s.

An interesting feature of aerosol deposition is reflected in the data of Figs. 3 and 11. There is a suggestion of a minimum deposition velocity at a radius of about 0.1 μm. It is well known in filtration theory, for example, that there exists a minimum in deposition rate resulting from the falloff of importance of inertial impaction, with an accompanying increase in diffusional deposition from increased Brownian motion. The "minimum" in these complimentary processes is expected to occur at about 0.1 μm.

There is very little direct information about the removal rate expected as a result of deposition on trees, buildings, or other objects at the earth's surface. The question is complicated, too, by the possible influence of electrical charge and reentrainment of particles after collection. Neither of the processes can be readily accounted for by theory at present, though both mechanisms have concerned investigators.

An interesting series of experiments was conducted in a wind tunnel on single conifer needles and conifer trees by Langer [50] and Rosinski and Nagamoto [51]. This is one of the few studies in the literature that gives information of the relative importance of inertial effects, electrical charging, and reentrainment. For single needles or leaves, electrical charging of $\sim 2\mu$m-diameter fluorescent ZnS dust with up to eight units of charge had no detectable effect at wind speeds of 1.2–1.6 m sec^{-1} [50]. The average collection efficiency was found to be $\sim 6\%$ for edgewise cedar or fir needles, with broadside values an order of magnitude lower. Bounce-off after striking the collector was not detected, but reentrainment could take place above ~ 2 m sec^{-1} wind speed. Tests on branches of cedar and fir by Rosinski and Nagamoto [51] suggested similar results as for single needles. The average scavenging efficiency was defined as the ratio of the number of particles deposited on the tree to the dose corresponding to the tree. This efficiency ran from 1 to 10% over the range 0.2–0.5 m sec^{-1}. Reentrainment was observed in early stages of a test, particularly on the upstream side of the trees, but it practically ceased after several hours of exposure. The average area coverage after several hours of exposure to 2.4-μm-diameter ZnS particles was about 0.4%.

One of the few studies of aerosol removal from the atmosphere by trees was reported a few years ago by Neuberger *et al.* [52]. They made measurements of ragweed pollen concentrations (large particles) and Aitken nuclei concentrations inside and outside a forested area. The data suggested that more than 80% of the pollen was removed in a dense coniferous forest. Parallel laboratory investigations showed that as high as 34% of the upstream concentration of Aitken nuclei would be removed by coniferous trees. Deciduous materials could only account for nearly a factor of two less removal of submicron particles.

In the absence of any better information, we must rely on the limited results of Neuberger *et al.* [52] and Rosinski and colleagues [50,51], in combination with data like Chamberlain's [49] for the deposition velocity over rough surfaces, to evaluate losses at the earth's surface.

Collision Processes in Aerial Suspensions. Without clouds or rain, collisions may take place between suspended particles in air by one of several mechanisms creating a relative velocity between the particles. These include: (a) Brownian motion, (b) turbulence, (c) gravitational settling (scavenging by fallout), (d) electrical attraction by species of different polar charge, and (e) attraction resulting from diffusiophoretic and thermophoretic forces. The first three are those most often considered as contributors to the transfer of particle from smaller size ranges to larger ones. However, collisions resulting from electrical and other phoretic forces also may be important factors in leading to agglomeration. Although relatively well understood in principle, these latter processes are more difficult in practice to evaluate in the atmosphere because of uncertainties in physical parameters like charge level and because of concentration–temperature gradients near suspended particles.

Collisions due to Brownian motion affect mainly the smallest particles in the atmosphere, whose thermal agitation is largest. The net loss by Brownian coagulation has been calculated theoretically in two extreme ranges of Knudsen number. In the absence of electrical forces, the loss rate is given by [2]

$$L_B = \frac{2}{3}\frac{kT}{\mu}\left[(v_1^{1/3} + v_2^{1/3})\left(\frac{1}{v_1^{1/3}} + \frac{1}{v_2^{1/3}}\right)\right]n(v_1)n(v_2), \qquad \text{Kn}\to 0 \qquad (37)$$

$$L_B = -(v_1^{2/3} + v_2^{2/3})\left[\frac{8kT}{\rho}\frac{(v_1 + v_2)}{v_1 v_2}\right]^{1/2}\Omega n(v_1)n(v_2), \qquad \text{Kn} \gtrsim 0.5 \qquad (38)$$

where

$$\Omega = \left\langle 1 - \frac{\pi}{6}\left(1 + \frac{\pi}{8}\right)\left(\frac{v_1 + v_2 m_G}{\rho v_1 v_2}\right)^{1/2}\frac{16 n_G(v_1^{2/3} v_2^{2/3})}{v_1 + v_2}\right.$$

$$\times \left\{\exp\left(-\gamma\frac{v_1^{1/3} + v_2^{1/3}}{\lambda}\right) + \frac{\gamma(v_1^{1/3} + v_2^{1/3})}{\lambda}\left[0.577 + \ln\frac{\gamma(v_1^{1/3} + v_2^{1/3})}{\lambda}\right.\right.$$

$$\left.\left.\left. + \sum_{\varepsilon=1}^{\infty}(-1)^{\varepsilon}\left(\gamma\frac{(v_1^{1/3} + v_2^{1/3})}{\lambda}\right)^{\varepsilon}\right]\right\}\right\rangle$$

The subscript G refers to the suspending gas and $\gamma = \text{const}\,(= \pi/4.8)$. There is a range of Knudsen number in which the detailed theory has not been developed. However, an estimate of the collisional behavior can be made as indicated in Fig. 12. Here, the rate of collision for equal-sized spheres is plotted as a function of Knudsen number relative to the rate at $\text{Kn} \to 0$. Evidently, there is a maximum in collision rate at $\text{Kn} \approx 5$. The main contribution to coagulation of atmospheric particles by Brownian motion takes place below 0.1 μm, or at $\text{Kn} > 1$.

For particles larger than about 0.1 μm radius, collisions in the atmosphere will be increasingly controlled with size by turbulence and by gravitational fallout. The theory for collision rates due to turbulent motion involves two mechanisms, one arising from the inability of particles to follow the rotational motion of small turbulent eddies in air, and the other from the velocity inhomogeneities associated with the turbulent atmosphere itself. The analytical model generally adopted is that of Saffman and Turner [53], for example, which assumes isotropic turbulence in the fine-scale structure of

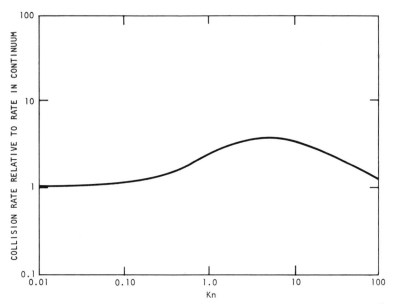

Fig. 12. Relative collision rate for equal-sized spheres undergoing Brownian motion [79].

atmospheric turbulence. For this case,

$$L_T = -1.3(\varepsilon/v)^{1/2}(v_1^{1/3} + v_2^{1/3})^3 n(v_1)n(v_2) \tag{39}$$

where ε is the dissipation rate of kinetic energy of turbulent air.

Calculation of the loss rate per unit volume from Eq. (39) suggests that turbulent coagulation has to be weak for all particle size ranges in the atmosphere compared with other mechanisms except under certain extreme conditions (see, e.g., Hidy [48]).

Collisions resulting from the capture of one particle falling through a cloud of other particles can only occur from unequal-sized particles. For falling spheres, the coagulation rate is given by

$$L_S = -\pi(v_1^{1/3} + v_2^{1/3})^2 E_p(v_1, v_2)\,|\,q_{s1} - q_{s2}\,|\,n(v_1)n(v_2) \tag{40}$$

where v_2 is the volume of the larger particle, E_p is the collision efficiency, and $|\,q_{s1} - q_{s2}\,|$ is the velocity difference between the two particles. The collision efficiency E_p is a complicated function of the particle radius ratio $(v_1/v_2)^{1/3}$.

For small particles with ratios of $v_1/v_2 \lesssim 0.1$, an approximation for E_p may be derived without consideration of the aerodynamic forces of interaction between particles. If the particles are spherical, Friedlander [54] calculated that

$$E_p \approx 1 - \frac{3}{2[1 + (v_1/v_2)^{1/3}]} + \frac{1}{2[1 + (v_1/v_2)^{1/3}]^3} \tag{41}$$

based on the theory of interception without Brownian diffusion.

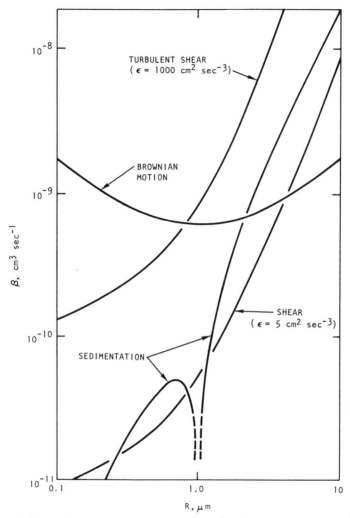

Fig. 13. Comparison of different collision mechanisms in tropospheric aerosols [54].

Using the idealized collision efficiencies for different encounter mechanisms, one can estimate the relative importance of these processes for various size classes of aerosols. An illustration of such a calculation is shown in Fig. 13. Here, the differences in the collision rate parameter

$$\beta = L_s/n(v_1)n(v_2)$$

are indicated for Brownian coagulation based on Eq. (37), turbulent coagulation based on Eq. (39), and coagulation by differential settling based on Eq. (41). The calculation was made after the method of Friedlander [54] for a

1-μm particle interacting with particles of radius R. The turbulence (shear) result was estimated for two different energy dissipation rates, the higher one being characteristic of air flow over the ground and the lower one being relevant to air near cloud base. From sample calculations of this type, one can readily appreciate that coagulation by Brownian motion may dominate collision processes of very small particles of $\lesssim 0.5$ μm radius, but turbulence

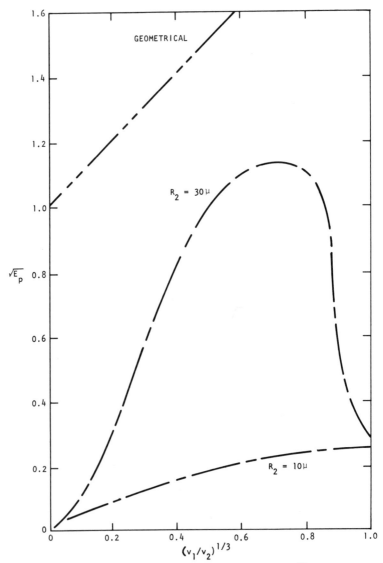

Fig. 14. Efficiency of collision for falling spheres [55].

and differential settling become overwhelmingly important at larger particle sizes.

Recent work has indicated that the collision efficiency between falling particles has to take into account the aerodynamic forces associated with the interaction of the bodies as they approach one another. This function has been evaluated to a reasonable accuracy only for spheres falling in a stagnant medium in the Stokes range of drag. Accounting for the aerodynamic forces in interaction between the two colliding particles and excluding electrical forces (see, e.g., Davis and Sartor [55], the collision efficiency has the form indicated in Fig. 14. The theory predicts a maximum in collision efficiency at $(v_1/v_2)^{1/3} \approx 0.6$ for particles larger than about 20 μm, with minimum values at the extremes of this ratio.

The net effect of collisions by different mechanisms is assumed to be additive, so that the influence of all mechanisms is considered to linearly depend on each of the processes.

The coagulation of aerosols in the atmosphere can be enhanced if bipolar electrical charging exists in the aerial suspension. Calculations [2] suggest that electrical forces will be most important for the smallest particles in the atmosphere, except in extreme cases near thunderstorms where electrical effects may be of significance for the larger particles, too. Recently, Sartor [56] has reported collision efficiencies for electrically charged hydrometeors in clouds. Such calculations could be extended to large aerosols with corrections for the electrical properties of the aerosol particles.

Diffusiophoresis and thermophoresis depend respectively on the strength of a vapor concentration gradient near an evaporating droplet or condensing droplet and on temperature gradients near particles. In the atmosphere, such gradients are relatively small, with the possible exception of those encountered near hydrometeors. In general, the diffusiophoretic forces and thermophoretic forces induced around condensing or evaporating hydrometeors are estimated to be small, except for extremely severe conditions of water vapor transport near the body. It is normally believed, then, based on theoretical evidence, that these processes contribute to only a small fraction of the removal of aerosols. However, the case against the importance of such forces in the atmosphere remains to be explored in further detail before discounting these effects completely, particularly in view of Slinn and Hales' [59] conclusions.

Rainout and Washout. Small suspended particles have been detected and studied in rain water for sometime (e.g., Junge [3]). Thus, it is known that aerosols, like trace gases, are removed efficiently by precipitation processes.

The estimation of the loss of particles by rainout involves consideration of the complicated and interacting factors that may contribute to scavenging of aerosols by cloud droplets or ice crystals. One mechanism for trapping particles comes about from the nucleation of the hydrometeor. Others involve migration of particles to the cloud drop by (a) phoretic forces during condensation or by evaporation, (b) electrical interactions, or (c) thermal agitation. Scavenging also will occur as the larger cloud particles fall downward through nuclei present in the surrounding air.

It is difficult to estimate the magnitude of the loss rate of aerosols by in-cloud processes. However, results such as those reported by Rosinski and Kerrigan [57] would place the removal efficiency large enough to yield 10^3–10^6 insoluble particles per cubic centimeter of precipitated water.

The minimum rate of removal by rainout will consist of the nucleation rate plus the loss by collisions during fallout as calculated by Eq. (40). Typically, clouds have 100 droplets cm^{-3} and last from 10–20 min. This would require a removal by nucleation of 0.1 cm sec^{-1} or higher. The rate of removal by collisions can be written in terms of the cloud droplet spectrum as

$$L_{CL} = -\omega n(v_1) = -n(v_1) \int_0^\infty \pi(v_1^{1/3} + v_p^{1/3})^2 q_{sp}(v_p) E_p(v_1/v_p) N(v_p) \, dv_p \qquad (42)$$

where the subscript p refers to the cloud droplets and $N(v_p) \, dv_p$ is the concentration of droplets in the size range v_p and $v_p + dv_p$. The collision efficiency E_p has been calculated for the extreme case of $(v_1/v_p) \to 0$ by Zebel [58] for submicron spheres diffusing by Brownian motion to spherical droplets in the presence of electrical forces. For cases of larger values of the radius ratio, aerodynamic forces have to be considered and curves such as those in Fig. 14 have to be used. Some estimates of the efficiency curves have been calculated incorporating electrical forces and these have been investigated by Sartor [56], for example.

Washout is somewhat easier to describe than rainout. The principal mechanism of removal here is the scavenging of particles during fallout of precipitating hydrometeors below cloud base. If the falling bodies are evaporating, a temperature gradient develops between the object and the surrounding air. Slinn and Hales [59] have estimated that this gradient may become sufficiently large that thermophoresis may have to be considered in the washout mechanism. For our purposes here, however, the only important mechanism is the gravitational collision process, following a loss relation like Eq. (42). Because the precipitating objects will have equivalent diameters exceeding 100 μm, the collision efficiency will differ from that estimated for the Stokes fluid dynamic range (Re$_p$ < 1). As an approximation for $v_1/v_p \ll 1$, the result for potential air flow around a sphere is used [2] with account of the interception effect, or

$$E_p \approx \frac{(Stk)^2}{(Stk + 0.25)^2} + \left(\frac{v_1}{v_p}\right)^2 + 2\left(\frac{v_1}{v_p}\right) \qquad (43)$$

where Stk $= 2q_{sp}\rho_1 v_1^{2/3}/9\mu v_p^{1/3}$ for captured particles having radius $\gtrsim 0.5$ μm.

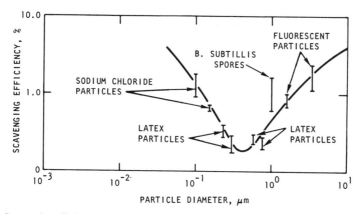

Fig. 15. Scavenging efficiency of 3-mm ice crystals. Vertical lines indicate range of experimental data [78].

For smaller particles undergoing Brownian motion, Slinn [60] has calculated the collision efficiency for colliding spheres to be

$$E_p \approx (4/Pe_p)(1 + 0.39Re_p^{1/2}Sc_p^{1/3}] \approx 2D/q_{sp}R_p \qquad (44)$$

Here, the Reynolds and Peclet numbers refer to the hydrometeor radius and Pe, Re, and Sc are defined in terms of the aerosol particle diffusivity and terminal velocity of the collecting sphere.

The scavenging efficiency of irregular hydrometeors such as ice crystals or snowflakes cannot be derived theoretically without great difficulty. Some experimental observations have been reported, however. An illustrative result is shown in Fig. 15. In this diagram the results of Sood and Jackson [78] are given for 3-mm ice crystals falling through aerosols composed of particles of different sizes and compositions. A minimum in the scavenging efficiency curve appears where diffusional deposition falls off with increasing particle size, but inertial deposition (impaction) also decreases with decreasing particle size. The main departure from a self-consistent experimental curve is seen for the *B. subtilis* spores. These particles are like short cylinders in shape rather than like spheres as were the other particles used in the experiment.

Condensation and Absorption of Vapors. Taking inventory of the quantities of condensable or absorbable vapors in the atmosphere, we find that this material can contribute in a substantial, perhaps dominant way to the growth and effective "removal" of particles from a given size range in the aerosol size spectrum. It is uncertain at present what mechanism is more common in the transition of condensable vapors to particles. In some cases, preferential formation of very tiny particles may take place by polymerization or other homogeneous reactions. In other cases, heterogeneous reactions involving preexisting particles or hydrometeors may dominate. From available evidence, it would appear that much of the SO_2 gas in the atmosphere evolves into particles by heterogeneous oxidation involving fog and cloud droplets. However, it is possible that much of the organic material producing aerosols involves homogeneous reactions forming new particles in the atmosphere.

The change in the aerosol size distribution associated with heterogeneous chemical reaction is difficult to calculate since no models have been established. However, one can make a crude estimate by the following argument.

The rate of "loss" of species in each size range by condensation or absorption of reactive gases may be estimated by considering a gas-diffusion-limited process. The fractional loss of particles then will be

$$\text{fractional loss} \cong m_A \Phi_A N_i / \rho N_i \langle v \rangle \qquad (45)$$

where $\langle v \rangle$ is the average volume of particles over the size range considered. The actual loss in number volume^{-1} second^{-1} is the fractional loss times the number of particles N_i.

Relative Significance of Removal Mechanisms. A kinetic model for the aerosol size distribution function taking into account all of the various mechanisms for production and removal constitutes a very complex relationship. Such an equation can be formulated, in principle, but there are sufficient uncertainties in the modeling of individual mechanisms that analysis of such a complete model equation would not be fruitful at this time. It is possible to gain insight into the relative importance of various identified mechanisms by making estimates of the magnitudes of loss rates from the relations listed above. As an illustration, suppose we consider the variation in loss rates as a function of altitude over an urban complex. At the ground, the total number concentration of aerosols is taken to be 10^5 as measured, say, by an Aitken nuclei counter. The decay of particles over ranges of size with altitude is taken from averaged data such as that in Fig. 8. For purposes of calculations, let us use three ranges of particle size with mean radius centered at 0.05, 0.5, and 5 μm, whose concentrations are N_i. The concentration of the smallest particles $N_{0.05}$ will be identified with the Aitken nuclei counts ($= N = 10^5$ cm^{-3}), the middle range $N_{0.5}$ ($= 10^2$ cm^{-3}) with an optical scattering instrument (roughly the visibility range), and the largest range $N_{5.0}$ ($= 1$ cm^{-3}), with an impactor or total mass filter.*

To estimate the significance of vapor condensation and absorption of gases, Eq. (45) is used. For comparison with other mechanisms, let us consider a typical situation where the condensable gas A is SO_2, with a concentration at the ground $\sim 10 \ \mu$g^{-3} and at cloud base and above $\sim 0.1 \ \mu$g^{-3}.

---

*Because of the shape of the size distribution curve, the total number concentration is strongly weighted toward tiny particles, but total mass is weighted toward the larger particles.

TABLE 4. Estimated Removal Rates of Aitken Nuclei in the Troposphere (particles lost $cm^{-3}\ sec^{-1}$)

Process	Height		
	Ground (urban) $N_i = 10^5\ cm^{-3}$	Near cloud base (2 km) $N_i = 10^3\ cm^{-3}$	In or above clouds (6 km) $N_i = 10^2\ cm^{-3}$
Sedimentation	10^{-6}	10^{-8}	10^{-9}
Inertial and diffusional deposition on obstacles at the Surface ($q_g = 0.1$ cm sec^{-1})	0.1	—	—
Convective diffusion ($D_T = 10^4$–$10^3\ cm^2\ sec^{-1}$)	10^{-1}–1	10^{-4}	10^{-6}
Condensation of vapors on particles	10^4	10^0	10^{-1}
Thermal coagulation	1	10^{-4}	10^{-6}
Scavenging by differential settling*† ($R_2 = 10\ \mu m$)	10^{-3}	10^{-7}	10^{-9}
Turbulent coagulation‡	10^{-3}	10^{-9}	10^{-11}
Washout by 1-mm spherical hydrometeors ($N_p = 10^{-3}\ cm^{-3}$)	10^{-3}	10^{-5}	—
Rainout by cloud processes (nucleation + collisions) ($R_p = 10\ \mu m$)**	—	10^{-2}§	10^{-4}

*Brownian diffusion to surface included.
†Calculated for $\rho_p = 1$ with 10-m particle concentration, $N_2 = 10^{-1}\ cm^{-3}$, $N_2 = 10^{-3}\ cm^{-3}$, and $N_2 = 10^{-4}\ cm^{-3}$, respectively.
‡Calculated for turbulence dissipation rate $\varepsilon = 10^3\ cm^2\ sec^{-3}$, $\varepsilon = 1\ cm^2\ sec^{-3}$, and $\varepsilon = 0.1$ $cm^2\ sec^{-3}$, respectively.
**Calculated for $N_p = 10^2\ cm^{-3}$, cloud base; $N_p = 10\ cm^{-3}$ at 6 km.
§Aitken nuclei are assumed too small to be a factor in cloud droplet nucleation; Brownian diffusion is included in scavenging.

It is assumed that about half of the gaseous SO_2 is removed by heterogeneous oxidation reaction to $SO_4^=$ involving "condensation" on particles.*

The rate of loss by sedimentation and diffusion may be evaluated from Eqs. (34) and (36). By analogy to sedimentation, the loss in the first meters of air to a rough surface and vegetation is taken proportional to the concentration gradient at the surface, and the deposition velocity is the proportionality coefficient. Losses by various wet and dry collision processes are calculated from Eqs. (37)–(43). The turbulent dissipation rates assumed are values taken from Fig. 4.1 of Lumley and Panofsky [61]. It is assumed for our purposes that dry scavenging by sedimentation involves 10-μm particles moving

---

*This is somewhat arbitrary based on Robbins and Robinson's [8] estimates for SO_2 removal.

through a cloud of smaller particles. Rainout is estimated assuming $R_p = 10\,\mu$m and $N_p = 10^2\,\text{cm}^{-3}$ at cloud base, but $10\,\text{cm}^{-3}$ at higher altitude, and assuming a nucleation rate of particles larger than $0.1\,\mu$m of $0.1\,\text{cm}^{-3}\,\text{sec}^{-1}$. Washout is estimated for $R_p = 1$-mm drops where concentration N_p is $10^{-3}\,\text{cm}^{-3}$. The droplet concentrations are taken as typical for cumulus and stratocumulus clouds as given by Fletcher [62]. In such an evaluation, we note that essentially all of the mechanisms operate continuously with varying degrees of efficiency except for mechanisms involving cloud development and precipitation. Cloud removal processes will be intermittent in nature, even though they represent an important factor in aerosol removal. There are estimates that, on the average, the earth is covered with clouds about half the time. Perhaps it rains (or precipitates) from cloud layers about 20% of the time. Thus, one may guess that about one-tenth to one-half the world aerosol population in the lower troposphere is influenced by washout and rainout at any one time. Washout, of course, will be of greatest interest near the ground, while rainout will be of importance at cloud base and above.

The results of the calculations are indicated in Tables 4–6. Bearing in mind the great deficiencies in the methods for making our estimations,

TABLE 5. Estimated Removal Rates of Large Particles ($R_1 = 0.5\,\mu m$) in the Troposphere (particles lost cm^{-3} sec^{-1})

Process	Height		
	Ground (urban) $N_i = 10^2\,\text{cm}^{-3}$	Near cloud base (2 km) $N_i = 1\,\text{cm}^{-3}$	In or above clouds (6 km) $N_i = 10^{-1}\,\text{cm}^{-3}$
Sedimentation	10^{-6}	10^{-7}–10^{-8}	10^{-9}
Inertial and diffusional deposition on obstacles at the Surface ($q_g = 0.01$ cm sec^{-1})	10^{-5}	—	—
Convective diffusion ($D_T = 10^5\,\text{cm}^2\,\text{sec}^{-1}$)	10^{-3}	10^{-5}	—
Condensation of vapors on particles	10^{-1}	10^{-5}	10^{-6}
Thermal coagulation	10^{-4}	10^{-7}	10^{-9}
Scavenging by differential settling* ($R_2 = 10\,\mu$m)	10^{-7}	10^{-11}	10^{-15}
Turbulent coagulation*	10^{-3}	10^{-9}	10^{-11}
Washout by 1-mm. spherical hydrometeors ($N_p = 10^{-3}\,\text{cm}^{-3}$)	10^{-8}	10^{-10}	—
Rainout by cloud process* ($R_p = 10\,\mu$m)	—	10^{-1}†	10^{-1}†

*Same values of N_2, ε, and N_p as used in Table 4.
†Assumed 0.1 particle cm^{-3} sec^{-1} nucleates.

TABLE 6. Estimated Removal Rates of Giant Particles (R_1 = 5.0 μm) in the Troposphere (particles lost $cm^{-3} sec^{-1}$)

Process	Height		
	Ground (urban) $N_i = 10^{-1} cm^{-3}$	Near cloud base (2 km) $N_i = 10^{-3} cm^{-3}$	In or above clouds (6 km) $N_i = 10^{-4} cm^{-3}$
Sedimentation	10^{-7}	10^{-11}	—
Inertial and diffusional deposition on obstacles at the surface (q_g = 0.1 cm sec^{-1})	10^{-6}	—	—
Convective diffusion ($D_T = 10^5 cm^2 sec^{-1}$	10^{-6}	10^{-8}	—
Condensation of vapors on particles	10^{-6}	10^{-10}	10^{-11}
Thermal coagulation	10^{-6}	10^{-10}	10^{-12}
Scavenging by differential settling* (R_2 = 10 μm)	10^{-8}	10^{-12}	10^{-10}
Turbulent coagulation*	10^{-4}	10^{-10}	10^{-12}
Washout by 1-mm spherical hydrometeors ($N_p = 10^{-3} cm^{-3}$)	10^{-7}	10^{-9}	—
Rainout by cloud processes (nucleation + collision)* (R_p = 10 μm)	—	10^{-3}†	10^{-4}†

*Same values of N_2, N_p, and ε as in Table 4.
†Assumed 0.1 particle $cm^{-3} sec^{-1}$ nucleates if nuclei are present.

establishing the relative importance of processes leads to some interesting implications. Condensation of vapor or attachment of ions, as noted previously, is of primary significance in the smallest particle fraction, but appears to be important over all size ranges at all altitudes. Aside from condensation of vapors in the layers in urban air near the ground, turbulent coagulation, with removal by impaction and sedimentation, becomes significant for the largest particles. The vast quantities of condensable or adsorbable gases present are significant in producing an effective "loss" by transfer of particles up the spectrum, and this process appears at least potentially as important if not more important than simultaneous processes of coagulation. It appears that convective diffusion and thermal coagulation are more important contributors to removal for the small particles, but essentially all mechanisms are weak for the larger particles. Even though turbulent coagulation is comparatively weak over all ranges of particle size, it may provide an essential mechanism to transport particles to larger particle sizes above $\sim 0.5 \mu m$ radius. It is likely that condensation of vapors is also an important process in this range, however.

Depending on the altitude of cloud base and the frequency of cloud formation, the wet processes leading to precipitation should play leading roles with diffusion in removing larger aerosols away from the ground. Coagulation by Brownian motion eventually will remain active in the loss of tiny particles only as long as their concentrations remains high.

At altitudes exceeding 6 km, intermittent rainout will be the major factor in aerosol removal. However, one could expect that high in the troposphere above cloud tops, say at ~ 10 km, all mechanisms would be rather weak in removing particles compared with processes below 5 km altitude.

From the considerations here, condensation of vapor and absorption of gases, including nitrogen oxides, sulfur-containing compounds, and (high-molecular-weight) hydrocarbons, emerges as an important factor in the effective growth and removal of particles as well as for the removal of the gases themselves. This has been recognized for some time, but still very little of the physical and chemical changes involved are understood at present. Thus, this aspect of aerosol behavior in the atmosphere remains largely an open question at this time.

As can be seen readily from such crude considerations as introduced above, the problem of evaluating quantitatively the removal of aerosols is very complicated. Indeed, such characteristic scales as a residence time may have complex and diffuse definitions for aerosols. Perhaps the simplest meaningful definition for the residence time of an aerosol particle in the atmosphere is the time it takes for any "tagged" suspended particle to reach the ground. This sidesteps the issue of the dispersion and physical and chemical transformations in the atmosphere. In achieving deposition, one need only consider the particle size, initial altitude, or an average altitude for particles. From the discussion of the deposition of particles as a function of size, there is expected to be a characteristic residence time such that the largest particles are rapidly lost by sedimentation and very small particles are removed quickly by diffusional attachment or rainout. However, particles in the 0.1–1-μm size range, which experience only weak sedimentation and small inertial and diffusional removal, should remain in the atmosphere for much longer times than either the smallest or largest particles.

Simple concepts such as mean residence times may not be useful in dealing with such suspended material. It is certainly true that rough estimates may be made of removal rates as well as mean residence times of particles in the troposphere. However, the many dynamical processes that contribute more or less randomly to transport, agglomeration, and loss of particles of a given size may produce statistical fluctuations in residence time that far exceed the mean value. Under such circumstances, the average times derived from a rather simplified interpretation of available observations may have little meaning when applied to reality. Certainly further research is needed in this important area of atmospheric chemistry to achieve meaningful statistical limits of such parameters as the mean residence time for aerosol particles.

Application to Modeling Size Distribution Functions. Perhaps the most sensitive parameter to loss and production mechanisms of atmospheric aerosols is the size distribution function. A possible consequence of the chain of events contributing to aerosol aging is the tendency for the tropospheric aerosol size spectrum to be similar in shape. The regularity in shape is most commonly detected in the large-size fraction of the size spectrum, beyond radii of about 0.5 μm. In this range, called the *Junge subrange*, a power-law behavior is often observed for large numbers of spectra averaged together. The Junge subrange is given empirically by*

$$n(v) \propto \phi v^{-2} \tag{46}$$

Here, ϕ is the volume fraction of particles, given by $\int_0^\infty vn(v)\,dv$.

Although this power-law form is frequently adopted for the upper end of the spectrum, there is no theoretical basis for it derived on dynamical arguments. In fact, Brock [63] has pointed out that the observations generally used to rationalize the applicability of Eq. (46) can be fit just as well by a lognormal distribution over the range of radii of interest.

Attempts to explain the achievement of regularity in shape of the size spectrum have developed along similar lines making use of a "simplified" kinetic equation. Assuming that the aerosol is chemically homogeneous and the atmosphere is stagnant and horizontally homogeneous in aerosol concentration, Eq. (31) can be rewritten more explicitly to serve as a model:

$$\frac{\partial n(v)}{\partial t} + \frac{\partial}{\partial v} Pn(v) = S(v, t) + \frac{1}{2} \int_0^{v_1} \beta(v_1, v_1 - v_2) n(v_1) n(v_2)\, dv_2$$

$$- \int_0^\infty \beta(v_2, v_1) n(v_1) n(v_2)\, dv_2 + q_s \frac{\partial n}{\partial y} + \frac{\partial}{\partial y}\left(D_T \frac{\partial n}{\partial y}\right) \tag{47}$$

where P is a condensation term, given for Maxwellian condensation as

$$3^{1/3}(4\pi)^{2/3} \left(\frac{\rho_A l_A^2 M_A}{\kappa \mathscr{R} T^2} + \frac{\rho_A \mathscr{R} T}{D_A M_A P_{AS}}\right)$$

with l_A the latent heat of A, κ the thermal conductivity of the medium, P_{AS} the saturation vapor pressure of component A, and $\mathscr{R}$ the gas constant. The second term on the left covers the net gain from absorption and condensation of trace gases [term C in Eq. (31)]. The first term on the right is the production rate by homogeneous gas-phase reactions [Term A, Eq. (31)]. The second and third terms on the right are the collision and scavenging terms [B, F, and G

---

*The Junge distribution is normally written in terms of the radius distribution function $n(r)$, giving an r^{-4} power law.

in Eq. (31)]; the fourth term on the right is the sedimentation term [D in Eq. (31)]; and the fifth term on the right denotes the loss by diffusional transport [E in Eq. (31)]. Loss by deposition is included in the lower boundary condition. These initial and boundary conditions may be written as

$$n \rightarrow 0 \quad \text{as} \quad v \rightarrow 0$$

$$n(v) \rightarrow 0 \quad \text{as} \quad y \rightarrow \infty, \quad t > 0$$

$$n(v) = n_0(v, y) \quad \text{at} \quad t = 0 \qquad (48)$$

$$\text{either} \quad D_T(\partial n/\partial y) + q_s n = 0 \quad \text{at} \quad y = 0, \quad t > 0,$$

$$\text{or} \quad n(v) = n_s(v), \quad t > 0$$

Even this highly idealized, simplified kinetic model is far too complicated to obtain mathematical solutions with current practice. Therefore, workers have so far explored only the possible applicability to aerosol behavior of solutions of even simpler kinetic equations.

Some of the earliest work of this nature was analytical, conducted by Smoluchowski at the turn of the century for particles undergoing Brownian motion in a continuum suspended in a cloud volume. This is equivalent to setting the left side of Eq. (31) equal only to terms in B on the right, with $\beta(v_1, v_2)$ given in terms of Eq. (37) for Kn $\rightarrow 0$. Smoluchowski's early work was extended over the years by several investigators [2,6]. An interesting result was deduced from the work of Hidy [64] and Wang and Friedlander [65] was that after long times, the distribution function predicted by such a model would become similar in shape to each other and independent of the initial distribution. These classes of asymptotic solutions are known as *self-preserving*. The upper end of the solutions given by these investigators dropped as $e^{-\Lambda v}$ ($\Lambda \equiv N/\phi$), which makes a size spectrum too narrow compared with the atmospheric distribution. Furthermore, the time scale to achieve self-preservation by coagulation alone is believed to be too great to be effective in the atmosphere, except under unusual conditions.

Recently, Lai et al. [66] found that a self-preserving distribution also may exist for free-molecular aerosol coagulations, where $\beta(v_1, v_2)$ is given in Eq. (38), with $\lambda/R > 10$. This calculation indicated that the shape of the upper portion of such a distribution is practically the same as that for a continuum case. This model for collisions is much more applicable to Brownian collisions between atmospheric aerosols, yet it fails to provide a broad size spectrum compatible with atmospheric aerosol measurements. Therefore, coagulation by Brownian motion alone cannot be responsible for the observed regularity in shape of the atmospheric spectrum.

The numerical studies in the 1950's of Junge and colleagues [3] indicated that only the region around the peak of the spectrum was sensitive to sources and coagulation (i.e., the region between 0.01 μm and 0.1 μm radius). They suggested that precipitation processes played a major role in shaping the aerosol spectrum, with a necessary termination at the extreme size ranges by sedimentation.

Later, a model was developed for steady-state conditions where small particles are fed upward by coagulation resulting from Brownian motion [54]. The input in the small size range is then balanced by removal via sedimentation. Using dimensional arguments, Friedlander was able to demonstrate that two power-law forms for the upper end of the spectrum are possible. The exponent for the coagulation subrange around 0.1–1.0 μm radius was found to be $-5/2$, while that for the sedimentation subrange for radii $> 10\,\mu$m was steeper, at $-19/4$.

Recently, Brock [63] has considered the possible asymptotic behavior of size spectra generated by different laws of condensation of particles. He found that a Junge subrange could be derived only under conditions where condensation followed a model where the rate of mass increase is proportional to the existing particle volume. Such a model does not appear to be compatible with current models for the physics of condensation.

Pich et al. [67] have examined a system of coagulating particles experiencing simultaneous condensation of vapor. They found that a self-preserving spectrum can develop under some circumstances, but its shape will be narrower than that predicted for coagulation alone.

Junge [68] hypothesized that the regularity in shape of the size spectrum of aerosols near the ground may be the result of the statistics of mixing particles from many sources. Hidy [48] also speculated that this may be the case in the troposphere even though a broad spectrum may stem from transport processes in the aerosol cloud in combination with collisions. One simple kinetic equation can be written if only the most important terms are chosen based on the scaling similar to those listed in Tables 4–6. Consideration of a "near-ground" model for a dry atmosphere involving a steady state with a balance of (a) coagulation by Brownian motion and turbulence, (b) sedimentation, and (c) eddy diffusion can lead to a power-law form in the extremely large size range, with the assumption of self-preservation. However, the Junge form can be achieved only if the volume fraction is constant with height in this model. At higher altitudes, the scaling in Table 6 would suggest that collision processes are weak but may be balanced by even weaker transport processes. In a steady state with such conditions, a Junge subrange can be expected over a portion of the size distribution.

Even the simplifications examined by Hidy [48] lead to an integro-differential equation like Eq. (46) for which complete solutions are not available as yet because of its complexity. The speculation that such a kinetic equation will lead to a broadened size distribution function more comparable with experiment than is obtained by considering coagulation alone must await verification by further numerical experiments.

4. SUMMARY AND CONCLUSIONS

The removal of different kinds of trace contaminants from the lower atmosphere proceeds by means of many different physical and chemical processes. In the case of gases, the cleansing of the atmosphere depends to a considerable extent on the chemical reactivity of the pollutant as well as its solubility in water. In the latter case, the development of water precipitation in the atmosphere plays a vital role in the removal process, as well as providing a necessary link in the hydrological cycle. Gases that are relatively inert and water-insoluble, such as CH_4 or N_2O, can remain in the air for many months, even years. Other more reactive materials, such as NH_3 and SO_2, are efficiently and rapidly depleted from the atmosphere so that their lifetimes appear to be only a few hours.

The action of sunlight plays a crucial part in providing the energy for chemical reactions of pollutant gases leading to their effective removal as other gases or as aerosols. Thus, the sun is a mixed blessing in these considerations in that it generates the extremely irritating photochemical smog but allows for the rapid removal of the more chemically reactive anthropogenic pollutants.

Aside from precipitation processes, the chain of events leading to the formation of aerosols from reactive gases represents a vital part of the removal mechanisms for such materials. Thus aerosols in the atmosphere may be intermediates in the evolution of gaseous pollutants as well as primary products of both anthropogenic and natural origin.

In general, the removal of aerosol particles from the air involves a larger variety of processes than is associated with gases. Again in the case of this class of contaminant, chemical reactions and precipitation from water clouds are important factors in the removal processes. However, mechanisms leading to deposition on obstacles such as diffusion and sedimentation also must be considered as well as collision processes such as those induced by Brownian motion.

Aerosol particles in the troposphere range in size from tens of angstroms to tens of microns. Because of this wide variation of size, the mechanisms important for removal may differ markedly in limited size ranges. Although virtually all of the removal mechanisms for aerosols have been identified, their quantification remains uncertain in some cases. A particularly important uncertainty concerns the rates of removal associated with in-cloud processes.

Removal processes of different kinds are believed to be major factors in determining chemical composition as a function of particle size as well as the size distribution of the atmospheric aerosol. Limited evidence suggests that condensation of vapors is important to particle growth, and probably competes strongly with coagulation by Brownian motion in dominating the effective removal of the smallest particles in air near the ground. Turbulent diffusion is important for all particles at this level, but sedimentation should be crucial in limiting the shape of the upper portion of the size spectrum. At higher altitudes, loss rates of aerosols in the absence of rain clouds decrease considerably, and "wet" processes involving precipitation of water are the dominant factor for all particles. Because of present difficulties in evaluation, the significance of electrical forces and other phoretic forces in aerosol removal remains uncertain. However, electrification is speculated to be an important factor leading to clearance of aerial suspensions.

Since the mechanisms for removal of aerosol particles vary markedly with particle size, there should be subsequent differences in residence times in the atmosphere. It is hypothesized that the smallest particles, those less than 0.1 μm in diameter, and the largest particles, those greater than 10 μm in diameter, are removed rapidly by attachment or rainout, and sedimentation, respectively. Particles in the range 0.1–10 μm in diameter, however, may have longer residence times since diffusional and inertial deposition mechanisms and sedimentation are weak in this range.

ACKNOWLEDGMENT

The effort resulting in this review was sponsored in part by the U. S. Atomic Energy Commission under Contract AT(04-3)-543.

REFERENCES

1. Study of Critical Environmental Problems (SCEP), "Man's Impact on the Global Environment," MIT Press, Cambridge, Mass., 1970, Chapter 2.
2. Hidy, G. M., and Brock, J. R., "The Dynamics of Aerocolloidal Systems," Pergamon Press, New York, 1970.
3. Junge, C. E., "Atmospheric Chemistry and Radioactivity," Academic Press, New York, 1963.
4. Bird, R. B., Stewart, W. E., and Lightfoot, E. N., "Transport Phenomena," Wiley, New York, 1960.
5. Renzetti, N. A., and Doyle, G. J., The chemical nature of the particulate in irradiated automobile exhaust, *J. Air Poll. Contr. Assoc.* **8**, 293–296 (1959).
6. Hidy, G. M., and Brock, J. R., An assessment of the global sources of aerosols, in "Proc. 2nd IUAPPA Clean Air Congress, Washington, D.C., December 1970," Academic Press, New York, 1971, pp. 1088–1097.
7. Hidy, G. M., and Friedlander, S. K., The nature of the Los Angeles aerosol, in "Proc. of 2nd IUAPPA Clean Air Congress, Washington, D.C.," Academic Press, New York, 1971, pp. 391–404.
8. Robinson, E., and Robbins, R., Sources, abundance, and fate of gaseous atmospheric pollutants, Stanford Research Institute Final Rept. Proj. PR-6755, Menlo Park, Calif., 1968; also supplemental report, 1969.
9. Brock, J. R., Attachment of trace substances on atmospheric aerosols, in AEC Symposium Series No. 22, U. S. Atomic Energy Commission, Oak Ridge, Tenn., 1971, pp. 373–380.
10. Souilleux, P. J., The measurement of the size spectrum and charge to total ratio of condensation nuclei having naturally occurring radon daughter products attached to them, *Health Phys.* **18**, 245–254 (1970).

174 Chapter 3

11. Corn, M, Montgomery, T. L., and Esmen, N. A., Suspended particulate matter: seasonal variation in specific surface areas and densities, *Environ. Sci. and Technol.* **5**, 155–157 (1971).
12. Rabinowitch, E. I., "Photosynthesis and Related Processes," Interscience, New York, 1945.
13. Chamberlain, A. C., Transport of gases to and from grass and grasslike surfaces, *Proc. Royal Soc. (London)* **A290**, 236–265 (1966).
14. Levich, V. G., "Physicochemical Hydrodynamics" (English transl.), Prentice-Hall, Englewood Cliffs, N.J., 1962, pp. 166–171.
15. Seiler, W., and Junge, C. E., Carbon monoxide in the atmosphere, *J. Geophys. Res.* **75**, 2217–2226 (1970).
16. Leighton, P. A., "Photochemistry of Air Pollution," Academic Press, New York, 1961.
17. Lemke, E. E., Thomas, G., and Zwiacker, W. E. Profile of air pollution control in Los Angeles County, Los Angeles County Air Pollution Control District, 1969.
18. Koyoma, T., Gaseous metabolism in lake sediments and paddy soils and the production of atmospheric methane and hydrogen, *J. Geophys. Res.* **68**, 3971–3974 (1963).
19. Bainbridge, A. E., and Heidt, L. E., Measurements of methane in the troposphere and lower stratosphere, *Tellus* **18**, 221–225 (1966).
20. Schütz, K., Junge, C., Beck, R., and Albrecht, B., Studies of Atmospheric N_2O, *J. Geophys. Res.* **75**, 2230–2246 (1970).
21. Bates, D. R., and Hayes, P. B., Atmospheric nitrous oxide, *Planet. Space Sci.* **15**, 189 (1967).
22. Georgii, H. W., *Geofis. pura e appl.* **47**, 155–171 (1960).
23. Acrivos, A., and Taylor, T. D., Heat and mass transfer for single spheres in stokes flow, *Phys. Fluids* **5**, 387–394 (1962).
24. Frössling, N., *Gerlands Beitr. Geophys.* **52**, 1970 (1938).
25. Dankwerts, P. V., Absorption by simultaneous diffusion and chemical reaction into particles of various shapes and into falling drops, *Trans. Faraday Soc.* **47**, 1014–1023 (1951).
26. Postma, A. K., Effect of solubilities of gases on their scavenging by raindrops, in U. S. AEC Symposium Series No. 22, U. S. Atomic Energy Commission, Oak Ridge, Tenn., 1970, pp. 247–259.
27. Beilke, S., and Georgii, H., Investigation on the incorporation of SO_2 into fog and rain droplets, *Tellus* **20**, 435–441 (1968).
28. Summers, P. W., Scavenging of SO_2 by convective storms, in AEC Symposium Series No. 22, U. S. Atomic Energy Commission, Oak Ridge, Tenn., 1970, pp. 305–318.
29. Andersson, T., Small-scale variations of the contamination of rain caused by washout from the low layers of the atmosphere, *Tellus* **21**, 685–692 (1967).
30. Georgii, H., and Wötzel, D., On the relation between droplet size and concentration of trace elements in rainwater, *J. Geophys. Res.* **75**, 1727–1731 (1967).
31. Urone, P., and Schroeder, W. H., SO_2 in the atmosphere: a wealth of monitoring data, but few reaction rate studies, *Environ. Sci. and Technol.* **3**, 436–445 (1969).
32. Schuck, E. A., and Doyle, G. J., Photooxidation of hydrocarbons in mixtures containing oxides of nitrogen and sulfur dioxide, Air Pollution Foundation (Los Angeles), Report No. 29, 1959.
33. Ogata, Y., Izawa, Y., and Tsuda, T., The photochemical sulfoxidation of *n*-hexane, *Tetrahedron* **21**, 1349–1356 (1965).
34. Doyle, G. J., Self-nucleation in the sulfuric acid–water system, *J. Chem. Phys.* **35**, 795–9 (1961).
35. Quon, J. E., Siegel, R. P., and Mulburt, H. M., Particle formation from photolysis of sulfur dioxide in air, in "Proc. 2nd IUAPPA Clean Air Congress, Washington, D.C., December 1970," Academic Press, New York, 1971, pp. 330–335.
36. Katz, M., and Gale, S. B., Mechanism of photooxidation of sulfur dioxide in the atmosphere, in "Proc. 2nd IUAPPA Clean Air Congress, Washington, D.C., December 1970," Academic Press, New York, 1970, pp. 336–343.
37. Harkins, J., and Nicksic, S. W., Studies in the role of sulfur dioxide in visibility reduction, *J. Air Poll. Control Assoc.* **15**, 218–221 (1965).

38. Stephens, E., Smog aerosol: Infrared spectra, *Science* **168**, 584–1586 (1970).
39. Johnston, H. F., and Conghanowr, D. R., Absorption of sulfur dioxide from air oxidation in drops containing dissolved catalyst, *Ind. Eng. Chem.* **50**, 1169–72 (1958).
40. Junge, C. E., and Ryan, T. G., Study of the SO_2 oxidation in solution and its role in atmospheric chemistry, *Quart. J. Roy. Meteor. Soc.* **84**, 46–56 (1958).
41. Van den Heuvel, A. P., and Mason, B. J., The formation of ammonium sulfate in water droplets exposed to gaseous sulfur dioxide and ammonia, *Quart. J. Roy. Meteor. Soc.* **89**, 271–275 (1963).
42. Scott, W. D., and Hobbs, P. V., The formation of sulfate in water droplets, *J. Atmos. Sci.* **24**, 54–57 (1967).
43. Terraglio, F. P., and Manganelli, R. M., The absorption of atmospheric sulfur dioxide by water solutions, *J. Air Poll. Control Assoc.* **17**, 403–406 (1967).
44. Miller, J. M., and de Pena, R. G., The rate of sulfate ion formation in water droplets in atmospheres with differential partial pressures of SO_2, in "Proc. of the 2nd IUAPPA Clean Air Congress, Washington, D.C., December 1970," Academic Press, New York, 1971, pp. 375–378.
45. Georgii, H. W., Contribution to the atmospheric sulfur budget, *J. Geophys. Res.* **75**, 2365–2372 (1970).
46. Urone, P., Schroeder, W. H., and Miller, S. R., Reactions of sulfur dioxide in air, in "Proc. 2nd IUAPPA Clean Air Congress, Washington, D.C., December 1970," Academic Press, New York, 1971, pp. 370–374.
47. Friedlander, S. K., The characterization of aerosols distributed with respect to size and chemical composition, *Aerosol Science* **1**, 295–307 (1970).
48. Hidy, G. M., The Dynamics of aerosols in the lower troposphere in assessment of airborne particles, "Proc. 3rd Univ. of Rochester Conf. on Environ. Toxicology," C. C. Thomas, Springfield, Ill., 1972, pp. 81–115.
49. Chamberlain, A. C., Radioactive aerosols and vapors, *Contemporary Phys.* **8**, 561–581 (1967).
50. Langer, G., Particle deposition and reentrainment from coniferous trees, Part II, *Kolloid Z.* **204**, 119–124 (1965).
51. Rosinski, J., and Nagamoto, C. T., Particle deposition on and reentrainment from coniferous trees, Part I, *Kolloid Z.* **204**, 111–119 (1965).
52. Neuberger, H., Hosler, C. L., and Kocmond, W. C., Vegetation as an aerosols filter, in "Biometeorology," Proc. 3rd Int. Biometeor. Congr., Pau, France, S. W. Tramp, and W. H. Weike, eds., 1967, part 2, pp. 693–702.
53. Saffman, P. G., and Turner, J. S., On the collision of drops in turbulent clouds, *J. Fluid Mech.* **1**, 16–30 (1956).
54. Friedlander, S. K., The similarity theory of the particle size distribution of the atmospheric aerosols, in "Aerosols, Phys. Chemistry and Appl." K. Spurny, ed., Czechoslovak Acad. Sci., Prague, 1964, pp. 115–130.
55. Davis, M. H., and Sartor, J. D., Theoretical collision efficiencies for small cloud droplets in stokes flow, *Nature* **215**, 1371–1372 (1967).
56. Sartor, J. D., Accretion rates of cloud drops, raindrops, and small hail in mature thunderstorms, *J. Geophys. Res.* **75**, 7547–7558 (1970).
57. Rosinski, J., and Kerrigan, T. C., The role of aerosol particles in the formation of raindrops and hailstones in severe thunderstorms, *J. Atmos. Sci.* **26**, 695–715 (1969).
58. Zebel, G., Capture of small particles by drops falling in electric fields, *J. Colloid and Interface Sci.* **27**, 294–304 (1968).
59. Slinn, W. G. N., and Hales, J. M., Thermophoretic influences, in "Precipitation Scavenging," AEC Symposium Series No. 22 U. S. Atomic Energy Commission, Oak Ridge, Tenn., 1970, pp. 411–424.
60. Slinn, W. G. N., Precipitation scavenging of submicron particles. Part A—Theory, in "Proc. USAEC Meteorological Information Meeting," AECL-2787, 527-520, Chalk River Laboratories, Chalk River, Ontario, 1967.

61. Lumley, J., and Panofsky, H., "The Structure of Atmospheric Turbulence," Wiley–Inter-science, New York, 1964, p. 123.
62. Fletcher, N. H., "The Physics of Rainclouds," Cambridge Univ. Press, 1966.
63. Brock, J. R., On size distributions of atmospheric aerosol, *Atmos. Environment* **5**, 833–842 (1971).
64. Hidy, G. M., On the theory of the coagulation of noninteracting particles in Brownian motion, *J. Colloid Sci.* **20**, 123 (1965).
65. Friedlander, S. K., and Wang, C. S., The self-preserving particle size distribution for coagulation by Brownian motion, *J. Colloid and Interface Sci.* **22**, 126–132 (1966).
66. Lai, F., Friedlander, S. K., Pich, J., and Hidy, G. M. The self-preserving particle size distribution for Brownian coagulation in the free-molecule regime, submitted to *J. Colloid and Interface Sci.* **39**, 395–405 (1972).
67. Pich, J., Friedlander, S. K., and Lai F. S. The self-preserving particle size distribution for coagulation by Brownian motion. III. Smoluchowski coagulation and simultaneous Maxwellian condensation, *Aerosol Sci.* **1**, 115–126 (1970).
68. Junge, C. E., Comments on "Concentration and size distribution measurements of atmospheric aerosols and a test of the theory of self-preserving size distributions," *J. Atmos. Sci.* **26**, 603–607 (1969).
69. Thiele, E. W., Relation between catalytic activity and size of particle, *Industr. Eng. Chem.* **31**, 916–920 (1939).
70. Möller, U., and Schumann, G., Mechanisms of transport from the atmosphere to the Earth's surface, *J. Geophys. Res.* **75**, 3013–3019 (1970).
71. Wagman, J., Lee, R. E., and Axt, C. J., Influence of some atmospheric variables on the concentration and particle size distribution of sulfate in urban air, *Atmos. Environ.* **1**, 479–489 (1967).
72. Ludwig, F. L., and Robinson, E., Variations in the size distributions of sulfur-containing compounds in ruban aerosols, *Atmos. Environ.* **2**, 13–23 (1968).
73. Lee, R. E., and Patterson, R. K., Size determination of atmospheric phosphate, nitrate, chloride, and ammonium particulate in several urban areas, *Atmos. Environ.* **3**, 249–255 (1969).
74. Blifford, I. H., Jr., and Ringer, L., The size and number distribution of aerosols in the continental troposphere, *J. Atmos. Sci.* **26**, 716–726 (1969).
75. Weickmann, H., Recent measurements of the vertical distribution of Aitken nuclei, in "Artificial Stimulation of Rain," Proc. 1st Conf. Phys. of Clouds and Precipitation Particles, H. Weickmann and W. Smith, eds., Pergamon Press, New York, 1955, pp. 81–88.
76. Selezneva, E. S., The main features of condensation nuclei distribution in the free atmosphere over the European territory of the USSR, *Tellus* **18**, 525–531 (1966).
77. Meszaros, A., Vertical profile of large and giant particles in the lower troposphere, in "Proc. 7th Intl. Conf. Condensation and Ice Nuclei," Prague–Vienna, 1969, pp. 364–368.
78. Sood, S. K., and Jackson, Y. M. R., Scavenging by snow and ice crystals, in "Precipitation Scavenging" AEC Symposium Series No. 22, U. S. Atomic Energy Commission, Oak Ridge, Tenn., 1970, pp. 121–136.
79. Hidy, G. M., and Brock, J. R., Some remarks about the coagulation of aerosol particles by Brownian motion, *J. Colloid Sci.* **20**, 477–491 (1965).
80. Hidy, G. M., "The Wind: The Origin and Behavior of Atmospheric Motion," Van Nostrand, New York, 1967.
81. Munn, R. E., "Descriptive Micrometeorology," Academic Press, New York, 1966.
82. Niki, H., Daby, E. E., and Weinstock, B., Mechanisms of smog reactions, to be published in "Advances in Chemistry," American Chemical Society, Washington, D.C., 1972.

Chapter 4

THE GLOBAL SULFUR CYCLE

James P. Friend

New York University, Department of Meteorology and Oceanography

1. INTRODUCTION

The chemistry of gaseous sulfur compounds in the atmosphere at concentrations less than 0.1 ppm depends upon nonuniform distributions of sources and sinks as well as the variable physical properties of the atmosphere itself. Because of natural variations in sunlight intensity, temperature, humidity, surface chemical characteristics, and atmospheric transport processes, the behavior of sulfur compounds, both gaseous and particulate, is complex, and may never be known in great detail. Similar statements might be made for other trace constituents of the atmosphere, notably nitrogen oxides and nitrates. In this chapter, however, we will be concerned only with sulfur compounds.

The study of the global cycles of trace atmospheric substances serves manifold purposes. It delimits the nature of the complex atmospheric reaction system; it provides a means of assessing the global impact of anthropogenic sources; and it clearly shows the relationship of the atmospheric chemical system to soil, water, and rock systems of the earth and to the biosphere.

Among materials of anthropogenic origin, sulfur oxides, carbon monoxide, and certain trace metals such as lead contribute substantially to the atmospheric concentrations of these materials. In contrast, the amounts of substances such as oxides of nitrogen, carbon dioxide, and hydrocarbons of anthropogenic origin are small fractions of the natural components of these materials in the atmosphere over the globe. The cycles of carbon monoxide and carbon dioxide are treated in Chapter 6. The nitrogen cycle has been studied by Robinson and Robbins[1], who found existing knowledge to be so sparse that a balanced cycle could not be given with a reasonable degree of certainty. (It must also be pointed out that an error in converting units of kilograms per hectare to tons per square meter renders much of the Robinson–Robbins nitrogen cycle invalid.)

There have been several formulations of the global sulfur cycle. These are listed separately in a bibliography at the end of this chapter. A brief review of the various cycles will be given at the end of the chapter.

This chapter presents a partially balanced global sulfur cycle including as much detail as warranted by current knowledge of the burdens and fluxes of sulfur among the atmosphere, biosphere, pedosphere (soil), hydrosphere, and lithosphere. We will draw upon the works of previous investigators in the subject as necessary. It will be seen that despite substantial uncertainties in various components of the global sulfur cycle, significant understanding of the role of sulfur in man's global environment may be gained.

2. THE NATURE OF THE SULFUR CYCLE

Reservoirs. The model of the global sulfur cycle consists basically of certain spheres or reservoirs between which transfer of sulfur may occur. The known paths of transfer are separately represented. The total amounts of sulfur in the spheres or subdivisions of spheres will be referred to as burdens. The rates of transfer of sulfur from one sphere to another (also between subdivisions of a single sphere) are globally integrated fluxes. They will often be referred to simply as fluxes.

Figure 1 is a schematic diagram showing the spheres of the model of the global sulfur cycle. The pedosphere is the portion of the earth comprising the soil. It will be noted that the biosphere is distributed between the hydrosphere (mostly oceans) and the pedosphere. Estimates of the amounts of sulfur in some of the spheres are also shown in Fig. 1. These will be discussed in more detail below. Most of the sulfur in the earth's crust resides in rocks and as dissolved sulfate in the oceans.

Transfer Rates and Box Models. It has been the practice in geochemistry to construct "box" models to describe the behavior of various substances. In these models, the boxes are considered to be reservoirs from which transfer is described by first-order kinetics, i.e., transfer from the ith box containing a total mass M_i to the jth box is given by

$$dM_i/dt = -\sum_j \alpha_{ij} M_i$$

where α_{ij} is the so-called transfer coefficient for the flux between the ith and jth boxes. These box models are often used to give quantitative estimates of

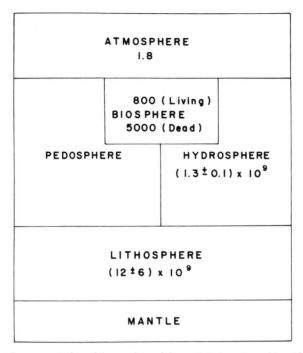

Fig. 1. Schematic representation of the portions of the earth (spheres) considered in the model of the sulfur cycle. Numbers show the amounts of sulfur in units of 10^{12} g (Tg) in portion.

changes that may occur. In effect, the transfer coefficients are parameters which are adjusted to make the model agree with observed data. If simple box models do not give satisfactory results, the complexity of the models can be increased by subdividing the boxes to create more reservoirs and more parameters. This type of modeling has met with varying degrees of success. (An example of a reasonably successful box model for describing radioactive fallout from nuclear weapons testing is that of Krey and Krajewski [2].)

In the present model of the sulfur cycle, the transfer rates are estimated from direct observations (e.g., deposition of $SO_4^=$ by precipitation) or by making reasonable assumptions concerning geochemical balances. Because of the multiplicity of processes that transfer sulfur across the air–earth interface, the complexity in distribution of sources and sinks, chemical transformations in the atmosphere, and the rapidity of these processes, representation of the atmospheric portion of the sulfur cycle in a box-model framework would be attended by large uncertainties. Accordingly, the reader is warned not to use the present sulfur cycle as a box model for making predictions of the effects of varying SO_2 emissions. Rather, the model should be viewed as an approximate representation of the state of the sulfur cycle as of 1970. Future observations will permit updating of the model.

Residence Times. Though we shun the use of first-order kinetics to construct the model or to describe it, the concept of residence time has proven to be very useful in atmospheric chemistry and in geochemistry. Strictly speaking, residence times (thought of as constants) can characterize a reservoir-transfer relationship only for first-order kinetics. The residence time τ_i of a substance in the ith box is determined in the following manner:

(a) Consider that the reservoir contains a mass of the substance M_i.

(b) According to the assumption of first-order kinetics, the rate of removal of material from the reservoir is

$$R_i = \beta_i M_i$$

where β_i is the net transfer coefficient for removal (the sum of all removal transfer coefficients).

(c) The residence time is defined as

$$\tau_i = M_i/R_i = 1/\beta_i$$

(d) In the case where the production rate of material P_i balances the removal, the reservoir is in secular equilibrium (steady state) so that

$$dM_i/dt = P_i - R_i = 0$$

In examining the atmospheric portion of the sulfur cycle, the residence time as described above is used to give important information on the time scale of the processes involved.

Units. The basic unit used for burdens in the various reservoirs is the teragram (Tg), equal to 10^{12} g or 10^6 metric tons. The unit for transfer is teragrams per year (Tg yr^{-1}). These units conform to international convention and they are convenient.

3. THE CHEMISTRY OF SULFUR IN THE GLOBAL ENVIRONMENT

Before we proceed to a detailed description of the features of the sulfur cycle, we will review briefly the chemical processes that govern the global behavior of sulfur. Several of these processes are discussed in more detail in Chapters 2 and 3. Much of what is given here follows material reviewed by Kellogg et al. [3].

The compounds of sulfur of primary interest in the atmospheric portion of the cycle are hydrogen sulfide (H_2S), sulfur dioxide (SO_2), sulfur trioxide (SO_3), and sulfate ($SO_4^=$). The first three forms are gaseous in the atmosphere, while the last form is liquid if it is sulfuric acid (H_2SO_4) or solid if in combination with cations other than hydrogen. Organic compounds such as mercaptans and sulfides might exist in the atmosphere, but they have not been studied in relation to the global sulfur cycle.

Hydrogen Sulfide. The source of hydrogen sulfide in the atmosphere is decaying vegetation. Its characteristic odor of rotten eggs may be detected in intertidal flats where bacteria in reducing muds decompose vegetation. Other swampy or boggy regions of continents are also sources of H_2S. The magnitude of this source has not been directly determined. It may be of considerable proportions. As will be seen, the transfer rate of H_2S is estimated by differences in the balanced cycle.

Hydrogen sulfide emitted into the atmosphere is apparently quite rapidly oxidized to SO_2. Ozone (O_3), oxygen atoms (O), and molecular oxygen (O_2) may all react with H_2S. The reaction between H_2S and O_3 is slow in a gaseous atmosphere though it may proceed at an appreciable rate in the presence of aerosol [4]. Atomic oxygen is generally present in very small concentrations in natural tropospheric air so that the step

$$H_2S + O \rightarrow OH + HS$$

may be quite slow. In polluted atmospheres under bright sunlight and in the upper atmosphere (above about 20 km), any H_2S present would be rapidly oxidized because of relatively high concentrations of O atoms. Hydrogen sulfide is partly soluble in water and the presence of dissolved O_2 and trace transition-metal ion catalysts may cause oxidation of the sulfide to sulfate. Such a process is known to occur in water bodies such as the Baltic Sea or the Black Sea where reducing bottom muds produce H_2S in solution. However, dissolved O_2 in the surface waters oxidizes the sulfide to sulfate and prevents the occurrence of detectable sulfide concentrations near the surface. The catalytic oxidation of H_2S in solution might proceed quite rapidly in the atmosphere in fog or cloud droplets, though little is known about this reaction system.

Sulfur Dioxide. The reactions of SO_2 in the atmosphere all lead to the formation of $SO_4^=$ and thus represent a removal mechanism for SO_2. These reactions are discussed in detail in Chapters 2 and 3 Photochemical mechanisms and the reaction of SO_2 with O atoms are too slow to be of significance in tropospheric air, although the latter reaction apparently accounts for the formation of stratospheric sulfate aerosol at altitudes above about 10 km. The reaction system which is probably of major importance for the global removal of SO_2 from the atmosphere is catalytic oxidation in solution in the presence of catalysts which may be NH_4^+ or trace quantities of transition metal ions. Schematically, the system may be represented as

$$SO_2 + (\tfrac{1}{2}O_2)_{dissolved} + H_2O \xrightarrow[\text{catalyst}]{} 2H^+ + SO_4^=$$

Sulfur dioxide may be removed from the atmosphere by this means in cloud droplets, fog, at moist surfaces of plants and soil, and at the surfaces of water bodies.

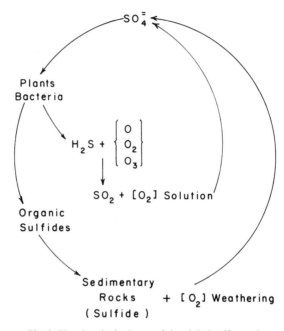

Fig. 2. The chemical scheme of the global sulfur cycle.

Crustal Sulfur. Sulfate in the oceans and in soils is converted by plants (perhaps with the aid of bacteria) to organic sulfide compounds. When the plants die, reductive decay releases H_2S. Oxidative decay probably produces SO_2 directly. Experiments conducted in Sweden and reported by Eriksson [5] have shown that under atmospheres with SO_2, plant-soil systems rich in sulfate tend to release H_2S while those poor in sulfate tended to gain in sulfate. This information is important in fortifying the assumption (to be made later) that the sulfur content of the pedosphere remains constant.

In the lithosphere, sulfur in igneous and metamorphic rocks is mostly sulfide and that in sedimentary rocks is about one-third sulfide (mostly pyrite) and two-thirds $SO_4^=$ (in evaporites) [6]. The process of weathering of rocks oxidizes the sulfides and results in the introduction of $SO_4^=$ into rivers (because the sulfur content of soil tends to remain steady). The sulfides in sedimentary rocks result from bacterial decay of vegetation in muds which later become rocks.

Figure 2 illustrates schematically the main chemical scheme of the global sulfur cycle and summarizes the above discussion.

4. CONCENTRATIONS OF SULFUR

4.1. The Atmosphere

Kellogg *et al.* [³] have reviewed most of the data bearing on concentrations of sulfur in clear air over continents and oceans. The data are sparse indeed. Below are listed ranges and estimates of the means for surface level air concentrations of the three main forms of atmospheric sulfur along with a comment on the data. Concentrations are expressed in units of micrograms per cubic meter or parts per billion by mass (ppbm). ($1 \, \mu g \, m^{-3}$ at STP is equivalent to 0.8 ppbm.)

H₂S. No reliable measurements have yet been made, though Junge [²] gives values of 2–20 $\mu g \, m^{-3}$ for Bedford, Mass. and New York City.

Robinson and Robbins [¹] estimate a mean concentration of 0.2 ppb. (It would appear that they use volumetric mixing ratios, but because of arithmetic errors, it is difficult to be certain. However, in the case of H_2S, it makes little difference since the conversion factor is the ratio of the molecular weight of air to the molecular weight of H_2S.) In this work, a global average mixing ratio of 0.2 ppbm for the troposphere is assumed. This assumption may be considered in error.

SO₂ Concentrations have been found in the range 0–3 ppbm in surface air over oceans, in Antarctica, and in Panama. Concentrations of 0.4–1.7 ppbm were found by Georgii [⁸] in aircraft flights over Colorado. Kellogg *et al.* [²] estimated an average tropospheric concentration of 0.2 ppbm for the Northern Hemisphere and 0.1 ppbm for the Southern Hemisphere. For simplicity, a concentration of 0.2 ppbm is used for the entire troposphere.

Sulfate. Measured concentrations over the land at the surface are in the range 0–20 $\mu g \, m^{-3}$. Typical values in nonurban regions [⁷] are of the order of 0.5 $\mu g \, m^{-3}$. Robinson and Robbins [¹] estimate a global average surface concentration of 2.0 $\mu g \, m^{-3}$. Georgii [⁸] found concentrations of 2–5 $\mu g \, m^{-3}$ over the Atlantic Ocean. Gillette and Blifford [⁹] in aircraft impactor examples found near-surface concentrations of 0.7 $\mu g \, m^{-3}$ over the Pacific Ocean offshore and 0.9 $\mu g \, m^{-3}$ over Death Valley. The concentrations diminished with altitude to about 0.2 $\mu g \, m^{-3}$ at 9 km height in both locations. In this work, a value of 1.5 $\mu g \, m^{-3}$ is used to represent sulfate concentrations in surface air over the continents.

The concentrations of the different forms of sulfur given above will be used later along with other information to estimate the atmospheric burden of sulfur, which in turn will be used to estimate lifetimes in the atmosphere. However, they are not used in formulating the transfer rates and in balancing the global cycle.

4.2. Crustal Sulfur

The average concentration of sulfate in the oceans is a well-established quantity equal to 2.650 mg kg^{-1}. The sulfur contents of the lithosphere has

TABLE 1. Sulfur Contents of Rocks (wt %)

Igneous rocks (sulfide and sulfate)	0.03
Metamorphic rocks (sulfide and sulfate)	0.03
Sedimentary rocks	
Sulfides	0.17
Sulfates	0.35

been reviewed by Holland [10] summarizing studies by Ricke [11] and Holser and Kaplan [6]. The data are much less certain than for the oceans. Table 1 is to be regarded as a set of crude estimates.

The concentrations of sulfate in river water are of significance in the global sulfur cycle. The processes of rock weathering, volcanism, soil leaching, and deposition in precipitation all contribute to the sulfate contents of rivers. Data analyzed by Livingstone [12] have given a widely accepted average sulfate concentration in the world's rivers of 11.2 ppm. Much of the data given by Livingstone were obtained prior to 1900. They undoubtedly are much less affected by pollution than are the present-day concentrations.

5. CONTENTS OF THE RESERVOIRS

The sulfur burdens of the various spheres of the cycle are generally obtained by multiplying the average concentrations as given in the previous section by the total mass of the medium in the reservoir. This is a straight-forward procedure and applies to all quantities considered except for atmospheric sulfate.

Atmospheric sulfate over the oceans comes primarily from the injection of sea salt nuclei into the surface air from which mixing transports some portion to higher altitudes. Maritime air moving over land can bring measurable quantities of oceanic sulfate into the interior regions of continents. Junge [7] summarizes aircraft sampling data which lead to the following estimates:

(a) The sea salt content in a column over the oceans is $10 \, \text{mg m}^{-2}$. Integrating this over the area of the oceans ($3.61 \times 10^{14} \, \text{m}^2$) and using the fraction of $SO_4^=$ in dissolved sea salt as the fraction of $SO_4^=$ in sea salt nuclei (namely 0.077) gives 0.28 Tg for the $SO_4^=$ content of air over the oceans.

(b) Byers (see Ref. 7) found an amount of sea salt over Chicago equivalent to an estimated $5 \, \text{mg m}^{-2}$ of surface. If we assume that half of this value is an average representative of all air over the continents, this leads to a total sulfate of oceanic origin over land of 0.026 Tg. (The land area of the earth is $1.49 \times 10^{14} \, \text{m}^2$).

Atmospheric sulfate of continental origin results mainly from gas-to-particle conversions via the chemical reactions discussed above. From

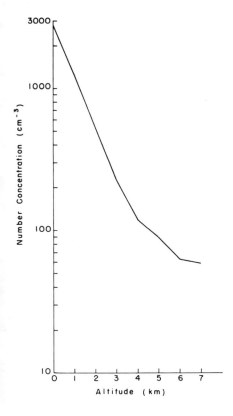

Fig. 3. The AFCRL model aerosol profile (clear).

aircraft samples [8,9], it is evident that the mass concentrations of sulfate diminish with height. It was previously estimated that the average concentrations of sulfate in surface air over the continents is 1.5 μg m^{-3}. An average vertical profile of sulfate is needed in order to arrive at an estimate of the total burden of continental sulfate. The AFCRL Model Aerosol Profile [13] shown in Fig. 3 is used in this work to represent the average vertical profile of sulfate. The following procedure provides the needed estimate:

(a) Adjust the surface value of the ordinate in Fig. 3 to unity.

(b) Scale the surface level concentration of 1.5 μg m^{-3} in the vertical by the adjusted profile.

(c) Integrate under the scales profile and multiply by the land area of the earth to obtain 2.90 mg m^{-2} × 1.49 × 10^{14} m^2 = 0.43 Tg of SO$_4^=$.

Table 2 summarizes all of the above results and lists the contents of all of the reservoirs in the global sulfur model with the exception of the mantle and the pedosphere. Figure 1 also illustrates the reservoirs and their contents. The pedosphere is thought to be in dynamic balance with the biosphere and the atmosphere, thus maintaining an invariant quantity of sulfur. Its burden is undoubtedly small compared to the sulfur in the hydrosphere and the

TABLE 2. Sulfur Contents of Reservoirs

Reservoir	Average concentration	Total sulfur in reservoir, Tg	Ref.
Atmosphere			
SO_2	0.2 ppbm	0.52	1
H_2S	0.2 ppbm	0.99	1,3
$SO_4^=$ (continental)	1.5 μg m^{-3} surface air	0.15	*
$SO_4^=$ (sea salt—oceans)	0.77 μg m^{-2}	0.093	*
$SO_4^=$ (sea salt—land)	0.19 μg m^{-2}	0.0087	*
Total $SO_4^=$	—	0.25	*
Total in atmosphere	—	1.8	*
Lithosphere			
Igneous rocks, S	0.03%	$(6 \pm 3) \times 10^9$	10,11
Metamorphic rocks, S	0.03%		10,11
Sedimentary rocks			
Sulfide S	0.17%	$(2.5 \pm 1.0) \times 10^9$	6,10
Sulfate S	0.35%	$(4 \pm 2) \times 10^9$	6,10
Hydrosphere			
Ocean $SO_4^=$	2.650 mg kg^{-1}	$(1.3 \pm 0.1) \times 10^9$	10

*The present work.

lithosphere. The composition of the mantle is subject to even more uncertainty than that of the lithosphere. The uncertainties shown for the quantities in the lithosphere are given by Holland [10] and he suggests that they may even be too small.

The sulfur burden of the atmosphere is indeed small, but as will be seen, is very important to the overall understanding of the sulfur cycle.

6. TRANSFER MECHANISMS AND RATES

The scheme of transfer rates or fluxes of sulfur from one reservoir to another constitutes the main part of the global sulfur cycle. Balance of portions of the cycle is obtained through assumptions that a burden in a given reservoir is neither increasing nor decreasing. In this presentation, emphasis is placed primarily on details of the atmospheric portion (the most active portion) of the cycle. In what follows, a brief description of each transfer mechanism is given. Tables and figures showing the scheme for each important section of the cycle are given so that comprehension of the final overall cycle will be easier.

6.1. The Atmosphere

The fluxes of sulfur from crust and mantle to the atmosphere are often called sources of atmospheric sulfur. The fluxes from the atmosphere to the

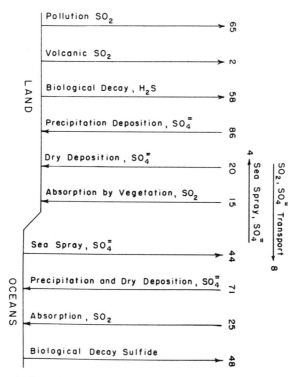

Fig. 4. The atmospheric portion of the sulfur cycle. Upward fluxes balance downward fluxes and atmospheric transport flux balances the atmosphere internally.

crust are known as sinks. It is of great concern to understand the role of pollution sulfur in the global sulfur cycle. As will be seen, most of the sulfur from anthropogenic sources is injected into the atmosphere, and from there makes its way through the cycle. The reader should refer to Fig. 4 in considering the following descriptions.

TABLE 3. Primary Global Anthropogenic Sources
of Atmospheric Sulfur*

Source	Emission rate, Tg yr^{-1}
Coal-burning	45.4
Oil-burning	5.5
Industrial (smelting and petroleum refining)	13.6
Transportation	0.4
Total	64.9

*Based on tables in SCEP [14].

6.1.1. Pollution Sulfur: 65 Tg yr^{-1}

These emissions are almost entirely as sulfur dioxide. The primary sources accounting for almost all of the emissions are given in Table 3 based on estimates given by the SCEP [14]. This reference gives data on U. S. sulfur emissions and on worldwide fuel consumption. Estimates given here of sulfur from oil- and coal-burning and transportation for the world are based on the assumption that the proportions of sulfur emitted are the same as for U. S. emissions. This estimate is very close to that of 64 Tg yr^{-1} given by Robinson and Robbins [1].

6.1.2. Volcanic Sulfur: 2 Tg yr^{-1}

This estimate was made by the author in connection with a study of sources of stratospheric sulfate aerosols. An outline of the calculation is as follows:

(a) Average annual volcanism yields $0.8 \, \text{km}^3$ of magma per year equivalent to 2.2×10^{15} g yr^{-1}.
(b) The composition of volcanic gases is roughly 95% H_2O, 4% CO_2, and 1% SO_2 (these are mole percentages).
(c) H_2O comprises about 5% of the mass of magma.
(d) Annual emission rate of S is 2 Tg yr^{-1}.

Most of the sulfur emissions from volcanoes undoubtedly remain in the troposphere. However, only a small fraction of the 2 Tg yr^{-1} would be needed to produce the stratospheric burden of sulfur estimated to be <0.1 Tg. The total volcanic sulfur estimated here is in reasonable agreement with that given by Kellogg et al. [3] of 1.5 Tg yr^{-1}.

6.1.3. Biological Sulfur (H_2S), Land: 58 Tg yr^{-1}

This flux is arrived at by difference through requiring the pedosphere to be balanced:

Biological decay, H_2S =	Atmospheric input to pedosphere	106	
	+ atmospheric input to land biosphere	15	
	+ fertilizer input to pedosphere	26	
	− river runoff from pedosphere	−89	
	= Total:	58	Tg yr^{-1}

The river runoff from the pedosphere is itself an amount estimated by difference as will be explained later.

6.1.4. Precipitation Deposition, Land : 86 Tg yr^{-1}

This flux was estimated by Kellogg et al. [3] based on chemical analyses of rainwater samples collected in the National Precipitation Sampling Network and other networks in Europe.

6.1.5. Dry Deposition, Land: 20 Tg yr^{-1}

Following the method of Robinson and Robbins [1], it is estimated that dry deposition of $SO_4^=$ is about 20% of the total deposition (106 Tg yr^{-1}). This is reasoned by analogy to nuclear fallout.

6.1.6. Absorption by Vegetation, SO$_2$: 15 Tg yr^{-1}

Kellogg et al. [3] estimated the flux in the constant-stress layer using a method based on relationships used for calculating fluxes of heat and momentum. This method appears to be more rational than using deposition velocities as had been done by Robinson and Robbins [1], who calculated 26 Tg yr^{-1} for this flux. However, details of Kellogg et al.'s calculations are not given.

6.1.7. Sea Spray, SO$_4^=$: 44 Tg yr^{-1}

This is a widely accepted estimate given in the work of Eriksson [15].

6.1.8. Precipitation and Dry Deposition, Oceans, SO$_4^=$: 71 Tg yr^{-1}

Junge [7] gives an estimate based on sparse data for deposition of $SO_4^=$ in rain over the oceans. This amounts to 57 Tg yr^{-1} in rain. An additional 14 Tg yr^{-1} of sulfur in dry fallout gives the total of 71 Tg yr^{-1}. Again the dry deposition is taken to be 20% of the total in analogy to nuclear fallout.

6.1.9. Absorption of SO$_2$, Oceans: 25 Tg yr^{-1}

Robinson and Robbins [1] estimate this flux using a deposition velocity of 0.9 cm sec^{-1} with a concentration of SO_2 of 0.2 ppb(v?). Kellogg et al. [3] essentially leave their cycle open by saying that it is difficult to say whether the oceans are a source or a sink (or both) of SO_2. However, significant concentrations of SO_2 have been reported over the Atlantic Ocean [16] measured on the research vessel *Meteor* in 1969. Concentrations ranged from 0 to 3 ppbm, peaking at about 50°N latitude and going to very small concentrations toward the North Pole and toward the Equator. In view of the relative importance of the catalytic oxidation of SO_2 in solution for removal of SO_2 in clouds and at various land surfaces, it seems quite reasonable to suppose that this mechanism operates in the sea surface waters. Laboratory investigations of the reaction all have shown quite rapid oxidation of SO_2 to sulfate with the rate being limited by pH as the solution becomes more acidic. The oceans, however, are not acidic enough to limit the reaction in this way. (See Chapter 3 for more details of this reaction.)

6.1.10. Biological Decay (Sulfide), Oceans: 48 Tg yr^{-1}

This is the flux necessary to balance the atmospheric portion of the cycle. There are only very sparse observations giving only a vague hint of an oceanic source of gaseous sulfur. Eriksson [5], Robinson and Robbins [1], and Junge [7] have all used the method of difference to suggest that the oceans may be a significant source of gaseous sulfur. The oxidation of H_2S would probably be so rapid in surface waters saturated with oxygen that no H_2S could escape to the atmosphere. A similar remark holds for SO_2. If there is such a source, it may be in the form of organic sulfides such as dimethyl sulfide.

6.1.11. Atmospheric Transport, Land ↔ Sea
6.1.11.1. Sea Spray, Sea → Land: 4 Tg yr^{-1}

Eriksson estimated that about 10% of the sea salt nuclei are transported to continental regions, based on analyses of Cl^- and $SO_4^=$ in rainwater samples. By assuming that neither Cl^- nor $SO_4^=$ from the ocean was lost, then "excess" $SO_4^=$ can be determined by subtracting the sea salt $SO_4^=$ from the total $SO_4^=$. Of course, the oceanic $SO_4^=$ is also determined.

6.1.11.2. SO_2, $SO_4^=$ Transport, Land → Sea: 8 Tg yr^{-1}

This flux balances the land and maritime portions of the atmosphere.

TABLE 4. Atmospheric Fluxes of Sulfur and Related Quantities in Various Sulfur Cycles (Tg yr^{-1})

Item	Eriksson [15]	Junge [7]	Robinson and Robbins [1]	Kellogg et al. [3]	This work
Pollution, SO_2	40	40	70	50	65
Biological decay, H_2S	110	70	68	90†	58
Precipitation disposition (land)	{65}	56*	70	86	86
Dry deposition (land)		15	20	10	20
Absorption by vegetation, SO_2	75	70	26	15	15
Volcanic, SO_2	—	—	—	1.5	2
Sea spray, $SO_4^=$	40	—*	44	43	44
Precipitation and dry deposition, SO_4 (oceans)	100	60*	71	72	71
Absorption, SO_2 (oceans)	100	70	25	—	25
Biological decay ($S^=$) (oceans)	170	160	30	—†	48
Atmospheric transport, SO_2, $SO_4^=$	− 10	− 30	+ 26	+ 5	+ 8
Sea spray transport	5	—*	4	4	4
Total atmospheric flux (input = output)	365	315	212	183	217
Fertilizer application	10	{25}	11	—	26
Rock weathering	15		14	—	42
Pedosphere → river runoff	55	70	48	—	89
Total river runoff	80	95	73	—	136

*Junge's model was for excess sulfur only, so the sea salt component was not included.
†Kellogg et al. estimate a total of 90 Tg yr^{-1} of sulfur from decay of land and oceanic biota.

Figure 4 shows schematically all of the above-discussed fluxes. Table 4 lists the same quantities along with those from the other portions of the cycle. The flux of sulfur into the atmosphere of 217 Tg yr^{-1} just balances the flux out of the atmosphere to the earth's surface. The transport term of 8 Tg yr^{-1} of SO_2 and $SO_4^=$ balances the atmosphere internally. These are essentially the two assumptions made concerning the atmospheric part of the global cycle. The result of the first assumption (balance of fluxes across the earth–air interface) is that a total source strength of sulfur of biological origin of 106 Tg yr^{-1} must be invoked. The strength of the land-based biological source (58 Tg yr^{-1} is estimated through balancing of the pedosphere. There are known biological sources of H_2S on land. However, there are no direct observations to support the estimate of the source strength (48 Tg yr^{-1}) due to biological decay in the oceans.

6.2. The Pedosphere

Figure 5 summarizes the pedospheric portion of the sulfur cycle. The pedosphere contains soil with sulfide and sulfate components.

6.2.1. Atmospheric Inputs: 121 Tg yr^{-1}

This flux is the sum of 6.1.4–6.1.6 of the atmospheric portion explained above.

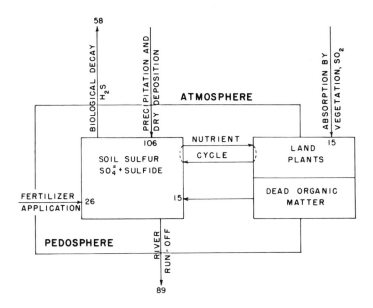

Fig. 5. The pedosphere portion of the sulfur cycle. The input fluxes balance the output fluxes.

6.2.2. Biological Decay, H_2S: 58 Tg yr^{-1}

Explained in 6.1.3. The sulfide is considered to be emitted from the soil by the action of anaerobic bacteria.

6.2.3. Plant $\leftrightarrow$ Soil Nutrient Cycle

This cycle is assumed to be in balance. Ultimately, the 15 Tg yr^{-1} taken up by plants from the atmosphere is transfused to the soil in dead organic matter.

6.2.4. Fertilizer Application: 26 Tg yr^{-1}

Eriksson [5] estimated a flux of 10 Tg yr^{-1} for 1960. By 1970, the global use of fertilizer nutrients had increased 2.6-fold from 25 Tg yr^{-1} in 1960 to 65.5 Tg yr^{-1} in 1970 [14]. It is assumed that the sulfate content of fertilizer nutrients has remained constant.

6.2.5. Pedosphere $\rightarrow$ River Runoff: 89 Tg yr^{-1}

This flux is a difference term between the known total river runoff and the estimated amounts for volcanic and rock-weathering inputs to rivers.

The annual circulation of sulfur through the pedosphere amounts to 147 Tg yr^{-1}. The assumption that the pedosphere is in dynamic equilibrium having a constant sulfur burden is consistent with experimental evidence discussed earlier in conjunction with the chemistry of sulfur.

6.3. The Hydrosphere

Figure 6 shows the scheme of fluxes related to the hydrosphere.

6.3.1. Atmosphere $\rightarrow$ Oceans: 4 Tg yr^{-1}

This is the net flux across the air–sea interface from 6.1.7–6.1.10.

6.3.2. Oceanic $SO_4^=$ $\rightarrow$ Marine Plants: 200 Tg yr^{-1}

Eriksson [5] estimated this flux using estimates of oceanic productivity and the rate of turnover of sulfur in anaerobic bottom sediments (blue mud). Eriksson suggests that this estimate may be too low, but he points out that it is of no consequence to the rest of the cycle. From Fig. 6, it can be seen that the flux from oceanic $SO_4^=$ to marine plants is essentially an internal transfer.

6.3.3. Sedimentation: 100 Tg yr^{-1}

Holser and Kaplan [6] estimate this flux based on an assumed steady state in the rate of sediment formation. With this assumption the rates of sulfide $(S^=)$ sedimentation and sulfate $(SO_4^=)$ sedimentation are in proportion to the abundance of those materials in sedimentary rocks. Thus the fluxes are: sedimentation, $S^=$, $36\,Tg\,yr^{-1}$; sedimentation, $SO_4^=$, $64\,Tg$ yr^{-1}.

6.3.4. Dead Organic Matter: $164\,Tg\,yr^{-1}$

This flux is the difference between the input to marine plants and the sedimentation of sulfides. The flux is further divided as follows: biological decay sulfide (atmosphere) $= 48\,Tg\,yr^{-1}$ (as explained above); oceanic $SO_4^= = 136\,Tg\,yr^{-1}$. This is yet another difference term. It is part of the internal transfer in the oceanic biosphere nutrient cycle.

6.3.5. River Runoff → Oceanic $SO_4^=$: $136\,Tg\,yr^{-1}$

The transport of sulfate to the oceans in rivers is comprised of components from the pedosphere (discussed above), the weathering of rocks, and volcanic emissions. These items and the entire river runoff transport term will be discussed in the next section.

6.3.6. Volcanism → Oceans: $5\,Tg\,yr^{-1}$

Explained in 6.5.

Examination of Fig. 6 and summing of the inputs and outputs for the hydrosphere show that this reservoir is not in balance. The burden in the oceans is increasing at the rate of $45\,Tg\,yr^{-1}$. In view of the uncertainties attendant to such terms as the atmospheric sulfide source from the oceans, The increase may not be real. At any rate, the calculated annual increase is such a small fraction of the total oceanic burden that it can have no measurable effect even if nearly correct.

6.4. River Runoff: $136\,Tg\,yr^{-1}$

Though the transport of sulfur in rivers is a flux term, all of which have been considered in conjunction with the reservoirs, it is an important component of the sulfur cycles. Transport in rivers is the main agency by which lithospheric sulfur enters the fluid spheres which contain the more active portions of the sulfur cycle. Figure 7 shows schematically the components of the river runoff flux.

The total flux was first based on the estimates made by Livingstone [12] of the $SO_4^=$ contents of the world's rivers and amounted to $120\,Tg\,yr^{-1}$. This was assumed by Eriksson [5] to reflect the fertilizer input of $10\,Tg\,yr^{-1}$ for 1960. In this work, it is estimated (as explained previously) that an additional $16\,Tg\,yr^{-1}$ of fertilizer input should be added to the total river runoff to reflect the current (1970) situation.

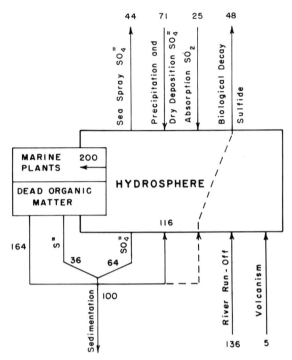

Fig. 6. The hydrospheric portion of the sulfur cycle. The net annual increase in sulfur content is 45 Tg yr^{-1}.

6.4.1. Volcanism (Land): 5 Tg yr^{-1}

Explained in 6.5.

6.4.2. Rock Weathering: 42 Tg yr^{-1}

Berner [17] estimated this flux using geochemical concepts relating to the weathering of sedimentary rocks which are represented primarily by pyrite and gypsum. By assuming that both minerals weather at the same rate, the Ca^{++} content of rivers can be used to estimate the weathering component of the total sulfate.

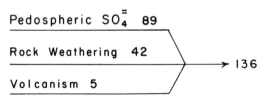

Fig. 7. Components of river runoff (Tg yr^{-1}).

6.4.3. From Pedosphere: 89 Tg yr^{-1}

This is the difference between the total flux and components 6.4.1. and 6.4.2.

6.5. Lithosphere

6.5.1. Sedimentation: 10 Tg yr^{-1}

Holser and Kaplan [6]. Explained in 6.3.

6.5.2. Weathering of Rocks: 42 Tg yr^{-1}

Estimated by Berner [12]. Explained in 6.4.

6.5.3. Crustal Degassing (Volcanism): 10 Tg yr^{-1}

This term is estimated by Holland [10] based on a geological argument that about 75% of the present sedimentary rocks have been recycled through the crust. The rate is approximately one-fourth of the present (steady-state) rate of deposition of sulfide sediments. An estimated uncertainty of 50% accompanies the flux.

Division of the total of 12 Tg yr^{-1} of sulfur in crustal and mantle degassing was somewhat arbitrarily accomplished. The estimate of the portion vented into the atmosphere through volcanism (2 Tg yr^{-1}) was discussed previously. The remaining 10 Tg yr^{-1} was divided evenly between land and oceanic volcanism. The portion going to land appears in river waters mainly in regions of geothermal activity.

No significance should be imparted to the lack of balance of the lithosphere. This model has used the term lithosphere to denote the reservoir-containing rocks. The present concept of the lithosphere is that it extends into the mantle, where there is presumably an exchange of mineral materials between what we call the crust and the mantle.

Furthermore, as Holland notes, the quantities associated with the geological sulfur cycle are very uncertain. No arguments depending critically upon them should be advanced and they should be used only for order-of-magnitude indications.

6.6. Mantle

6.6.1. Mantle Degassing (Volcanism): 2 Tg yr^{-1}

Holland [10] based this estimate on the steady-state formation of all the sulfur in the oceans and sedimentary rocks plus one-half that in igneous and metamorphic rocks.

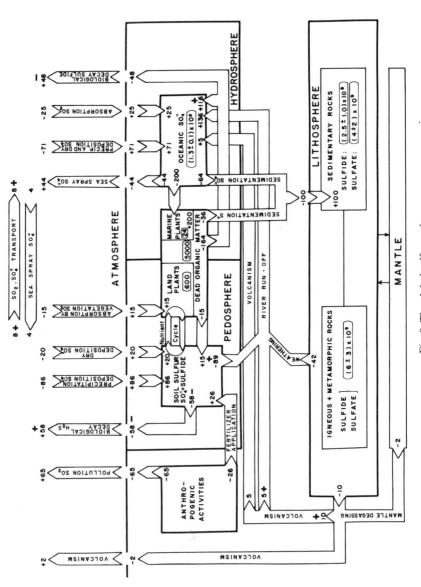

Fig. 8. The global sulfur cycle.

7. THE GLOBAL SULFUR CYCLE

Figure 8 is a diagram of the overall global sulfur cycle combining all of the burdens and fluxes that have been discussed individually. Table 4 lists all of the burden and fluxes along with references.

A review of all the components of the sulfur cycle reveals several areas in which estimates are based on sparse data and so may be in error by an unknown amount. The fluxes to the atmosphere of sulfide sulfur from biological decay are particularly shakey estimates, as discussed above. To investigate the implications of this uncertainty, imagine a change in the flux denoted by "biological decay, H_2S" by a positive increment $+\Delta$. This is indicated in Fig. 8 by a large plus or minus sign. In order to rebalance the atmospheric and pedospheric portions of the cycle without changing the fluxes based on the data, the adjustments indicated by large plus signs and minus signs must be made. The reader may judge for himself whether or not a large increment may be permitted on the basis of present knowledge.

Residence Time of Atmospheric Sulfur. The residence time of sulfur in the atmosphere may be calculated according to (see introduction)

$$M/R = 1.8 \text{ Tg}/217 \text{ Tg yr}^{-1} = 0.0083 \text{ yr} = 2.7 \text{ days}$$

where M is the atmospheric burden of sulfur, or 1.8 Tg (Table 2), and R is the total flux of sulfur from the atmosphere, or 217 Tg yr^{-1} (Table 4). It is thus seen that the atmosphere overturns its burden on the order of 100 times each year. Because of uncertainties in estimating both M and R and because the atmosphere does not generally behave as a reservoir with first-order kinetics, the residence time calculated above must be used in a circumspect manner. In general, it may be seen that the atmospheric portion of the sulfur cycle is very active indeed. Residence times for various sulfur emissions would be expected to vary according to atmospheric and surface conditions relating to the chemical and physical mechanisms of transformation and removal. The magnitudes of the residence times are of the order of a few days.

8. DISCUSSION

It is a curious matter that of the prior investigators of the global sulfur cycle, atmospheric chemists generally tended to ignore one or more of the important geological processes of sedimentation, rock weathering, or volcanism, and geochemists tended to ignore most of the biological and anthropogenic processes that are important to the atmospheric and river-runoff portions of the cycle. In this work, an attempt has been made to use present-day knowledge of atmospheric and geological processes so as to obtain a more comprehensive and detailed view of the global sulfur cycle than has been previously presented. It is not our purpose here to delineate the differences between the present work and the previous ones. The interested reader is urged to consult the bibliography on the sulfur cycle listed at the end of the chapter. This discussion is primarily concerned with atmospheric and pollution-related aspects of the cycle.

8.1. Comparison with the Cycles

The only fluxes connected with the atmospheric portions of the sulfur cycle that are directly supported by measurements are pollution SO_2, river runoff, and deposition by precipitation. The other fluxes are based on the same data plus assumptions about physical and chemical processes and balances of various reservoirs.

Table 4 is a comparison of fluxes and associated quantities of five models of the sulfur cycle, including the present work. The main differences between the model developed in this chapter and the other recently proposed models (atmospheric portions) arise because of the new estimates of the components of river runoff from fertilizer sulfate and rock weathering. The increased flux in river runoff in turn results in an estimated flux of H_2S from biological decay that is smaller than in other models. Generally, there is rough equality in the estimates of the total of terrestrial and marine biological emissions in the present work and those of Robinson and Robbins [1] and Kellogg et al. [3]. Estimates of the other fluxes vary by lesser absolute amounts partly because of the similarity in the approaches of the various investigators. Each flux in the present model other than pollution SO_2 and those mentioned above is taken from one or another of the previously presented cycles.

8.2. Background Concentrations of Atmospheric Sulfur

The body of data on the temporal and spatial distributions of concentrations of sulfur compounds in the unpolluted atmosphere is very small. In fact, it is too small for a reliable estimate of the atmospheric sulfur burden to be made. Furthermore, the sparseness of data does not permit determination of secular variations in background concentrations (as has been so well done for CO_2). Precise determinations of the atmospheric burden and secular trends in sulfur concentrations await the establishment of a proper program of global-scale sampling and analysis.

It should be clear to the reader that the model of the sulfur cycle cannot be used as a basis for estimating changes in burdens or concentrations in the atmosphere resulting from projected future emissions of pollution SO_2.

8.3. Mobilization of Sulfur by Man

It is of interest to compare the amounts of sulfur released into the environment by man's activities with the amount released by natural sources. Such releases have been termed mobilization by Bertine and Goldberg [18], who estimated the amounts of trace elements released into the atmosphere

TABLE 5. Estimates of Present Rate of Sulfur Mobilization (Tg yr^{-1})

Source	Amount
Pollution SO_2 (air)	65
Fertilizer application (rivers)	26
Total pollution sulfur	91
Volcanism (air + rivers)	7
Weathering of rocks (rivers)	42
Total natural sulfur	49

by burning of fossil fuels. (Because of various errors, their estimates of the amounts of sulfur released from natural and anthropogenic sources are incorrect. This was pointed out by Friend [19].)

The comparison may be made directly by considering the model of the sulfur cycle. It is shown in Table 5. Evidently, man releases almost twice as much sulfur to the environment as does nature. Judging from recent trends in fossil fuel and fertilizer usages, it appears that the ratio is increasing and may approach 10:1 in the near future.

8.4. Pollution Sulfur in River Waters

The concentrations of sulfate in the world's river waters reflect quite directly the deposition on land of atmospheric sulfur compounds. They must then in some way reflect the flux of pollution sulfur into the atmosphere (and also contain a component from fertilizer application). An estimate of the amount of pollution sulfur in river runoff may be made using the sulfur cycle and *additional* assumptions. Such an estimate was recently made by Berner [17], who calculated that approximately 28% of the total river runoff was of pollution origin (pollution SO_2 plus fertilizer $SO_4^=$).

The following sequence represents how an estimate of the pollution component of river runoff can be made:

(a) Because of the short residence time of SO_2 in the atmosphere, it may be assumed that most of the pollution SO_2 is deposited as sulfate on the land. Let us assume 60 Tg yr^{-1} of the 65 Tg yr^{-1} of the sulfur pollution SO_2 is deposited on the land.

(b) From (a), the total input of pollution sulfur to the pedosphere is 86 Tg yr^{-1} (26 Tg yr^{-1} from fertilizer).

(c) Assume pedospheric sulfur is uniformly mixed with respect to the outputs to the atmosphere (H_2S) and river runoff.

(d) Assume no accumulation or depletion of pedospheric sulfur. (This is an assumption already made to balance the cycle in the model.)

(e) The fraction of pollution sulfur in the pedospheric component of river runoff is

$$86/(89 + 58) = 0.58$$

(f) Total pollution sulfur from the pedosphere entering river runoff is

$$0.58 \times 89 \text{ Tg yr}^{-1} = 52 \text{ Tg yr}^{-1}$$

(g) The fraction of pollution sulfur in the world's river runoff is

$$53/136 = 0.38$$

It is estimated here that about 40% of the sulfur contained in the river waters of the world is from anthropogenic sources. If we had used a balanced sulfur cycle with Eriksson's [5] estimate of the flux from fertilizer sulfate (10 Tg yr^{-1}), the estimated fraction would have been in excellent agreement with Berner's calculations.

The calculations performed here and in 8.3 illustrate how the information contained in the model of the sulfur cycle may be used. The discussion in 8.2 contains a caveat on how it may not be used. As in all modeling applications, the user must beware of drawing conclusions that depend in a sensitive manner on the assumptions made.

REFERENCES

1. Robinson, E., and Robbins, R. C., Sources, abundance, and fate of gaseous atmospheric pollutants, Stanford Research Institute Final Rept. Proj. PR-6755, Menlo Park, Calif. (1968).
2. Krey, P. W., and Krajewski, B., Comparison of atmospheric transport model calculations with observations of radioactive debris, J. Geophys. Res. 75, 2901 (1968).
3. Kellogg, W. W., Cadle, R. D., Allen, E. R., Lazrus, A. L., and Martell, E. A., The sulfur cycle, Science, 175, 587 (1972).
4. Cadle, R. D., and Ledford, M., Int. J. Air Water Pollut. 10, 25 (1966).
5. Erickson, E., The yearly circulation of sulfur in nature, J. Geophys. Res. 68, 4001 (1963).
6. Holser, W. T., and Kaplan, I. R., Isotope geochemistry of sedimentary sulfates, Chem. Geol. 1, 93 (1966).
7. Junge, C. E., "Atmospheric Chemistry and Radioactivity," Academic Press, New York, 1963.
8. Georgii, H.-W., Contribution to the atmospheric sulfur budget, J. Geophys. Res. 75, 2365 (1970).
9. Gillette, D. A., and Blifford, Jr., I. H., Composition of tropospheric aerosols as a function of altitude, J. Atmos. Sci. 28, 1199 (1971).
10. Holland, H. D., "The Chemistry and Chemical Evolution of the Atmosphere and Ocean," in preparation.
11. Ricke, W., Cosmochim. Acta 21, 35 (1960).
12. Livingstone, D. A., Chemical composition of rivers and lakes, U. S. Geol. Surv. Prof. Pap. 440-G, 1963.

13. McClatchey, R. A., Fenn, R. W., Selby, J. E. A., Volz, F. E., and Garing, J. S., Optical properties of the atmosphere (Revised), Air Force Cambridge Research Laboratories Report No. AFCRL-71-0279, Environmental Research Paper No. 354, 1971.
14. Study of Critical Environmental Problems, "Man's Impact on the Global Environment," MIT Press, Cambridge, Mass. 1970.
15. Eriksson, E., The Yearly circulation of chloride and sulfur in nature; meteorological, geochemical, and pedological implications, 2, *Tellus* **12**, 63 (1960).
16. Study of Man's Impact on Climate (SMIC), "Inadvertent Climate Modification," MIT Press, Cambridge, Mass., 1971.
17. Berner, R. A., Worldwide sulfur pollution of rivers, *J. Geophys. Res.* **76**, 6597 (1971).
18. Bertine, K. K., and Goldberg, E. D., Fossil fuel combustion and the major sedimentary cycle, *Science* **173**, 233 1971.
19. Friend, J. P., Technical Comment, *Science*, **175**, 1278 (1972).

BIBLIOGRAPHY OF SULFUR CYCLES

Eriksson, E., The yearly circulation of sulfur in nature, *J. Geophys. Res.* **68**, 4001 (1963).
Holser, W. T., and Kaplan, I. R., Isotope geochemistry of sedimentary sulfates, *Chem. Geol.* **1**, 93 (1966).
Junge, C. E., "Atmospheric Chemistry and Radioactivity," Academic Press, New York (1963).
Kellogg, W. W., Cadle, R. D., Allen, E. R., Lazrus, A. L., and Martell, E. A., The sulfur cycle, *Science*, **175**, 587 (1972).
Robinson, E., and Robbins, R. C., Sources, abundance, and fate of gaseous atmospheric pollutants, Stanford Research Institute Final Rept. PR-6755, Menlo Park, Calif, 1968.

Chapter 5

THE CHEMICAL BASIS FOR CLIMATE CHANGE

STEPHEN H. SCHNEIDER*
AND
WILLIAM W. KELLOGG

National Center for Atmospheric Research,† Boulder, Colorado

1. INTRODUCTION

It is obvious that the climate of the planet earth has undergone many rather drastic changes in the past, and there is every reason to believe that there will be other changes in the future. The question is not whether our climate will change, but rather: What causes it to change? If we knew the answer, perhaps we would be able to foresee the future climates in store for us.

There is now another factor to be taken into account in any speculation about climate changes of the future, for man has altered the face of the earth and the composition of the atmosphere on such a large scale that his influence can no longer be ignored relative to nature's. There is the haunting realization that mankind may have caused climate changes already, though we do not yet know to what extent [1].

This realization has prompted atmospheric scientists and others to assess what we know about man's impact on regional and global climates. Two major studies devoted to this broad question were the Study of Critical

---

*Work begun while Resident Research Associate at Institute for Space Studies, Goddard Space Flight Center, NASA, New York, New York.
†The National Center for Atmospheric Research (NCAR) is sponsored by the National Science Foundation.

Environmental Problems (SCEP), held in Williamstown, Massachusetts, in the summer of 1970 [2], and the Study of Man's Impact on the Climate (SMIC), held at Wijk, near Stockholm, Sweden, in the summer of 1971 [3]. Both were organized by Professors Carroll L. Wilson and William H. Matthews of the Massachusetts Institute of Technology, and the present authors played active parts in SMIC—one of us, Kellogg, was also a group leader in SCEP. The problem of climate change and the influence of man's activities is, by its very nature, so complex that these studies had to be inter-disciplinary, and therefore drew on scientists from many fields, who collabo-rated in the preparation of the SCEP and SMIC reports.

This chapter is based on the SMIC report, and is a kind of distillation of the findings of SMIC with emphasis on the role of atmospheric chemistry. We briefly review the records of past fluctuations of the climate, outline the theory of the general circulation of the atmosphere–ocean system and show how this system can be simulated by mathematical models, indicate where man may be influencing atmospheric composition, and indicate a few of the areas that need more research, particularly those related to atmospheric chemistry. The interested reader is encouraged to turn to the SCEP and SMIC reports for more details, and also to the collection of papers prepared by the SCEP team edited by Matthews et al. [4].

2. LESSONS TO BE LEARNED FROM THE PAST

The history of the planet earth goes back nearly five billion years, when the solar system was created. However, we cannot decipher the history of the climate with much certainty beyond the last 500 million years. This history is written in the sediments of the oceans, the rocks, the shapes of mountains and ocean bottoms, (for the more recent periods) the layers of ice in the glaciers and ice sheets of Greenland and the Antarctic, and in tree rings. An example of such a glacier record is shown in Fig. 1, from the work of Dansgaard et al. [5]. The record is not always very clear, and many unanswered questions remain, but the outlines of the story are emerging.

One of the most striking facts that emerges from a study of the climates of this 500-million-year period, from the beginning of the Cambrian, is that for about 90% of this time the earth had poles that were not more or less permanently frozen or "glaciated." Since both poles are now covered with ice and snow the year around, it is fair to say that we are living in a kind of ice age. If the poles were free of this ice cover in summer, we would have a very different climate everywhere, as will be pointed out below [6,7].

There was a widespread glaciation lasting for 30–50 million years during the late Paleozoic (300–250 million years ago), followed by a warm period with no glaciation at the poles, and then a gradual cooling trend took place that led to the beginning of the present Antarctic glaciers about five million

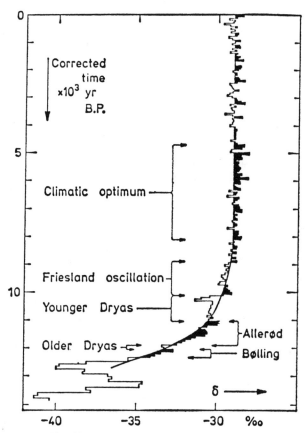

Fig. 1. Variations of the $^{18}O/^{16}O$ oxygen isotope ratio with depth below the surface (converted to age of sample) in an ice core taken from the Greenland ice sheet at Camp Century. An increase of this ratio accompanies an increase in temperature at middle and high latitudes in the Northern Hemisphere, and the values exceeding the smoothed curve (shaded in black) would therefore correspond to periods warmer than the long-term trend. The climatic periods shown refer to the European sequence [5].

years ago. The Northern Hemisphere ice covers of Greenland and the Arctic Ocean were probably established later, about two million years ago.

During this period of polar glaciation in which we live there have been at least four or five periods with dramatic increases in glaciation interspersed with relatively warm periods. These have been marked by the formation of ice sheets over much of North American and Eurasia, each one building up over a period of some 100,000 years and terminating rather quickly, in about 10,000 years. The last such "ice age," the Wisconsin, seems to have reached its maximum in both hemispheres about 20,000 years ago, and the continental ice sheets (except for those of Antarctica and Greenland) disappeared about 10,000 years ago.

For the last 8000 years, the climate of the earth has been relatively constant compared to the glacial–interglacial transitions just described, but there have been many smaller fluctuations as indicated in Fig. 1. These smaller fluctuations, amounting to 1 or 2°C change in mean annual temperature at middle latitudes (probably at least twice that in polar regions), have had serious consequences for life in marginal parts of the earth, i.e., where small degradations in the growing conditions can have a disastrous effect on agriculture. The story of the Viking exploration during the warm period prior to A.D. 1200, and the loss of the Norse colonies in Greenland as the Atlantic ice advanced, is a well-known example of the effect of climate change on the affairs of men. The period A.D 1550–1700 has been called "the Little Ice Age" and history has clearly recorded the suffering it caused in many parts of Europe. In the last century, with more adequate records of temperature being kept at many stations throughout the world, it is possible to document a rise of about 0.8°C in the global mean temperature between 1850 and 1940, as shown in Fig. 2, prepared by Mitchell [8]. In polar regions above 70° latitude the fall of temperature in the past decade has been about 1°C, several times greater than the global average (see Fig. 3.8 in SMIC [3]). The November, 1972 issue of *Quaternary Research* contains a valuable collection of papers relating to very-long-term climate change, and presents a complete set of references on this subject.

These are some of the fluctuations that have occurred in the past, and they could hardly (except possibly those of the last few decades) have been caused by any of man's activities. A number of theories have been advanced to account for such changes in the climate, and they tend to fall into the following categories:

1. Fluctuations in solar emission.

2. Influence of variations of the orbital parameters of the earth, such as tilt of the earth's axis and ellipticity of the earth's orbit around the sun.

3. Volcanic activity and the effect of volcanic dust on the radiation balance of the atmosphere.

4. Continental drift and the presence or absence of land masses at the poles that favor glaciation.

5. Quasiperiodic changes in ocean circulation, perhaps linked with the presence or absence of ice in the Arctic Ocean, perhaps linked with the behavior of the Antarctic ice sheet.

6. Almost-intransitivity of the land–ocean–atmosphere system, a theoretical concept in which the system appears to have more than one stable climate regime, where the shift from one climatic condition to another is accomplished without there necessarily being any changes in the external forcing functions or boundary conditions (such as solar input, continental drift, or atmospheric pollution). Rather, these changes are merely natural fluctuations inherent in the complex system of nonlinearly interacting processes that determine the climatic regime. Although the system possesses

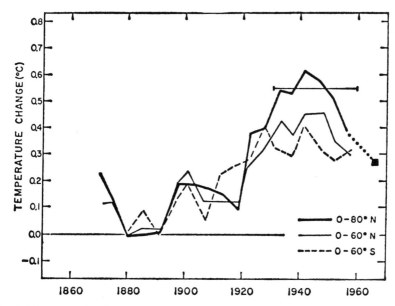

Fig. 2. Mean annual surface temperatures for various latitude bands. The method used by Mitchell [8] to extend the graph beyond 1960 is explained in SMIC [3]. The horizontal bar shows the mean value of temperatures in the 0–80°N band for 1931 to 1960.

only one stable climate when statistics are taken over an infinite period, the natural fluctuations could be observed as climatic changes over a finite interval of time [9].

Notice that we have roughly ranked these from the most deterministic theory, in the sense that the changes in climate are accounted for by a completely external fluctuation (in solar output), to the least deterministic, in the sense that climate changes are seen as random events in a naturally fluctuating system with no single stable condition over a finite period of time. One cannot dismiss any of these theories outright, and the final answer may very well be in terms of several of them together.

The approach to an explanation of climate change must be by at least two avenues: a better knowledge of past climates and changes in the conditions of the land–atmosphere–ocean system, and better mathematical (or physical) models that simulate this system. Such models must eventually be complete enough to include all the important physical and chemical factors that govern the heat balance of the atmosphere, oceans, and land surfaces. Then we will be able to answer some of the questions that we have raised by doing experiments with these models. In a very real sense, such models together with the observations that are needed to verify their predictions must be the key to an adequate theory of climate change, and without such a theory, we can hardly expect to predict man's impact on the climate with any certainty.

3. APPROACH TO A THEORY OF CLIMATE: INTERACTIONS AMONG ATMOSPHERIC CHEMISTRY, RADIATION, AND DYNAMICS

From the description of various different means by which man's activities may affect atmospheric composition and chemistry given in the following section, it will be apparent that such influences on the climate are not generally direct ones, but arise from the indirect effects connected with numerous interacting processes. In order to understand the consequences of changes associated with man's activities, therefore, we shall now examine the main factors and processes that govern all climatic changes, natural or man-induced. These processes can be classified under the subdivisions of radiative processes, atmosphere–ocean transport processes, and hydrological processes.

Some physical and mathematical aspects of a theory of climate will be presented in this section, with emphasis on the effects on the radiation balance that result from changes in atmospheric composition. This will be followed by a discussion of some possible indirect climatic effects (called "feedback mechanisms") resulting from such changes in the radiation balance. A short description of the hierarchy of mathematical models of climate is then given, with emphasis on the kinds of feedback mechanisms that can be treated by the various models.

3.1. Summary of Factors Affecting the Climate

The main factors that determine the climate of the earth–atmosphere system (taken to mean the atmosphere, oceans, and land and ice surfaces combined) are the input of solar radiation, the earth's rotation, the character of the earth's surface, and the chemical constituents of the earth-atmosphere system. Ultimately, it must be changes in the heat balance of this system that cause climate changes, and the key to this balance is provided by the many radiative processes that determine the distribution of the energy that drives the atmospheric "heat engine" [10].

3.1.1. Radiative Processes

3.1.1.1. Incoming Solar Radiation

The sun radiates energy corresponding roughly to an ideal (black) radiator with a temperature of about 5800°K [11]. This implies that 90% of the radiant energy lies in the interval with wavelengths from 0.4 to 4 μm, with a maximum intensity in the green portion of the visible spectrum at 0.48 μm.

About 30% [12] of the incoming solar energy is reflected back to space and is unavailable to warm the earth. This reflected fraction is called the "planetary albedo." Reflection occurs from the clouds, the earth's surface, and from molecules and particles present in the atmosphere. The clouds contribute the largest part of the albedo, reflecting about 25% [13] of the incoming radiation when averaged over a long period of time, but due to the natural variability of cloudiness over the globe, the earth's albedo can change substantially from day to day and also season to season [14].

The cloudless part of the earth comprises the remaining 5% of the global albedo. The albedo of the cloudless part of the earth is determined by the surface albedo and by reflection from atmospheric molecules and suspended particles. The latter, though contributing at most a few per cent to the total albedo, can be of great practical importance, since such particles are a factor that can be biased in one direction by man's activities.

Of the incoming radiation, about 25% is absorbed by gases, clouds, and particles in the atmosphere, 30% is reflected to space as discussed above, and the remainder is absorbed at the earth's surface [13]. This identifies another factor that can be affected by man's activities. Since man has significantly altered the character of the earth's surface, he has indirectly affected the climate (at least in limited regions) by disturbing the heat budget through changes in the character and albedo of the surface (see Chapter 7 of Ref. 3).

3.1.1.2. Outgoing Infrared Radiation

The incoming solar energy that is absorbed by the earth–atmosphere system must be balanced by an equal amount of outgoing radiant energy. Otherwise, the temperature of the earth would undergo a continuous change until the "energy balance" is restored. The earth emits radiant energy, as do all physical things, in proportion to its absolute temperature. But, since the wavelength of maximum radiant energy is inversely proportional to the temperature of the radiator, the earth emits radiation primarily in the long-wave or infrared region, with most of the energy residing in the wavelengths from 4 to 100 μm. If we calculate the total solar energy absorbed in the earth–atmosphere system and equate this to the escaping infrared radiation, then we can determine an "effective radiation temperature" of the planet from the Stefan–Boltzmann law relating the flux of radiant energy to temperature. This gives a numerical value of about $-20°C$ for the effective temperature of the earth, whereas we know the average *surface* temperature to be about $+14°C$.

The difference in the two temperature values is, of course, due to the presence of our atmosphere. The optically active gases, principally water vapor, carbon dioxide, and ozone, absorb and reemit infrared radiation in selective "bands" of the infrared spectrum. Clouds and particles also affect the infrared radiation, with the clouds (except thin cirrus clouds) absorbing nearly all the infrared radiation they receive throughout the infrared spectrum, and the particles absorbing or scattering *relatively* little infrared radiation, depending upon the character of the particulate material. (See, for example, the texts by Goody [15], Craig [16], or Kondratyev [17] for a detailed treatment of atmospheric radiation.)

The average surface temperature is higher than the effective radiative temperature primarily because the atmosphere is semitransparent to solar radiation but nearly opaque to infrared radiation as a result of absorbing gases and clouds. Thus, the surface, which absorbs much of the solar radiation, becomes a heat source for the lower atmosphere, which on the average cools steadily with increasing altitude to about 10 km. This part of the atmosphere is called the troposphere. The average tropospheric vertical temperature "lapse rate," $-\partial T/\partial z \sim 6.5°K\,km^{-1}$, is affected by both radiative heating and vertical convective processes [18].

The warm surface layer emits infrared radiation, most of which is intercepted by optically active atmospheric gases, clouds, and particles. These constituents reemit radiation both up to space and back down to the surface, the latter reducing the net loss of heat from the surface. Since the atmospheric emitters are colder than the surface, they emit proportionally less radiant energy. Because of this, the total outgoing infrared radiation from the earth–atmosphere system is less than the radiant energy emitted by the surface alone, and the effective radiation temperature of the earth is influenced more by the temperature of the colder atmospheric gases and cloud tops (which emit radiation roughly like a black body with the temperature of the atmosphere at the cloud tops) than the warmer surface below. This phenomenon has often been called the "greenhouse effect". Should the amount of an infrared-absorbing gas in the atmosphere be increased, it would then intercept a larger fraction of the infrared energy coming upward from the warm layers near the surface. Thus, the outgoing infrared flux to space would be reduced by adding infrared-absorbing gases to the atmosphere. Furthermore, this would also increase the downward infrared flux in the lower atmosphere, further warming the surface. The net result of the greenhouse effect is that an increase in the concentration of infrared absorbers in the atmosphere would lead to a rise in the surface temperature. This would be required in order to maintain constant infrared emission to space by the earth–atmosphere system, assuming that the planetary albedo remained unchanged.

Factors in the heat balance that could be significantly altered by man's activities are the amount of CO_2 or particles in the atmosphere. If the latter could also have an effect on cloud formation processes, then additional modifications to the heat budget might result. These influences will be identified more quantitatively in Section 3.2.

3.1.2. Atmosphere–Ocean Transport Processes

Although the amount of solar radiation absorbed in the earth–atmosphere system must be equal, over a sufficiently long time, to the outgoing infrared radiation for the earth as a whole, this is not necessarily true locally.

The input of solar radiation in the equatorial latitudes exceeds the outgoing infrared flux. This condition is referred to as a "positive" radiation balance. In middle latitudes, the incoming and outgoing radiant energies are of comparable magnitude, while at the polar latitudes, the outgoing infrared flux exceeds the absorption of incoming solar radiation by a large margin [12,13]. The latter results both from the relatively small values of average incoming solar energy at the poles and from the relatively high values of albedo of the polar ice caps. This unequal or differential solar heating of the globe occurs primarily in zones (latitude belts) and, when coupled with the rotation of the earth, is the driving force behind the motions we recognize as winds and ocean currents. These motions, both horizontal and vertical, regulate the distribution of temperature, cloudiness (which also requires the presence of suitable cloud nuclei), and precipitation over the globe.

The motions of the system result in a transport of heat from areas of positive radiation balance (the equatorial regions) to areas of negative radiation balance (the polar regions). In this process, the jet streams, trade winds, eastward winds of the mid latitudes, westward winds of the polar latitudes, and migratory large-scale weather systems or eddies are all generated in the atmosphere [19]. When the north–south temperature gradient is large, the circulation system becomes more unstable, generating large-scale transient eddies (storm systems) which transport additional heat poleward, tending to reduce some of the equator-to-pole temperature difference [20]. (This identifies a "negative" climatic feedback mechanism discussed further at the end of Section 3.2.)

The atmosphere carries heat in two forms: sensible and latent. When, for example, warm air is *directly* transported to a cold region, this can be thought of as a transport of "sensible" heat. Water vapor which is evaporated at the earth's surface can also be transported by the atmosphere. When it undergoes cooling in the presence of suitable nuclei (particles), the water vapor may condense into drops, thereby releasing the "latent" heat that was needed originally to change it from liquid water to water vapor. The process of evaporation, transport of water vapor, condensation, precipitation, and reevaporation, which is called the hydrological cycle, is responsible for about one-third of the net heat transported across the 30°N and S latitude circles. At this latitude, the sensible heat transport by the atmosphere accounts for another third of the total heat transported, and the oceans carry the remaining third of the total heat flowing poleward [21], through ocean currents, such as the Gulf Stream in the Atlantic. The ocean surface temperatures, which are controlled by turbulent mixing processes in the upper hundred meters or so of the oceans [22], play an important role in the exchange of sensible and latent heat with the atmosphere. Unfortunately, our knowledge of these processes is still too inadequate to permit satisfactory quantitative determination of the important role of the oceans in the global heat budget.

The possibility of climate modification by *direct* human interference with the motions of the earth–atmosphere system is less likely than the chance of man's activities modifying the radiation balance or cloud nucleation processes *indirectly*. However, by diversion of rivers or ocean currents, for example, man might trigger additional mechanisms which could ultimately have an impact on climate. Man is also affecting the level of the natural underground reservoirs of water by his mining of ground water for domestic use and in irrigation projects. This influences the hydrological cycle and could have an effect on the climate through changes in evaporation rates and the surface albedo. See Chapter 7 of the SMIC Report [3] for further details.

3.2. Physical and Mathematical Formulations of the Theory of Climate

In order to evaluate the effects of changes in atmospheric composition and chemistry on the climate, we must first identify the physical principles that govern the climate and then translate these physical statements into mathematical equations. Solution of these equations would provide, in principle, a theory of the climate. Following this approach, we then would summarize the effects of man's activities by additional mathematical statements which would be solved together with the equations of the general theory of climate for the purpose of evaluating man's impact on climate.

This section will present the principal mathematical equations involved in climate theory and indicate which terms are most affected by changes in atmospheric chemistry. Although a detailed treatment of the approximations that are often made in attempts to solve the equations is beyond the scope of this chapter, a discussion of climatic feedback mechanisms and climate models is presented at the end of this section.

3.2.1. General Theory

The factors governing climate (solar radiation, ocean circulation, ice cover, winds, etc.) that were discussed briefly in the previous section must all be related by the equations of the general climate theory. The principles of conservation of mass, momentum, and energy, taken together with the thermodynamic and chemical laws governing the change in material composition of the land, sea, and air, comprise the fundamental theoretical basis for the theory of climate. These form a coupled, nonlinear, three-dimensional set of partial differential equations, which are to be solved subject to the external input of solar radiation and for a given initial state of the land–sea–air system.

Each variable is related to the others in such a way that changes in one invoke simultaneous variations in others, which, in turn, can have a feedback effect on the original variable. For example, part of the solar radiation absorbed by the oceans is used to evaporate water, which can then form clouds that reduce the solar radiation reaching the surface, which in turn reduces evaporation (a negative feedback in this case).

The motions in the system take place on a variety of time and space scales. While features of the general circulation with a scale of 1000 km persist for days, small-scale quasirandom motions in the atmosphere and the oceans lasting for seconds and occupying a few meters of space are of crucial importance in the turbulent transport of heat, momentum, and atmospheric constituents. For practical applications, these are usually treated statistically and their effects are often related to average conditions over periods of the order of a day. The process of relating the statistical effects of very small-scale motions to conditions on a much larger scale—one that can more easily be computed—is called parameterization.

To attempt to solve the equations of climate theory with the knowledge and tools available to us in the foreseeable future, it is necessary to ignore the details of small-scale motions occurring on a continuous spectrum of scales and treat the system at a discrete number of points in space (called the "grid") and in time (called the "time step"). Then, all motions occurring with scales smaller than the grid size must be treated by parameterization or be completely ignored. The technique of selecting appropriate grid size, time step size, and parameterization technique for use in the construction of approximate theories of climate is the art of "modeling," and the particular choice of these elements determines the model. Of course, different models also vary with regard to the number and choice of physical and chemical factors they can take into account. A summary of various types of climate models is given at the end of this section. For a thorough collection of references and a more detailed discussion of modeling techniques, see Chapter 6 of the SMIC Report [3].

3.2.1.1. Primitive Equations

The following set of equations which relate the motions of the atmosphere to the input of radiant energy, and which are dependent upon the chemical composition of the atmosphere and physical and chemical conditions of the bounding earth's surface, provide an "in principle" answer to the question of the effect of changes in atmospheric chemistry on the variables that describe the state of the atmosphere. These atmospheric variables are eastward and northward components of the wind (relative to the earth's surface) u and v; vertical velocity w; pressure p; density and temperature ρ and T; and fraction by mass of atmospheric chemical constituents q_i. Taken together, they determine the climate, when statistics of these variables are retained over a sufficiently long period of time.

The first of the set of equations that we must solve are the equations of motion, which can be written

$$\frac{\partial u}{\partial t} + u\frac{\partial u}{\partial x} + v\frac{\partial u}{\partial y} + w\frac{\partial u}{\partial z} = 2\Omega(v\sin\varphi - w\cos\varphi) - \frac{1}{\rho}\frac{\partial p}{\partial x} + F_x \quad (1)$$

$$\frac{\partial v}{\partial t} + u\frac{\partial v}{\partial x} + v\frac{\partial v}{\partial y} + w\frac{\partial v}{\partial z} = 2\Omega u\sin\varphi - \frac{1}{\rho}\frac{\partial p}{\partial y} + F_y \quad (2)$$

$$\frac{\partial w}{\partial t} + u\frac{\partial w}{\partial x} + v\frac{\partial w}{\partial y} + w\frac{\partial w}{\partial z} = 2\Omega u \cos\varphi - g - \frac{1}{\rho}\frac{\partial p}{\partial z} + F_z \tag{3}$$

where F is the frictional force, Ω is the earth's angular velocity, φ is the latitude angle, and g is the acceleration of gravity. [Equation (3) is usually used in large-scale models in a highly simplified form neglecting all terms involving w, Ω, or F_z; then it is known as the hydrostatic equation.] The next equations to be solved simultaneously are the equation of continuity

$$\frac{\partial\rho}{\partial t} + u\frac{\partial\rho}{\partial x} + v\frac{\partial\rho}{\partial y} + w\frac{\partial\rho}{\partial z} = -\rho\left(\frac{\partial u}{\partial x} + \frac{\partial v}{\partial y} + \frac{\partial w}{\partial z}\right) \tag{4}$$

and equation of state

$$p = \rho RT \tag{5}$$

where R is the gas constant for air (a weakly dependent function of the local water vapor content of the atmosphere). Taken together, these five equations involve six unknowns: u, v, w, p, ρ, and T. The sixth equation is the energy equation, derived from the laws of thermodynamics that relate the changes in thermodynamic variables to the rate of radiative heat input $\dot{Q}$:

$$C_p\frac{dT}{dt} - \frac{1}{\rho}\frac{dp}{dt} = \dot{Q} - L\frac{dq_w}{dt} - C_l l\frac{dT}{dt} \tag{6}$$

where C_p is the specific heat of air at constant pressure, $\dot{Q}$ is the diabatic heating (from radiative processes), L is the latent heat of vaporization released when the specific humidity q_w is changed, and C_l is the specific heat of the liquid water content l. These equations, together with the additional relationships needed to determine the radiation field heating rate $\dot{Q}$ and the moisture field q_w provide a complete description of the state of the atmosphere. The derivative d/dt is equivalently $[(\partial/\partial t) + \mathbf{V} \cdot \nabla]$.

The form of the primitive equations described here is not that usually used in practical models of the atmospheric circulation, since the dependent variables in Eqs. (1)–(6) describe the "instantaneous" state of the atmosphere, including phenomena occurring with scales smaller than the grid size used in practical models. Therefore, it is common to subdivide the instantaneous properties into a mean value and fluctuating component, for example:

$$u = \bar{u} + u' \tag{7}$$

where the mean value component $\bar{u}$ is averaged over a suitable time interval Δt (large compared with small-scale turbulent motions and small in comparison to large-scale synoptic phenomena):

$$\bar{u} = \frac{1}{\Delta t}\int_{t-(\Delta t/2)}^{t+(\Delta t/2)} u\, dt \tag{8}$$

Thus, for a given time interval Δt, the mean of the fluctuating component (u') is zero.

Applying (7) to (1) and after some manipulation, the incompressible x-momentum equation can be obtained in the following form:

$$\frac{\partial \bar{u}}{\partial t} + \bar{u}\frac{\partial \bar{u}}{\partial x} + \bar{v}\frac{\partial \bar{u}}{\partial y} + \bar{w}\frac{\partial \bar{u}}{\partial z} = +f\bar{v} - \frac{1}{\rho}\frac{\partial \bar{p}}{\partial x}$$

$$- \frac{1}{\rho}\left[\frac{\partial(\overline{\rho u'u'})}{\partial x} + \frac{\partial(\overline{\rho u'v'})}{\partial y} + \frac{\partial(\overline{\rho u'w'})}{\partial z}\right] \quad (9)$$

where f is the Coriolis factor $(2\Omega \sin \varphi)$ and the vertical component of the Coriolis acceleration has been neglected. Here, the three "eddy stress" terms appear on the right-hand side and represent the transfer of momentum caused by small-scale transport of turbulent eddies. The viscous stress term F_x is neglected in (9) since it is several orders of magnitude smaller than the eddy stress term for most applications to general circulation models.

In practice, it is frequently assumed (by analogy to viscous stresses) that the eddy stresses are proportional to a coefficient times the wind shear. For example,

$$-\overline{\rho u'w'} = K_{xz} \, \partial\bar{u}/\partial z \quad (10)$$

where K_{xz} is called the coefficient of eddy viscosity or "*Austauch*" coefficient for the diffusion of momentum.

The derivation of these equations and the approximation techniques necessary for their solution are well beyond the scope of this chapter (see, for example, Haltiner [23], Hess [24], or Thompson [20]). Nevertheless, the interdependence of the atmospheric variables can clearly be seen merely by listing this set of coupled nonlinear partial differential equations.

The role of atmospheric chemical composition, though weakly coupled to the mechanical variables (through the change in atmospheric density and gas constant associated with variation in moisture content), is nevertheless strongly involved in the atmospheric processes of radiative heating and latent heating, since the former depends upon the presence in the atmosphere of optically active chemical constituents, and since the release of latent heat through the condensation of water vapor in the atmosphere always occurs on a suitable particle nucleus. The importance of changes in the chemical composition of the lower atmosphere on the radiative heating will be discussed below, whereas the role of particles in cloud and precipitation process (which are strongly, though indirectly, related to the radiation field) is presented by Pruppacher in Chapter 1.

Figure 3 shows the well-known blackbody emission curves for solar radiation (6000°K) and planetary infrared radiation (245°K) as a function of wavelength. Below these curves the total atmospheric absorption spectrum is given, indicating the effect of various trace gas constituents on the radiation

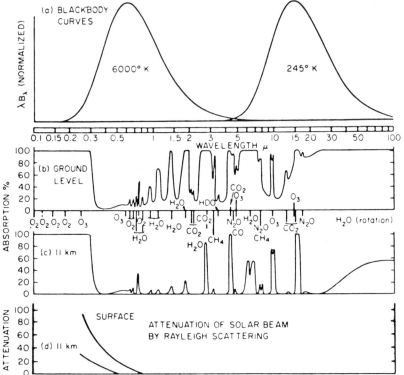

Fig. 3. (a) Blackbody emission for 6000°K and 245°K, being approximate emission spectra of the sun and earth, respectively (since inward and outward radiation must balance, the curves have been drawn with equal areas, though in fact 30% of solar radiation is reflected unchanged); (b) atmospheric absorption spectrum for a solar beam reaching the ground; (c) the same for a beam reaching the tropopause in temperate latitudes; (d) attenuation of the solar beam by Rayleigh scattering at the ground and at the temperate tropopause [10].

balance. The absorption of radiation in the lower atmosphere [below 11 km, which is the difference between curves (b) and (c)] is mainly due to the presence of several water vapor absorption bands and the 15-μm CO_2 absorption band, which occurs near the center of the infrared blackbody energy spectrum. Trace gases such as N_2O, CO, or CH_4 have only a minor effect on the atmospheric radiation field, and are therefore not usually considered to be among those chemical constituents having an important influence on climate. However, gases such as SO_2, H_2S, NH_3, and NO_2 can have a significant indirect or delayed effect on the radiation field since these substances can be photochemically converted into atmospheric particles, as discussed in Chapters 2, 3, and 4. The role of particles in scattering and absorbing radiation is discussed in some detail below.

3.2.1. Radiative Transfer—Incoming Solar Radiation

The incoming solar radiation, represented schematically in Fig. 3(a), is scattered and absorbed by the earth's surface and the atmosphere as described qualitatively in previous sections. To express the attenuation of solar radiation by atmospheric constituents, we consider the change dI_λ in intensity I_λ of a parallel monochromatic beam as given by Beer's law

$$dI_\lambda = I_\lambda \sigma_\lambda \rho \, ds \tag{11}$$

where ρ is the density of the attenuating medium, σ_λ is the extinction coefficient per unit mass, and ds is the path length. Integrating (11) yields

$$I_\lambda = I_{0\lambda} \exp \left(-\int_0^s \sigma_\lambda \rho \, ds \right) \tag{12}$$

where $I_{0\lambda}$ is the intensity of radiation at the point $s = 0$. The exponential expression in (12) is called the "monochromatic transmission function" or "transmittance" and the exponent $\int_0^s \sigma_\lambda \rho \, ds$ is the "monochromatic optical thickness" of the medium along the path length $(0, s)$.

Extinction of the direct beam is due to the combined effect of absorption and scattering. If only scattering occurs, all the energy present in the original beam is maintained in the radiation field, but some of this is redistributed in all directions. When the scattered and transmitted energies integrated over all angles are less than the incident beam energy, then absorption has occurred. The absorption and scattering coefficients depend upon the wavelength of radiation and the optical properties of the medium. For a layer of particles, for example, the "optical properties" refer to the real part of the index of refraction η_r of the particles and to the index of absorption η_i, or imaginary part of the index of refraction. Together with the size distribution and individual shapes of the particles, the optical properties determine the extinction coefficient for the particle layer [25]. This can be a highly variable function of wavelength. Since the index of refraction and particle size depend strongly on the chemical composition of the particles, changes in the number, size, and chemical composition of suspended atmospheric particles can therefore significantly affect the radiation field.

Part of the solar beam is scattered by air molecules. The average radius of air molecules is small with respect to the median wavelengths of solar radiation, and the scattering coefficient is approximately that given by Rayleigh [25]. This type of scattering has a simple angular distribution; that is, the intensity of radiation scattered in directions about the initial path is symmetric with respect to the initial direction and is also symmetric with respect to a perpendicular to the initial direction. Thus, an equal amount of energy is scattered in the "forward" direction $(0 \leq \theta \leq \pi/2)$ as is "back-scattered" $(\pi/2 \leq \theta \leq \pi)$. The coefficient of Rayleigh scattering is inversely proportional to the fourth power of the wavelength. Thus, air molecules scatter a large fraction of the short-wavelength solar spectrum, especially in the ultraviolet and blue regions (Rayleigh scattering therefore accounts for the blue sky color), while Rayleigh scattering from air molecules has little influence on infrared radiation, as can be seen in Fig. 3(d).

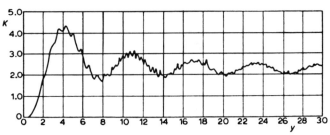

Fig. 4. The Mie scattering cross section k as a function of the size parameter y and for $\eta_r = 1.50$ [11].

However, as the radius of the scattering particle becomes comparable to the wavelength of radiation, the scattering is no longer the $\sim \lambda^{-4}$ Rayleigh scattering. When the radius r of spherical scattering particles is in the range $0.1\lambda \leq r \leq 25\lambda$, true for much of the suspended atmospheric dust for wavelengths in the visible region, it is necessary to use the more complicated theory of Mie [25]. The Mie scattering cross section S is then given by

$$S = \pi r^2 k(y, \eta) \tag{13}$$

where k is a function that depends upon the real part of the refractive index η_r and the size parameter $y = 2\pi r/\lambda$. Figure 4 shows the Mie scattering cross section $k(y, \eta)$, as a function of y and for $\eta_r = 1.5$. From the figure and the definition of parameter y, we can deduce the particle radius that has a maximum scattering effect on visible radiation (with $\lambda = 0.55\,\mu$m). Since the scattering cross section is a maximum at about $y = 4$, then the particle radius of maximum scattering of visible radiation is about $r_m = (\lambda/2\pi)y = (0.55/2\pi)$ $4 \approx 0.35\,\mu$m. (Figure 4 is for particles with the same radius. When the particle sizes obey a distribution law, the Mie cross-section curve may change shape somewhat.) Particles of this size are abundantly produced by man's activities, as described later on. The "phase function," or angular scattering distribution resulting from Mie scattering by particles, is no longer symmetric with respect to a perpendicular to the direction of incidence, as can be seen in Fig. 5, and is also dependent on the parameter y. As the particle size increases, the preponderance of forward scattering also increases.

For a typical atmospheric aerosol (suspended layer of particles in the atmosphere), there is a distribution of particle sizes given by Junge [26] in the form

$$n(r) \propto r^{-b} \tag{14}$$

where $n(r)$ is the number of particles with radius r. In relatively unpolluted air, and for the particle size range $0.1 \leq r \leq 10\,\mu$m, which is the important size range for Mie scattering of the visible part of the solar beam, b is roughly equal to 4. That is there are relatively more suspended particles with radii near 0.1 than 10 μm (see Chapter 3 on particle removal processes). One consequence of this size distribution is the fact that such an aerosol layer has

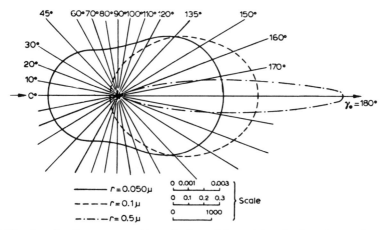

Fig. 5. The phase function or scattered intensity due to Mie particles of various sizes as a function of scattering angle γ_0 (in the text, $\theta = \pi - \gamma_0$) for green light [11].

a larger effect on scattering visible radiation than on infrared radiation [27]. However, in local regions of heavy pollution where deviations from the background Junge size distribution may occur, particles may also have significant effect on infrared radiation [28].

In addition to scattering radiation the particles can also absorb energy. This can have a consequence on the climate in several possible ways: by changing the albedo of the earth; by altering the atmospheric heating rate $\dot{Q}$; and also by decreasing the solar energy reaching the earth's surface. We consider first the effect of a particle layer in the lower atmosphere on changing the effective albedo of the underlying surface. Figure 6 shows an aerosol layer above the earth's surface with surface albedo α_S. Let α_{BS} be the fraction of sunlight (i.e., solar flux integrated over wavelength) back-scattered to space from the aerosol layer and let α be the fraction of flux absorbed in the aerosol layer. The transmission of radiation through the particle layer is $t = 1 - \alpha_{BS} - \alpha$. Then α_S of the transmitted beam is reflected from the surface (neglecting directional effects of the particles in changing α_S), with t times this amount escaping to space through the aerosol layer. Thus, the total or effective albedo of the combined aerosol–surface system is obtained by summing the power series [29]

$$\alpha_E = \alpha_{BS} + [t^2\alpha_S/(1 - \alpha_S\alpha_{BS})] \qquad (15)$$

The significance of changes in absorption and back-scatter fractions on the effective albedo can be seen in Table 1, computed using Eq. (15). The table shows that, except for local cases of very large aerosol absorption-to-back-scatter fractions or very large surface albedo, the effect of an aerosol is to increase the albedo of the earth's surface. Since the albedo of the cloudless fraction of the earth is about 0.14 on a global average [13], the global effect of

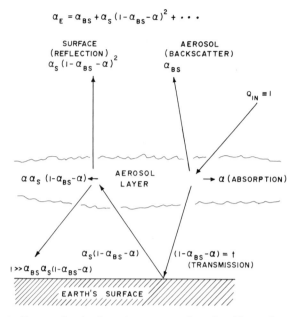

$$\alpha_E = \alpha_{BS} + \alpha_S (1 - \alpha_{BS} - \alpha)^2 + \cdots$$

Fig. 6. Schematic diagram showing how the presence of an absorbing and scattering aerosol layer in the lower atmosphere can change the effective albedo of the earth's surface [29].

an aerosol layer (with absorption equal to back-scatter and concentrated primarily below the cloud tops) would appear to be an increase in the albedo of the cloudless fraction of the earth. An increase in the entire earth's albedo of 1 % corresponds roughly to a global average surface temperature decrease of 2°C [30].

Multiple-scattering calculations have been performed to determine quantitatively the parameters α_{BS} and α (Rasool and Schneider [27]; Yamamoto and Tanaka [31], described in Chapter 8 of the SMIC Report [3]).

TABLE 1. Computed Values of Effective Albedo
of the Aerosol–Surface System for Various
Values of α_S, α_{BS}, and α

α_S	α_{BS}	α	α_E	$\alpha_E - \alpha_S$
0.1	0.01	0.01	0.1062	+0.0062
0.1	0.05	0.00	0.1408	+0.0408
0.1	0.05	0.05	0.1314	+0.0314
0.1	0.05	0.10	0.1226	+0.0226
0.05	0.05	0.05	0.0961	+0.0461
0.15	0.05	0.05	0.1622	+0.0122
0.28	0.05	0.05	0.2800	0.0000
0.50	0.05	0.05	0.4654	−0.0346

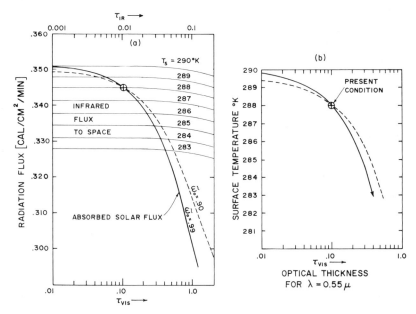

Fig. 7. In (a), both the global average values of absorbed solar radiation flux and terrestrial infrared radiation flux to space are plotted as a function of increasing accumulation of aerosols in the atmosphere. The absorbed solar flux decreases with increasing optical depth for both values of the single scattering albedo parameter $\tilde{\omega}_0$. The infrared flux to space, calculated for several values of surface temperature, is practically unaffected by increasing aerosols, as explained in the text. The intersections of infrared and visible flux values determine the equilibrium global surface temperature for a given optical thickness τ_{VIS} and are cross-plotted in (b) [27].

In these analyses, it was necessary to assume values for the optical properties of aerosols, the size distribution of the particles, and the effect of the particles on the part of the earth covered by clouds. In both studies, the background Junge size distribution [Eq. (14) with $b = 4$] was used and the particles were assumed to have no effect on either the amount of cover, height, or albedo of the cloudy half of the earth. In these independent studies, a real index of refraction $\eta_r = 1.5$ was assumed, and two values of the index of absorption η_i were given, $\eta_i \approx 0.00$ and $\eta_i \approx 0.01$, the latter implying aerosol absorption comparable to back-scatter for a median visible wavelength of 0.55 μm. For the infrared, a median wavelength of 10 μm, and an absorption index of $\eta_i = 0.1$ were used by Rasool and Schneider; Yamamoto and Tanaka neglected infrared effects.

In the results of Rasool and Schneider given in Fig. 7, the infrared flux to space in the presence of the aerosol layer is computed, but the effect of an increase in aerosol optical thickness on reducing the infrared radiation ($\lambda = 10$ μm) to space is found to be relatively less significant than the resulting increase in global albedo (for $\lambda = 0.55$ μm) for both cases of η_i. This is a consequence of the size distribution as mentioned previously. The global average surface albedo was taken to be 0.10. The computed increase in the albedo of the cloudless part of the earth results in a decrease of absorbed solar flux in the earth–atmosphere system. The intersection of the absorbed solar flux curves with outgoing infrared flux values determines the global average equilibrium surface temperature as a function of aerosol optical thickness, which is cross-plotted on Fig. 7(b). The single scattering albedo $\tilde{\omega}_0$, which depends upon η_i, is given in the figure for the two cases of index of absorption assumed in the calculations, $\eta_i \approx 0.00$ ($\tilde{\omega}_0 = 0.99$) and $\eta_i \approx 0.01$ ($\tilde{\omega}_0 = 0.90$). The single scattering albedo (described further, for example, by Hansen and Pollack [32]) is a measure of the ratio of light scattered to the sum of light both scattered and absorbed in a *single* encounter with a particle.

The computations of Yamamoto and Tanaka agree with those of Rasool and Schneider in predicting a significant cooling of the earth from an increase in aerosols. However, both these results depend, of course, on the major assumptions of the models: global average horizontal distribution of model variables; assumed values for the aerosol optical properties (particularly η_i); and finally, on the assumption of the noninteraction of the particles with cloudiness and the atmospheric motions, as discussed by Charlson et al. [33] and Rasool and Schneider [34]. At present, it is impossible to determine even the direction of coupled effects of aerosols and cloudiness. Mitchell [35] has argued that particles might indirectly decrease cloudiness, thereby offsetting any increase in global albedo, whereas Twomey [36] (see Fig. 8.9 of the SMIC Report [3]) has shown that particles could increase the albedo of clouds by directly altering cloud microstructure. As for possible coupled interactions between aerosol-induced changes in the radiation balance or vertical atmospheric radiative heating rates and a response by the general circulation or vertical atmospheric motions, the understanding of the effects of these kinds of "feedback mechanisms" requires further modeling by large-scale numerical models, preferably including ocean–atmosphere coupling as discussed later (see also Chapter 6 of the SMIC Report [3]).

Nevertheless, these global average models do indicate that changes in atmospheric chemistry can significantly affect climate. This suggests that programs for the monitoring of the geographic distribution of particle concentrations and optical properties be implemented, along with the improvement of mathematical models, incorporating as many coupled effects or feedback mechanisms as can be included with the available computing machinery.

3.2.1.3. Radiative Transfer—Outgoing Infrared Radiation

In the previous section, we discussed the effect of changes in some of the chemical constituents in the lower atmosphere on the absorption and scattering of the incoming solar beam, stressing in particular the role of particles. Now the influence of some atmospheric constituents on the outgoing infrared or planetary radiation will be considered, with particular emphasis on the selective absorption of infrared radiation by gases such as carbon dioxide and water vapor. The relative transparency of these gases to solar radiation compared to their opaqueness to infrared radiation leads to the well-known greenhouse effect, described qualitatively earlier in this chapter. We will consider briefly the equations of radiative transfer that describe the process of selective absorption of infrared radiation and show why an increase in atmospheric CO_2 will lead (in the absence of other coupled effects) to a warming of the climate.

The fundamental equation that describes infrared radiative transfer, which is obtained by using Kirchhoff's law and is valid in the lower atmosphere to a sufficient degree of approximation (see Goody [15]), is

$$dI_v/ds = -\rho K_v(I_v - B_v) \tag{16}$$

where I_v is the intensity of monochromatic radiation of wave number v ($= 1/\lambda$) traveling along the path ds, B_v is the intensity of blackbody or Planckian emission, ρ is the absorber density, and K_v is the absorption coefficient. For the presence of two or more absorbers, the product ρK_v should be replaced by the sum of the products for each respective absorber. For a horizontally homogeneous atmosphere, (16) can be integrated from the bottom of the atmosphere ($z = 0$) to the top ($z = H$) and over all zenith angles ($0 \leq \theta \leq \pi/2$) in the form of an upward $F^\uparrow$ and downward $F^\downarrow$ radiation flux passing through the level $z = \bar{z}$.

The net radiative flux in the atmosphere at level $z = \bar{z}$, F_{NET}, is the difference between the upward and downward fluxes, $F^\uparrow - F^\downarrow$. The vertical divergence of F_{NET} is proportional to the radiative atmospheric heating rate $\dot{Q}$, from Eq. (6). The upward flux at $z = \bar{z}$ for radiation in some frequency interval $(v, v + \Delta v)$ (in cm^{-1}) is comprised of two terms,

$$F_v^\uparrow(\bar{z}) = B_v(T(o))\tau_v(\bar{z}, o) + \int_0^{\bar{z}} B_v(T(z))(d/dz)\tau_v(\bar{z}, z)\,dz \tag{17}$$

The first term on the right-hand side of (17) is the contribution to the upward flux from the blackbody radiation (Planckian function, B_v) in the frequency interval $(v, v + \Delta v)$ from the earth's surface, with temperature $T = T(z = 0)$. This surface flux, $B_v(T(o))$ is multiplied by the transmittance of the atmosphere τ_v [for the frequency interval $(v, v + \Delta v)$ and for the atmosphere between the surface, $z = 0$, and the atmospheric level $z = \bar{z}$] to determine the surface contribution to $F^\uparrow(\bar{z})$, the upward flux at $z = \bar{z}$.

The second term in (17) represents the contribution to the total upward flux from emission of infrared radiation by optically active gases below the level $z = \bar{z}$. Unlike the surface flux, which is ideal blackbody radiation, the atmospheric emission is highly frequency-dependent because of the selective absorption of CO_2 or H_2O, and the integral has a strong v dependence through the transmission function τ_v.

Similarly, the downward infrared radiation flux at $z = \bar{z}$ arriving from above $\bar{z}$ is given by

$$F_v^\downarrow(\bar{z}) = \int_{\bar{z}}^H B_v(T(z))(d/dz)\tau_v(\bar{z}, z)\, dz \tag{18}$$

Thus, the downward infrared flux results only from the atmospheric infrared emitters, since the incoming planetary infrared radiation at $z = H$ is essentially zero.

The transmission of the atmosphere to infrared radiation in the frequency interval $(v, v + \Delta v)$ and between atmospheric altitudes $z = z_1$ and $z = z_2$ can be expressed in terms of the optical thickness of the atmosphere $u_v(z_2, z_1)$ by

$$\tau_v(z_2, z_1) = (2/\Delta v) \int_v^{v + \Delta v} E_{i3}[u_v(z_2, z_1)]\, dv \tag{19}$$

Here, the monochromatic optical thickness is

$$u_v(z_2, z_1) = \int_{z_1}^{z_2} (\rho_1 K_1 + \rho_2 K_2 + \ldots)\, dz \tag{20}$$

and E_{i3} is the third-order exponential integral which accounts for the integration of radiation intensity over all zenith angles. The optical thickness is proportional to the product of the densities of the optically active atmospheric constituents, ρ_i, and their respective absorption coefficients K_i. For many practical radiative transfer problems, (19) can adequately be approximated by (see pp. 134–144 of Kondratyev [37])

$$\tau_v(z_2, z_1) \propto \exp[-\beta u_v(z_2, z_1)] \tag{21}$$

where $\beta = \sec \bar{\theta} \approx 1.66$, which indicates an average zenith angle $\bar{\theta}$ of about 53°. See Refs. 15–17 and 37 for details of the derivations and approximations involved in the equations of the transfer of infrared radiation through the lower atmosphere.

The downward flux of infrared radiation warms the lower layer of the atmosphere and is in part responsible for the greenhouse effect—where the earth–atmosphere system as a whole emits infrared radiation to space at an effective temperature near the colder temperatures of the middle layers of the troposphere (the effective radiating temperature of the earth is about 253°K), rather than the warmer surface layers (the average surface temperature is about 287°K). By integrating (17) over frequency for clear sky conditions, it can be shown that only about 30% of the surface radiation flux, $\int B_v(T(o))\, dv$, is able to escape directly through the atmosphere to space. Most of the outgoing infrared flux at the top of the atmosphere originates in the colder middle layers of the troposphere.

Thus, increasing the infrared optical thickness of the atmosphere by increasing the concentration of absorbing constituents such as CO_2 or H_2O

will affect the infrared radiation fluxes. First of all, the transmission of the atmosphere to the warmer surface radiation will be decreased. At the same time, the increased opacity of the lower atmospheric layers will augment the downward infrared radiation, warming the surface layers. Second, the height of layers in the middle troposphere that effectively emit radiation to space will be slightly raised due to the increased opacity of the lower layers from augmented CO_2, thereby lowering the effective planetary radiation temperature. Thus, the outgoing infrared flux will be reduced. Assuming that increased CO_2 in the atmosphere would have no effect on the albedo of the earth–atmosphere system, then the increased CO_2 will not alter the amount of solar energy absorbed by the earth–atmosphere system. However, since the infrared flux to space would be decreased by the addition of CO_2 to the atmosphere, the tropospheric temperature would be forced to rise—until radiation balance of the outgoing infrared flux with the unchanged amount of absorbed solar energy would be reestablished. In order to compute quantitatively the effect of increases in CO_2 on tropospheric temperature [by use of (17) and (18)], it is necessary to know the vertical atmospheric temperature profile in order to evaluate the integrals on the right-hand sides of these equations. A standard assumption used in global average climate models for such an application is to assume that the atmospheric lapse rate never exceeds the observed "critical" lapse rate for convective stability: $-6.5°K\ km^{-1}$. This sort of model, called a "radiative–convective" model by Manabe and Wetherald [18], has been used to predict an average surface temperature increase of about 2°K for a doubling of atmosphere CO_2 from the current 300 ppm to 600 ppm. At the same time, the model predicts a stratospheric cooling of several degrees kelvin. The model accounts for the additional infrared greenhouse effect from increased atmospheric water vapor that might be expected from increased tropospheric temperatures. That is, CO_2 warming would increase the absolute amount of atmospheric water vapor under the assumption of constant relative humidity [38]. However, in Manabe and Wetherald's model, the increased water vapor is also allowed to *lower* the planetary albedo, further increasing the amount of absorbed solar energy in the earth–atmosphere system through the mechanism of increased atmospheric absorption of sunlight by the water vapor in the near-infrared. (This identifies a "positive feedback mechanism" between changes in the radiation balance arising from changes in the amount of CO_2 and the general response of the earth–atmosphere system.) Climatological averages for global cloudiness were adopted for the calculation and the cloudiness was assumed to be *unaffected* by the changes in radiative heating induced by increasing CO_2. As for the question of very large increases in CO_2, Rasool and Schneider [27] noted that the surface temperature increase levels off with large increases in CO_2. This is because the surface temperature increases logarithmically with CO_2 amount.

Thus, although these models clearly show that the initial consequence of an increase in CO_2 is a warming of the climate, the magnitude of these tentative predictions must await further modeling studies that include the feedback mechanisms relating changes in CO_2 amount to changes in atmospheric water vapor and cloudiness, changes in the lapse rate, alteration to the global circulation, and, possibly most important, shifts in the extent of polar ice.

The next section will identify some of the important feedback mechanisms often referred to in this chapter and will outline the kinds of mathematical models needed to elucidate these effects.

3.2.2. Feedback Mechanisms That Govern Climate

So far, we have shown that changes in the chemical composition of the lower atmosphere can affect climate, primarily through changes in the radiation balance. Although existing climate models are already able to predict that these changes are potentially significant, their conclusions must await improved models that account for other coupled effects or feedback mechanisms. In this section, we list some of these feedback effects and try to show how they could be incorporated into different types of mathematical models. The interested reader should also see Chapter 6 of the SMIC Report [3].

3.2.2.1. Assorted Feedback Mechanisms in the Earth–Atmosphere System

In order to study the sensitivity of the climate to changes in the concentration of a certain pollutant, it is first necessary to determine the immediate effect of the pollutant on the radiation balance. After this, one must establish how other coupled or interacting processes (often called feedback mechanisms) might act to dampen (negative feedback) or amplify (positive feedback) that effect on the climate.

Some of the clearly identifiable feedback mechanisms that govern climate are as follows:

Temperature–Radiation Coupling. Since all physical things radiate energy in some direct proportion to their absolute temperature, an increase in the temperature of a body also results in an increase in the amount of energy radiated from the body (with all other factors, such as chemical composition, remaining unchanged). Thus, an increase in the input of heat to a body, which would raise its equilibrium temperature, would also lead to an increase in the radiation leaving the body. This process tends to limit or stabilize the temperature response of a body to changes in energy input. Therefore, temperature–radiation coupling can be identified as a stabilizing or *negative* feedback mechanism.

Water Vapor–Greenhouse Effect Coupling. The atmosphere is believed to maintain a somewhat uniform distribution of *relative humidity* over a large range of lower atmospheric temperatures (as mentioned earlier in this chapter) even though the *absolute* amount of water vapor in the air varies strongly with atmospheric temperature. The absolute amount of water vapor in the atmosphere determines, to a large extent, the opacity of the lower atmosphere to infrared radiation as described earlier. Thus, increases in atmospheric temperature at constant relative humidity would lead to increased greenhouse effect, which gives rise to further increases in temperature. Therefore, water vapor–greenhouse effect coupling is a destabilizing or *positive* feedback mechanism.

Snow and Ice Cover–Albedo–Temperature Coupling. The high reflectivity of snow and ice as compared to water and land surfaces is a dominant factor in the climate of polar regions. However, the extent of the snow and ice cover of the earth's surface strongly depends upon the surface temperature. Thus, lowering of the planetary temperature would be expected to lead to a longer-lasting snow and ice cover, which would increase the planetary albedo— causing a decrease in solar energy absorbed in the earth–atmosphere system —thereby further lowering the temperature. This *positive* feedback mechanism has caused many scientists (see the SMIC Report [3]) to consider the question of the permanency or stability of the polar ice caps, since changes from the present extent of ice coverage would appear to "feed" on themselves.

Changes in the planetary temperature of greater than a few degrees or variations in energy input to the earth larger than a few per cent would appear, if sustained over a sufficiently long period (see the discussion of semiempirical climate models in the next section) to be sufficient to result in either a rapid expansion or complete disappearance of the polar ice sheets. However, this snow-and-ice cover–albedo–temperature coupling positive feedback mechanism must also be viewed in the light of other coupled processes that might modify these conclusions. For example, hydrological processes should also be included in the feedback loop, since ice and snow are merely the solid phase of water. Thus, the strong positive link between the extent of snow and ice cover and the local temperature assumed in the preceding discussion will be effective insofar as there is an appropriate amount of precipitation—to build up continental glaciers, for example. This case is a good example of the general complexity of the land–ocean–atmosphere system, since the hydrological processes depend in part upon the condition of the earth's surface, the dynamic state of the atmosphere, and the global cloudiness. Some of these other coupled interactions referred to above are discussed next.

Cloudiness–Surface Temperature Coupling. Most clouds are both excellent absorbers of infrared radiation and good reflectors of solar energy [13]. Thus, changes in cloudiness of the order of even a few per cent can have an effect on the radiation that is quite significant. Using global average models (discussed in the next section), we can show that changes in cloudiness could be an important climatic feedback mechanism, but we are as yet unable to determine even the direction of feedback control possibly exerted by clouds on the surface temperature. Because of the competing effects of decreased solar energy absorption and reduced planetary infrared emission with increasing cloud cover, we must consider the geographic distribution of cloud amounts, cloud heights, and cloud optical properties in order to calculate the effect of changes in any of these cloud parameters on the local radiation balance—and ultimately on the surface temperature. In addition, the formation and optical properties of clouds depend upon the kind and availability of suitable particles to act as cloud droplet nuclei and upon the local relative humidity and atmospheric stability (which are consequences of the hydrological and dynamic states of the atmosphere). Thus, the possible climatic feedback effect of cloudiness is not yet clear and remains to be determined through further observational and modeling studies in which the models must include the coupled effects of dynamic, hydrological and radiative processes [39].

Radiative–Dynamic Coupling. In the preceding discussion, the interactive nature of the land–ocean–atmosphere system has been emphasized. Thus, changes in the radiation balance initially caused, for example, by an increase in CO_2 could result in a dynamic response that might reorder the atmospheric motions in such a way as to either offset or accelerate any climate changes linked initially to the increase in CO_2. The direction of any such feedback coupling would have to be determined by specific model experiments using models of sufficient complexity to include all important feedback mechanisms.

Ocean–Atmosphere Coupling. Although the intrinsic link between the oceans, the atmosphere, and the climate has been implied in the preceding discussion, it is of sufficient importance to list this coupling separately. In addition to the obvious role of the oceans by acting as a source of water for the hydrological cycle, the dynamic coupling of atmospheric winds and temperatures with ocean circulation and sea surface temperatures plays a major role in determining our climate [40–42].

Also, the great heat storage capacity of the oceans acts to limit the extremes of seasonal climate that would otherwise be experienced in the middle and polar latitudes were it not for the presence of the ocean. This thermal "flywheel" effect of the oceans also increases the response time of the climate to changes in energy input [43]. Thus, the interactive role of the atmosphere and oceans in shaping the climate over the long term cannot be overstressed [9].

3.2.2.2. Hierarchy of Mathematical Models for Climate Simulation

Global Average Models. The simplest types of mathematical climate models use global average values of the physical variables. For example, although the solar radiation incident at a particular place on earth is, in actuality, a function of the time of year and the location of the place on the globe, in these models the sun's radiation is assumed to be uniformly distributed over the globe. Thus, the horizontal redistribution of energy over the planet by motions in the oceans and atmosphere, driven by differential solar heating of the globe, cannot be included in a global average model. That is, all but the first two feedback mechanisms just discussed are excluded. However, the effect of changes in the chemical composition of the atmosphere on the radiation balance can be computed to a first approximation by using a global average model in which solar radiation, planetary albedo, surface temperature, atmospheric temperature, and the vertical distribution of optically active atmospheric constituents are all averaged horizontally over the earth's surface. But the average vertical distribution of the atmospheric parameters can be considered explicitly. Thus, global average models are useful in simulating certain aspects of global climate and man's impact on it. They are particularly valuable in the determination of the *initial* effect of a pollutant on the radiation balance on a global scale.

For example, global average models with realistic vertical distribution of atmospheric absorbers and scatterers can study the relative magnitude of the effect of increases in CO_2 or aerosols on the radiation balance [18,27,31]. In fact, temperature–radiation coupling and water vapor–greenhouse effect coupling have already been incorporated in these global average models.

To determine the ultimate effect of a certain pollutant on climate, more realistic models including additional feedback mechanisms and with better spatial and temporal resolution should be used. But whereas large-scale models are difficult to interpret and costly to run, global average models, which require comparatively little computer time, are a necessary first step in determining if a specific pollutant might have an impact on climate.

Semiempirical Climate Models. Another approach to climate modeling is to parameterize the dynamic equations of climate theory by replacing the variables of the atmospheric motions with empirical relationships. For example, the net transport of heat into a zonal (latitude) belt can be parameterized by assuming the net heat flux added to the zone by the "circulation system" equal to an empirical coefficient times the difference between the zonal temperature and the planetary mean [30]. Alternatively, the dynamics can be parameterized in terms of several eddy coefficients [see Eq. (10)] and a mean meridional wind velocity, as has been done by Sellers [44], where the numerical values of these zonal parameters are selected to fit empirically derived data of the latitudinal values of heat transport (such as the data of Sellers [45] or Rasool and Prabhakara [46]).

Another relation based on empirical data for outgoing infrared radiation flux can be constructed which computes infrared flux to space as a function of surface temperature and cloud cover and includes the effect of the water vapor–greenhouse effect feedback mechanism [30]. When used in the heat balance equations [which are basically finite-difference formulations of the energy equation (6)] for the earth–atmosphere system with the earth divided into zonal belts, these parameterizations form semiempirical climate models.

Such models can be used to investigate qualitatively the effects of certain feedback mechanisms.

Using empirical determinations for the albedo of the ice and the temperature at which it persists, it is possible to use semiempirical models to study climate stability. The first results show that changes in the mean planetary temperature of the order of 1°C may be sufficient to permit unstable growth of the polar ice or a permanent melting, depending on the sign of the change in mean temperature [21,44]. More recent results of Schneider and Gal-Chen (to be published) show that, although the models of Budyko and Sellers are extremely sensitive to perturbations in external factors (such as the solar constant), they are relatively insensitive to changes in internal conditions (such as the initial temperature distribution or ice cover).

These empirical models, while convenient to use, do not adequately include the effects of radiative–dynamic coupling, where changes in the atmospheric motions and transports of heat and momentum could occur as a result of changes in the radiation balance. More specifically, in these models, the coefficients that parameterize the dynamic effects of the atmosphere are only functions of latitude and are based on yearly average data.

However, this parameterization might be improved by making the coefficients (eddy diffusion, etc.) functions of the model variables (e.g., zonal-average temperatures and meridional temperature gradients), where more general climate models might be used for determining this improved parameterization. The purpose of this approach would be the use of very complex and comprehensive models (requiring a great deal of computer time) to provide the semiempirical factors that are required by the simpler climate models and then to conduct experiments with these more economical models. Of course, one would like to use data from the real atmosphere to derive these parameterizations, but nature is seldom cooperative enough, and climate changes cannot be affected by our will as can be done in a model.

Although these models do not yet adequately represent the damping effect exerted by the large heat capacity of the oceans and large glaciers, all the important effects of the general circulation, the influence of hydrological processes, or possible changes in cloudiness, their results are sufficiently convincing to compel us to proceed with the development of more realistic large-scale models in which these other important feedback mechanisms can be more properly included.

Statistical Dynamical Models of Climate. By averaging variables with respect to longitude and assuming that the most important nonlinear interactions are those between the mean and fluctuating motions, one may formulate a set of averaged "primitive" equations. These become a complete system of equations in which the variables are statistics (zonal averages of the mean motions) of the climatic elements that describe the detailed state of the atmosphere, such as average temperature, average wind velocity, net transport of heat, momentum transport, variance of temperature, and eddy kinetic energy —all defined at several discrete levels in the atmosphere and considered functions of latitude and time [47, 48]. These types of models, which are more detailed than either the global average or the semiempirical models, yet require considerably less computer time than large-scale dynamic models, have provided some important insight into how various parts of the earth–atmosphere system interact with each other. They should be developed further, with attention given to determining their sensitivity to variations in some of their parameters.

The kinds of models discussed so far are simulation models, where the sources and sinks of energy and momentum are computed simultaneously with the mean states of the model-generated variables. A more physical (or mechanistic) modeling approach, where the sources and sinks of energy and momentum are prescribed, has been used by Dickinson [49] to study the response of the mean circulation to various types of thermal and eddy forcing.

Large-Scale Dynamic Models. In the general theory of climate described earlier, the instantaneous states of the atmosphere, oceans, and land surfaces are represented by continuous three-dimensional fields of temperature, pressure, velocity, and chemical constituents with their sources and sinks, each of which is related to the others by a series of physical laws or equations (see 3.2.1.1.). However, in a practical climate model, these continuous fields are represented on a finite-sized grid and all subgrid-sized motions are either parameterized or ignored while the large-scale dynamics are treated explicitly. Since the computing time increases about eight times when the grid size is halved, the resolution and consequent accuracy of the model depends upon the size and speed of available computers and on the reliability of the subgrid parameterization. The equations are integrated numerically over the grid at each time step, advancing from one time step to the next as many times as is needed to cover the time period of study.

For numerical weather prediction the model focuses on the atmosphere alone by specifying solar radiation and sea surface temperatures as external boundary conditions. The total integration period for such a prediction model is of the order of a week of simulated time or less, in which the coupled effect of the atmosphere on the state of the oceans is small and radiation influences can be treated simply. The grid size is as small as is practical (100–500 km on a side) and the distribution of clouds is usually an empirical parameterization based upon the local relative humidity and atmospheric stability, or may be only a climatological distribution, since the effect of small changes in cloudiness on the global circulation is also small for this short integration time. Many of these simulations have already been able to predict reasonably a number of the features of the global circulation, precipitation, temperature, and pressure patterns. Their ability to predict large-scale atmospheric phenomena is well beyond our own intuition. Of course, all small-scale (less than ~ 100 km) details are excluded without finer grid size and improved parameterization [20,23,50-53].

However, for climate simulation, we need statistics over periods far in excess of one week, and in long-range considerations, interaction of the ocean with the atmosphere becomes of paramount importance. Therefore, using the same computers (the best available), we must also model the oceans and couple the changes in the oceans to the atmosphere and vice versa. In addition, we must integrate the equations for such a joint ocean–atmosphere model for orders-of-magnitude longer periods of simulated time than for the numerical weather prediction experiment. We are then faced with the practical necessity of choosing judiciously among larger grid size, larger time step, fewer physical details, or more parameterization.

Additional difficulties are encountered in ocean modeling. Whereas for the case of the atmosphere we can gain confidence in our model by comparing its statistics with the volumes of collected meteorological data, no similar extensive knowledge of the time-dependent features of the oceans is readily available to help us verify the accuracy of our oceanic model. Furthermore, the small-scale turbulent mixing processes in the oceans are even more poorly understood than they are for the atmosphere, and subgrid parameterization becomes all the more difficult. While "predicting" climate change may be beyond our capabilities for some time to come, it may soon be possible to evaluate how large a perturbation (possibly man-made) to the environmental system could be tolerated before the system might respond by a change in climatic regime.

Nonetheless, a simplified joint ocean–atmosphere model has been developed [43] which is already able to simulate many of the features of the atmosphere and several of the large-scale properties of the oceans discussed in the previous section. Many of the important feedback mechanisms are included.

With the advent of the upcoming generation of computers, better data on the state and chemical composition of the oceans and the atmosphere, and increased cooperation among a broad spectrum of scientists and mathematicians, the reliability of climate modeling is certain to improve. Despite the problems, present and unforeseen, the use of climate models is essential for the continuing study of climate theory and the impact of changes in atmospheric chemical constituents on climate.

In the next section, we discuss how atmospheric chemistry is changing and try to assess the role of man's activities in effecting such changes.

4. NATURAL AND MAN-MADE INFLUENCES ON ATMOSPHERIC COMPOSITION

This chapter has, up till now, laid the groundwork for a discussion of the possible influence that man might have on atmospheric chemistry and composition and the consequent changes in regional or global climate. Since the previous chapters have dealt in some detail with the various man-made chemical components and the trends in their concentrations, this section will be in the form of an overview of the situation.

It is obvious that the air over our large cities has been markedly changed by the contaminants that pour from smokestacks and exhaust pipes, but to assess the larger-scale changes in radiation balance or climate of a region we must look at something considerably larger than a city, and then man-made contaminants are diluted and modified or partly removed from the air. For the atmosphere on a global scale man's activities become considerably less obvious, and it appears that we can concentrate on just three main subjects in the discussion that follows: (1) the particles that man introduces directly, or indirectly from gaseous contaminants that subsequently turn into particles; (2) the carbon dioxide content of the atmosphere, which has been increasing due to the burning of fossil fuels; and (3) possible influences on the stratosphere from high-flying aircraft.

Those familiar with the subject will note that we have chosen to ignore some very important aspects of the general question of man's impact on the climate, such as the extensive changes he has already made to the land surface, the redistribution of water as he dams rivers and irrigates deserts, and the heat that he releases into the air–ocean system by his demands for usable energy. These and other factors must certainly be taken into account in any comprehensive discussion, and the reader is referred to the SMIC Report [3] for a more complete treatment.

4.1. Carbon Dioxide from Fossil Fuels

The vast amount of carbon stored in the ground in the form of coal, petroleum, and natural gas has been used to generate heat to power our factories, machines, and automobiles, to heat our dwellings, and to generate electrical power. The products of combustion, mainly H_2O and CO_2, go into the atmosphere. The amount of water vapor released in this way is trivial compared to that already in the atmospheric system, but the case for CO_2 is a different matter.

There are two ways by which CO_2 can be removed from the atmosphere, and both are slow. The first is by solution into the oceans. It is estimated that the total amount of CO_2 in the oceans is about 60 times that in the atmosphere. This suggests that the ocean is a very large reservoir, and could therefore be a large "sink" for the CO_2. This is true, except that the deep ocean reservoir takes an exceedingly long time to respond to any change in the atmosphere because of its slow turnover time, measured in centuries. The CO_2 in the atmosphere is in equilibrium with only the top few hundred meters of the ocean when measured on a human time scale of a few decades.

Thus, since the beginning of the century, when fossil-fuel burning began to escalate rapidly, only about one-quarter of the CO_2 added to the atmosphere has had a chance to be absorbed by the oceans [2,3]. See the chapter by Keeling for the facts in the case.

The other sink for added CO_2 is the global "biomass," the aggregate of all living things. An increase of CO_2 concentration, as all greenhouse operators know, will generally result in faster photosynthesis by plants, and this will result in an incorporation of part of the extra CO_2, mostly in the forests. A complication here is that this incorporation of the added CO_2 is to some extent temporary, because after the trees finish growing they die and decay (or are burned) and some of their carbon returns to the air as CO_2. While the quantities involved in this cycle are not well enough known to permit us to make an accurate global budget, the best estimate is that another quarter of the CO_2 added during this century has been taken up by the biomass [2,3].

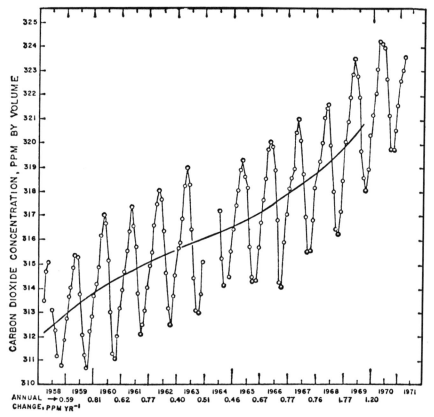

Fig. 8. Mean monthly values of CO_2 concentration at Mauna Loa, Hawaii for the period 1958–1971. The data are those of D. Keeling and his colleagues, as presented in the SMIC Report [3].

That leaves about 50 % of the CO_2 added by man still in the atmosphere, and when one compares the total tonnage of CO_2 remaining in the atmosphere corresponding to the increase in concentration shown in Fig. 8 with the estimated CO_2 production by man, the fraction is indeed about half. On the assumption that this partition will continue at about the same rate, and accepting projections of the energy generation and other fossil fuel requirements for the next few decades, one can predict a further increase in CO_2 concentration in the atmosphere of 20–25 % by A.D. 2000 [3].

The implications of such an increase in CO_2 for climate change have already been discussed in Section 3.2.1, where the effect of an increase in an infrared-absorbing gas was shown to result in an increase in the temperature at the ground and a cooling of the stratosphere. It is not known exactly what change the 20–25 % increase in CO_2 by A.D. 2000 would produce, but it is probably in the neighborhood of 0.5°C in midlatitudes (perhaps more at the poles) [2,3,18], and this is comparable to the changes in the last century, changes presumably from natural causes.

One cannot help being struck by the fact that Figs. 2 and 8 seem to imply a contradiction for the period from 1940 to the present, since the rise in CO_2 *should* have resulted in a global warming rather than a cooling. However, the temperature in fact fell by about 0.3°C, so there must have been some other factor at work that overwhelmed the effect of the CO_2. In the next section, we will discuss one such other possibility, the change in radiation balance due to dust and smoke in the atmosphere.

4.2. Particles in the Atmosphere

4.2.1. Sources, Sinks, and Residence Times

There is no way to obtain a precise measure of the various sources of particles that are always present in the lower atmosphere, but we can estimate them, and the largest source is probably soil and rock debris picked up by the wind. Other major contributors are sea salt (from windblown droplets that evaporate), forest fires and other open fires, and conversion of certain gases to particles in the atmosphere.

Table 2, taken from the SMIC Report [3], assigns estimates of contributions from these various sources, and it will be seen that the man-made contributions can account for from 5 to 50 % of the total *mass* of suspended particles. The spread in this ratio is an indication of the uncertainty. At midlatitudes in the Northern Hemisphere, where most of the human activity that produces these emissions occurs, the man-made fraction must be very much greater, and it is likely that in this large belt the man-made particle production roughly matches nature's.

In making this comparison, we must take account of those particles that started out as gases and were subsequently converted to particles, the two

TABLE 2. Estimates of the Weight of Particles Smaller
Than 20 μm Radius Emitted into or Formed in the
Atmosphere (10^6 metric tons yr^{-1})

Natural	
Soil and rock debris*	100–500
Forest fires and slash-burning debris*	3–150
Sea salt	(300)
Volcanic debris	25–150
Particles formed from gaseous emissions:	
Sulfate from H_2S	130–200
Ammonium salts from NH_3	80–270
Nitrate from NO_x	60–430
Hydrocarbons from plant exudations	75–200
Subtotal	773–2200
Man-made	
Particles (direct emissions)	10–90
Particles formed from gaseous emissions:	
Sulfate from SO_2	130–200
Nitrate from NO_x	30–35
Hydrocarbons	15–90
Subtotal	185–415
Total	958–2615

*Includes unknown amounts of indirect man-made
contributions.

main factors in this process being the sulfur compounds and certain hydro-carbons. This will be discussed in the next section. Notice that most of the man-made particles are formed in this manner.

All particles, whether windblown dust, sea salts, sulfate salts, smoke, smog, or haze, are ultimately removed from the air by one of four mechanisms: (1) dry sedimentation or settling out; (2) impaction against an object on or near the ground; (3) rainout (or snowout), in which the particles are attached or incorporated in cloud droplets by diffusion and are subsequently carried to the ground when the droplets grow to falling drops (or fall as snowflakes); and (4) washout, in which falling rain (or snow) scavenges the particles below the cloud. While these processes are extremely variable and obviously depend on meteorological conditions, on the whole their relative effective-ness in the lower troposphere is in the order listed, the least important being sedimentation and the most important being washout. These processes are treated further in the chapters by Cadle, Pruppacher, and especially Hidy.

The net result of all these removal processes is that in midlatitudes the average residence time of a particle in the lower troposphere is less than a week, in the upper troposphere it is of the order of a month or more, and in the stratosphere, it may range from less than 1 yr just above the tropo-pause to more than 5 yr at the stratopause (about 50 km) [54,55].

This variation in the residence time means that the particle content of the lowest layers depends on what is immediately upwind, and it changes considerably over distances of 500–1000 km when the air passes from land to sea or vice versa. In contrast to this, the particles in the upper troposphere can travel several times around the world, on the average, and there is little difference between the character of the particles found at, for example, 6–10 km over the U. S continent and over the Pacific at the same altitude [56]. The still longer residence times in the stratosphere are significant because any pollutant continuously injected there will accumulate, and this is the subject of Section 4.3.

There has been a gradual upward trend in the particulate loading of the atmosphere at most nonurban Northern Hemisphere stations where observations have been taken for more than a decade, as measured by the attenuation of the direct solar beam by particle scattering—the degree of "turbidity"— and by the quantity of material collected on filters [57] (see the lower portion of Fig. 9). An exception is the station on top of Mauna Loa, which seems to be more influenced by turbidity changes in the stratosphere than in the troposphere [58] (Fig. 10). In many cities, the mass concentration of particles has decreased *locally* as a result of air quality control actions, as shown in the upper part of Fig. 9. It is almost certain that the entire Northern Hemisphere reflects the ever-increasing addition of particles from industry, power plants, automobiles, and slash-and-burn agricultural practices (though the last source is largely confined to the tropics and subtropics).

So far, we have discussed the atmospheric particles themselves: their sources, sinks, and residence times, and also the upward trend in total Northern Hemisphere turbidity and particle concentration due to man's activities. (We will deal with volcanoes and their influence on the stratospheric turbidity in the next section.) The purpose of all this has been to relate the effect of particles to the radiation balance of the atmosphere and the possible leverage that man may have on the climate through his increasing production of particles. This increase in atmospheric particle loading in the Northern Hemisphere has apparently taken place in spite of efforts to reduce smoke emission in the cities [57].

In connection with the last point, the reduction in smoke from large coal-burning power plants and steel mills has been achieved by precipitating the larger particles that are the most obvious to the eye, but the smaller particles of submicron size still largely escape. The larger ones would have been removed first from the atmosphere, so the escaping particles are just those that last longer and contribute most to the hemispheric turbidity trend wherever they are. This is borne out by Fig. 9, where the urban mass concentrations of particles in the U. S. has gone down as a result of air pollution control efforts, while it has gone up at the nonurban stations. Since climatic effects are dependent on the global energy balance, which is an area-weighted

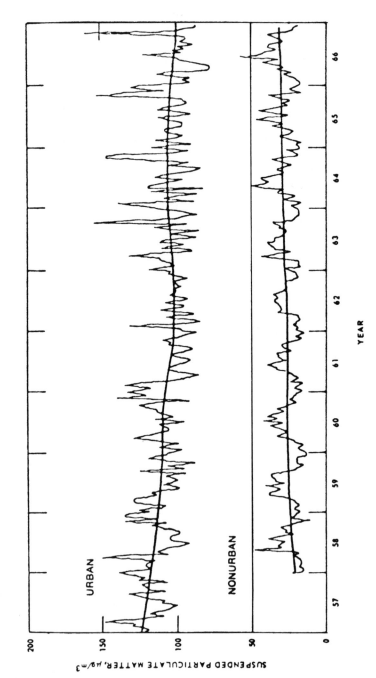

Fig. 9. Long-term trends at 58 urban and 20 nonurban sites. The jagged curves are the averages for each sampling interval; the smoother trend curves were calculated using a technique that combined weighted measurements of curve fit and smoothness. The average increase at the nonurban sites during this period is a steady 3–4% per year [57].

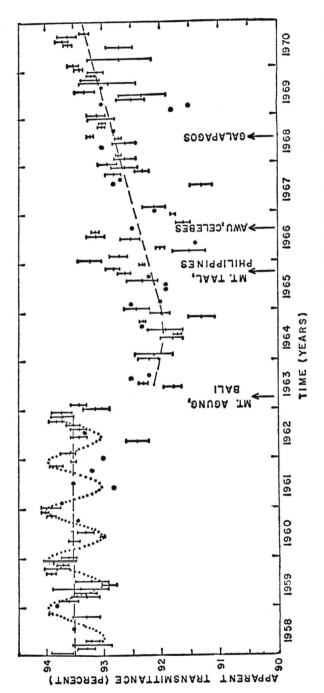

Fig. 10. Transmittance of normal-incidence solar radiation at Mauna Loa, Hawaii (19°N). The dotted line between 1958 and 1962 is a suggested seasonal variation in transmittance. The dashed lines are best-fit curves to the transmittance data displayed as either bar graphs giving range of uncertainty or as heavy dots. Analyses are confined to morning observations. Times of major volcanic eruptions appear along the bottom of the figure [58].

phenomena, the rural and oceanic particle loadings are most significant for climatic influences.

4.2.2. Particles Formed from Gases

It can be seen from Table 2 that roughly half the particles in the atmosphere are formed from gaseous emissions, both natural and man-made. This rather surprising fact has emerged in recent years partly as a result of studies of photochemical smog and partly as a result of extensive assays of the composition of aerosols in various parts of the world. This is a subject that is dealt with in some detail in the chapters by Cadle and Friend, so we will merely touch on it here.

A collection of atmospheric particles anywhere in the world will contain a certain amount of sulfate, since this substance seems to be virtually ubiquitous. An analysis of the sources of this sulfate [59,60] shows that, on a world-wide basis, about half comes from natural biological processes that produce hydrogen sulfide (H_2S) or sulfur dioxide (SO_2) and more than one-quarter comes from the burning of fossil fuels that contain sulfur, especially coal, producing SO_2. The largest other source is sea spray that leaves sea salt particles, including sulfates, suspended in the air; and volcanoes are another, smaller contributor.

The gaseous sulfur compounds H_2S and SO_2 are both oxidized in a matter of hours or days to sulfur trioxide (SO_3), and this immediately hydrolyzes to form sulfuric acid (H_2SO_4). The rate of this oxidation process is greatly speeded when these gases are dissolved in water droplets containing traces of heavy metal ions that act as catalysts. The H_2SO_4, being very hygroscopic, forms droplets by accreting water vapor from the air when the humidity is moderately high, and any positive ions in the air such as ammonia or free metals will combine in the droplet to form sulfate salts. The net result is a continual production of sulfate in the air from the gaseous sulfur compounds.

The formation of organic particles from volatile terpenes (natural, coming from vegetation) and unburned gasoline (man-made) is the result of a more complex process involving reactions with oxides of nitrogen, water vapor, and oxidants in the presence of solar ultraviolet radiation—the so-called "smog reactions." The "blue haze" observed in many parts of the world is usually caused by such natural organic particles [61] and can be distinguished from smog by the latter's brownish color.

The nitric acid (HNO_3) formed from the further oxidation of the oxides of nitrogen in the presence of water vapor is also a gas, but in the lower atmosphere it combines quite rapidly with ammonia or metallic ions to form nitrate salts. These salts are usually found attached to other particles.

A special situation exists in the stratosphere, where the low humidity and lack of terrestrial dust particles inhibit some of these particle-forming processes and permits others. That will be treated in the next section.

It is significant, returning once more to Table 2, that *most* of the man-made particles start out as gaseous emissions—this is true even if we consider forest fires and slash–burn tactics as man-made contributions. Particles formed in this way are initially very small and grow by coagulation over a period of time (measured in days) to provide a spectrum of particle sizes that follow the Junge distribution given by Eq. (14) (see the chapter by Hidy). These smaller particles are especially effective (due to their large numbers) in scattering (and absorbing) solar radiation, so they play a role in controlling the radiation balance of the atmosphere that is larger than their relative mass would suggest. By preventing large soot particles from leaving our smoke stacks, we are improving the appearance of the atmosphere close to the source, but the smaller particles and some of the invisible gases, especially SO_2, will be important in determining the turbidity of the region downwind. This is borne out by Fig. 9.

4.3. Changes in the Stratosphere

The stratosphere, which starts at an altitude of about 10 or 11 km in the polar regions and about 15 km in the tropics (the transition from troposphere to stratosphere being called the "tropopause"), is characterized by an extremely low water vapor concentration and a high ozone concentration.

The stratosphere is also more stable than the troposphere, since the temperature above the tropopause is either constant or increases with height. (This results primarily from the absorption of solar energy by ozone in the upper stratosphere.) This stability suppresses convection and vertical mixing. This, combined with the absence of any precipitation processes to cleanse the air, means that any trace substances introduced into the stratosphere will remain there for a much longer time than in the troposphere. The residence time of a trace gas or aerosol introduced just above the tropopause varies from about six months at high latitudes to 1 yr at low latitudes; at 20–30 km, the residence time is 2 or 3 yr, increasing to about 5 yr at 50 km [54,55]. These residence times have been deduced to a large extent by observations of the removal rates of radioactive tracers introduced by nuclear tests.

The extreme stability of the stratosphere, coupled with the low density of the air there, implies that any contamination of the stratosphere will be more long-lasting and significant than a corresponding contamination of the troposphere, where effective removal processes are constantly at work. There are two cases in point: contamination by large volcanic eruptions and contamination by high-flying jet aircraft. The first is a well-recognized natural phenomenon that has occurred many times, while the latter is only now a source of possible concern. Since we can apply the lessons learned by observing the stratospheric effects of eruptions to the jet aircraft question, we will discuss these lessons first.

There have been a number of well-recorded explosive volcanic eruptions in the past century [8,62,63], the most notable being that of Krakatoa in 1883,

Santa Maria in 1902–04, and Agung in 1963, all in the equatorial zone. Large eruptions that have recently occurred at higher latitudes were Hekla in 1947, Mt. Spurr in 1953, Bezymyannaya in 1956, and Surtsey in 1963–65. The tropical eruptions cited above certainly sent large quantities of particles and gases well into the stratosphere [8,62] and the gold and purple sunsets that the stratospheric particles produced were seen all over the world for two or three years following each event. With the establishment of radiation-monitoring stations, it was possible to observe the decrease in direct and total solar radiation reaching the ground, and reductions of the *direct* radiation have ranged from 5 to 20 % [30]. The reduction of *total* radiation was about one-sixth as much, but this is still an appreciable effect. It is quite possible that past periods of prolonged volcanic activity have resulted in appreciable cooling of the lower atmosphere, producing a true climate change.

While there was little doubt that these eruptions injected particles into the stratosphere, their identity was unknown until 1959–61, when C. Junge and his colleagues at the Air Force Cambridge Research Laboratory sampled them and discovered that they were predominantly sulfate particles, most likely droplets of sulfuric acid. There was a peak in the sulfate particle concentration at 18–20 km altitude [64]. A continuing stratospheric sampling program by means of high-flying RB57-F aircraft was begun in 1968 by the National Center for Atmospheric Research in collaboration with the Air Weather Service and the Atomic Energy Commission, and the composition of the stratospheric aerosols has been studied in great detail [65-67]. Between 1968 and 1970, the amount of filterable material at about 20 km was 30–40 $\times$ 10^{-4} ppm by mass, which was some five or six times more than the concentration observed by Junge in 1960–61. The predominance of sulfate was confirmed, and the other constituents such as silicon, sodium, chlorine, etc., were generally a factor of ten or more lower in concentration. Oddly enough, the cellulose filters used in these aircraft flights (IPC filters) showed roughly as much nitrate as sulfate, but it has been concluded that the nitrate was probably nitric acid *vapor* that was adsorbed by the cellulose fibers and was not in particulate form. The concentrations of nitric acid vapor deduced in this way are consistent with those determined spectroscopically by the University of Denver group [68].

Consistent with the predominantly volcanic origin postulated for these particles, there was a decrease in stratospheric particle concentration from 1970 to 1971 to values approximating those observed by Junge in 1960–61, before the large series of tropical eruptions—recall that the last large eruption was that of Fernandina (Galapagos) in 1968, and that the residence time of particles at 20 km is about 2 yr. This decrease has been documented by a number of other particle-observing techniques, including ground-based lidar. It also showed up as a trend of decreasing turbidity observed at the high-altitude observatory on Mauna Loa, Hawaii, shown in Fig. 10, which presumably reflected the changes in the total particle burden above the station.

The largest of the eruptions of the 1960's was that of Mt. Agung (Bali) in March 1963. There was a remarkable rise in stratospheric temperatures by as much as 6°C at heights of 20 km and above at equatorial upper-air sounding stations following this eruption, attributed by Newell [69] to the Agung dust cloud. While there may be some doubt that the temperature rise was purely an effect of the dust cloud, there being simultaneous dynamical changes in the circulation of the stratosphere, the close coincidence in time does strongly suggest a relation between the two events. Thus, this eruption may have caused a significant change in the heat balance and circulation of the stratosphere—though we should add that no corresponding change occurred in the troposphere that has been attributed to the same cause.

Now we turn to the question of the high-flying jet aircraft, particularly the fleets of supersonic transports of the future (SST's) and the possibility that they might affect the stratosphere in one way or another. This possibility was raised in the SCEP Report [2], and was hotly debated in the U. S. during the Congressional hearings on the SST appropriation throughout the spring of 1971. The SMIC Report [3] contains a summary of the atmospheric factors in the controversy, and the interested reader is also referred to the paper by Johnston [71].

If there were large numbers of SST's operating in the stratosphere—the number generally assumed has been 500 aircraft, each flying 7 hr per day— their engine exhaust would add a substantial amount of water vapor, CO_2, NO, SO_2, unburned hydrocarbons, and soot particles to the stratosphere at about 20 km altitude. Assuming the numbers of aircraft mentioned above and taking account of the 2-yr average residence time of any substance introduced at this altitude, the SCEP Report concluded that the CO_2 would not cause any noticeable change, but the H_2O added would be sufficient to cause a global average increase of about 10 % in the stratosphere, and considerably more along the heavily traveled airways. The NO added would be appreciable, but unfortunately we do not know the present concentration in the stratosphere because it has not been adequately measured. The SO_2, as discussed in the previous section, would be oxidized to form sulfate particles, and this together with the soot would add an amount that could be comparable to the particles from the Agung eruption. (This depends, however, on the sulfur content of the jet fuel used, which could be reduced if necessary.) The unburned hydrocarbons would probably end up as particles also. In short, it was predicted that the SST's could affect the stratospheric composition on a global scale.

The next question to ask is, of course: What is the significance of such a change in the stratosphere? There are two kinds of effects that have been studied, namely changes of stratospheric temperature and changes in the amount of ozone that exists naturally throughout the stratosphere. We will briefly summarize the present thinking with regard to both influences.

There does not seem to be any precise way of calculating the direct influence of the particles on the heat balance, but by analogy to the strato-

spheric temperature change that followed the Agung eruption, it was sur-
mised by the SCEP Report that the SST-produced particles might cause an
increase in stratospheric temperature of a few degrees. The added water
vapor would cause a very small cooling of the stratosphere due to its radiative
effects (see Section 3.2.1.), thereby partly counteracting the effects of the
particles. It is not yet clear whether there would be a significant change in the
climate at the surface, but this obviously deserves further study. Included in
such a study should be the possibility that the increased water vapor might
cause some cirrus or other high clouds to form that would not otherwise be
there, and these, too, could influence the heat balance.

The second influence, the interaction of the exhaust products with stratospheric ozone, could be more significant to man and has therefore become
more of a *cause célèbre*. Although many of the chemical and photochemical
reaction rates that enter into the complex interactions between ozone, water
vapor, and oxides of nitrogen are poorly known, it seems likely that the net
result of adding H_2O and NO will be a *decrease* in the total ozone content.
The SCEP Report, basing its conclusions largely on the early work of Park
[70] (with J. London) that considered only the reactions of O_3 with H_2O,
concluded that the worldwide effect might be a decrease of about 1 %.
Subsequent studies by London and Park as well as by Johnston [71], Crutzen
[72], and Nicolet and Vergison [73], taking account of the reactions of NO
with O_3 as well as those involving H_2O, have suggested that the effect of
SST's on the total ozone might be somewhat greater than the SCEP estimate,
but the SMIC Report in summarizing all the arguments comes to the cautious
conclusion that we really cannot make any firm estimate until (a) we have
better determinations of reaction rates and (b) we take proper account of the
very important dynamic effects of circulations and mixing processes along
with the photochemistry.

The implications of a decrease in total O_3 in the stratosphere are not in
climatic terms in the usual sense, but rather the effect that such a decrease
would have on the solar ultraviolet radiation reaching the ground. Since O_3
absorbs ultraviolet starting at around 0.32 μm and essentially screens it
completely from the surface for wavelengths below 0.30 μm, any decrease in
total atmospheric (stratospheric) O_3 will let more ultraviolet in this wavelength interval through. These are just the wavelengths that cause sunburn,
and, in some cases, skin cancer for light-skinned people, so the significance of
any decrease of O_3 lies in its effects on the skin of every person who goes out
in the sunlight—and that is just about everyone.

There is no clear resolution to this argument at the present time, as
pointed out by SMIC. The picture is further confused by the observation
that during the past 6 yr or more the water vapor in the stratosphere at about
20 km has increased by about 50 % (over one station, Anacostia, Md., which
is the only place where adequate measurements of this quantity have been
made) [74], while at a number of stations throughout the world, the total

ozone has also risen by 5 or 6 % [75]. These trends are shown in Figs. 11 and 12. Notice that this is in direct contradiction to the prediction, based on photochemistry alone, that an increase in H_2O should cause a *decrease* in O_3. Clearly, there are other factors at work in the real atmosphere that we do not yet understand, the most important one probably being those connected with the transport and turbulent diffusion of these trace gases in the stratosphere.

In spite of the practical and theoretical difficulties involved, the question of the possible influence of a fleet of SST's on the stratosphere is being vigorously pursued by a number of groups at this time. [This entire subject is being discussed in a series of monographs sponsored by the U.S. Department of Transportation's Climatic Impact Assessment Program (CIAP). These monographs should be available by early 1974.] We have an opportunity, rare in the history of modern man, to determine the environmental effect of a new technological development *before* it actually takes place, while there is still time to forestall the development if it turns out that it would cause an unacceptable hardship on mankind.

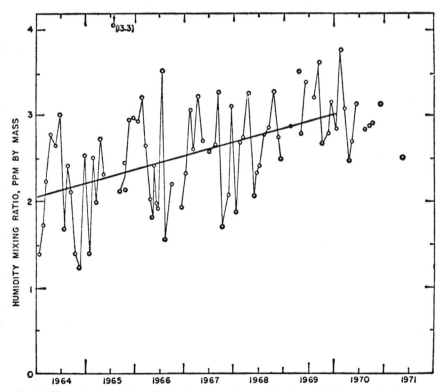

Fig. 11. Water vapor mixing ratios in the stratosphere at 50 mb or 20.6 km near Washington, D. C., United States. The straight line is a best-fit line between 1964 and 1969. Circles in adjacent months have been connected by straight lines. The upward trend between 1964 and 1969 appears to have leveled off after early 1970 [3,74].

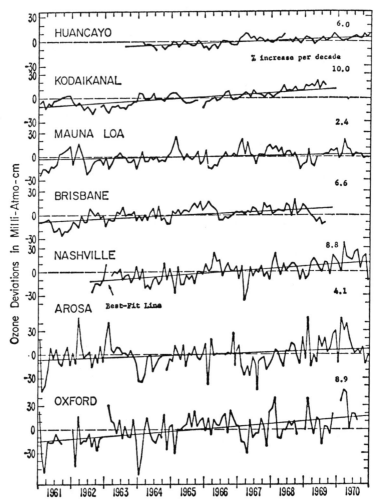

Fig. 12. Total ozone trends for selected world stations expressed as departures from normal between 1958 and 1970. Other stations show an increase similar to that in this figure, while some exhibit no apparent trends or decreasing trends. Before 1960, when fewer stations were in operation, both increasing and decreasing total ozone trends of about the same magnitude as in the figure were observed [75].

ACKNOWLEDGMENT

The authors wish to express their appreciation for a thoughtful reading of this chapter and helpful discussions to J. Murrey Mitchell, Jr.

REFERENCES

1. Kellogg, W. W., Predicting the climate, in "Man's Impact on the Climate," W. H. Matthews, W. W. Kellogg, and G. D. Robinson, eds., MIT Press, Cambridge, Mass., 1971, Chapter 4.

2. Study of Critical Environmental Problems (SCEP), "Man's Impact on the Global Environment," MIT Press, Cambridge, Mass., 1970.
3. Study of Man's Impact on Climate (SMIC), "Inadvertent Climate Modification," MIT Press, Cambridge, Mass., 1971.
4. Matthews, W. H., Kellogg, W. W., and Robinson, G. D. (ed.), "Man's Impact on the Climate," MIT Press, Cambridge, Mass., 1971.
5. Dansgaard, W., Johnsen, S. J., Clausen, H. B., and Langway, C. C., in "The Late Cenozoic Glacial Ages," symposium edited by K. K. Turekian, Yale Univ. Press, New Haven, Conn., 1971.
6. Flohn, H., *Geologische Rundschae* **54**, 504–575 (1964).
7. Lamb, H. H., "The Changing Climate," Methuen, London, 1966.
8. Mitchell, J. M., Jr., Summary of the problem of air pollution effects on the climate, in "Man's Impact on the Climate," W. H. Matthews, W. W. Kellogg, and G. D. Robinson, eds., MIT Press, Cambridge, Mass., 1971, Chapter 8.
9. Lorenz, E. N., Climatic change as a mathematical problem, *J. Appl. Meteorology* **9**, 325–329 (1970); reprinted as Chapter 9 in "Man's Impact on the Climate," W. H. Matthews, W. W. Kellogg, and G. D. Robinson, eds., MIT Press, Cambridge, Mass., 1971.
10. Robinson, G. D., Some meteorological aspects of radiation and radiation measurement, in "Advances in Geophysics," H. E. Landberg, and J. Van Miegham, eds., Academic Press, London, 1970.
11. Robinson, N., "Solar Radiation," Elsevier, Amsterdam, 1966, Chapter 3.
12. Vonder Haar, T. H., and Suomi, V. E., Measurements of the earth's radiation budget from satellites during a five-year period, *J. Atmos. Sci.* **28**, 305–314 (1971).
13. London, J., and Sasamori, T., Radiative energy budget of the atmosphere, *Space Research XI*, 639–649 (1970); reprinted as Chapter 6 in "Man's Impact on Climate," H. Matthews, W. W. Kellogg, and G. D. Robinson, eds., MIT Press, Cambridge, Mass., 1971.
14. Winston, J. S., The annual course of zonal mean albedo as derived from ESSA 3 and 5 digitized picture data, *Monthly Weather Review* **99**, 818–827 (1971).
15. Goody, R. M., "Atmospheric Radiation," Oxford Univ. Press, London, 1964.
16. Craig, R., "The Upper Atmosphere, Meteorology and Physics" Academic Press, New York and London, 1965.
17. Kondratyev, K. Ya., "Radiation in the Atmosphere," Academic Press, New York and London, 1969.
18. Manabe, S., and Wetherald, R. T., Thermal equilibrium of the atmosphere with a given distribution of relative humidity, *J. Atmos. Sci.* **24**, 241–259 (1967).
19. Lorenz, E., "The Nature and Theory of the General Circulation of the Atmosphere," World Meteorological Organization, Geneva, 1967.
20. Thompson, P. D., "Numerical Weather Analysis and Prediction," New York, 1961, p. 118.
21. Budyko, M. I., "Climate and Life," Hydrological Publishing House, Leningrad, 1971; also, The future climate, *EOS, Trans. Am. Geophys. Union* **53**, 868–874 (1972).
22. Neumann, G., and Pierson, W. J., Jr., "Principles of Physical Oceanography," Prentice–Hall, Englewood Cliffs, New Jersey, 1966, Chapters 13 and 14.
23. Haltiner, G. J., "Numerical Weather Prediction," Wiley, New York, 1971.
24. Hess, S., "Introduction to Theoretical Meteorology," Holt, Rinehart, and Winston, New York, 1959.
25. Van de Hulst, H. C., "Light Scattering by Small Particles," Wiley, New York, 1957.
26. Junge, C., "Atmospheric Chemistry and Radioactivity," Academic Press, New York, 1963.
27. Rasool, S. I., and Schneider, S. H., Atmospheric carbon dioxide and aerosols: effects of large increases on the global climate, *Science* **173**, 138–141 (1971).
28. Atwater, M. A., Radiative effects of pollutants in the atmospheric boundary layer, *J. Atmos. Sci.* **28**, 1367–1373 (1971).
29. Schneider, S. H., A comment on climate: the influence of aerosols, *J. Appl. Meteorology* **10**, 840–841 (1971).

30. Budyko, M. I., The effect of solar radiation variations on the climate of the earth, *Tellus* **21**, 611–619 (1969).
31. Yamamoto, G., and Tanaka, M., Increase in global albedo due to air pollution, *J. Atmos. Sci.* **29**, 1405–1412 (1972).
32. Hansen, J. E., and Pollack, J. B., Near-infrared light scattering by terrestrial clouds, *J. Atmos. Sci.* **27**, 265–281 (1970).
33. Charlson, R. J., Harrison, H., and Witt, G., Technical comment on Rasool and Schneider (1971), *Science*, **175**, 95–96 (1972).
34. Rasool, S. I., and Schneider, S. H., Reply to Charlson, Harrison, and Witt (1972), *Science*, **175**, 96 (1972).
35. Mitchell, J. M., Jr., The effect of atmospheric aerosols on climate with special reference to temperature near the earth's surface, *J. Appl. Meteorology* **10**, 703–714 (1971).
36. Twomey, S., The influence of atmospheric particles on cloud and planetary albedo, in the "Proceedings of the International Conference on Weather Modification, Canberra, Australia," 265–266, American Meteorological Society, Boston, 1971.
37. Kondratyev, K. Ya., "Radiative Heat Exchange in the Atmosphere," Pergamon Press, London, 1965, pp. 134–144.
38. Möller, F., On the influence of changes in CO_2 concentration in air on the radiative balance of the earth's surface and on the climate, *J. Geophys. Res.* **68**, 3877–3886 (1963).
39. Schneider, S. H., Cloudiness as a global climatic feedback mechanism: The effect on the radiation balance and surface temperature of variations in cloudiness, *J. Atmos. Sci.* **29**, 1413–1422 (1972).
40. Namias, J., Seasonal interactions between the north Pacific ocean and the atmosphere during the 1960's, *Monthly Weather Review* **97**, 173–192 (1969).
41. Newell, R. E., Vincent, D. G., Dopplick, T. G., Ferruzza, D., and Kidson, J. W., The energy balance of the global atmosphere, in "The Global Circulation of the Atmosphere," G. A. Corby, ed., Royal Meteorological Society, London, 1969, pp. 42–90.
42. Palmén, E., and Newton, C. W., "Atmospheric Circulation Systems," Academic Press, New York and London, 1969.
43. Manabe, S., and Bryan, K., Climate calculations with a combined ocean–atmosphere model, *J. Atmos. Sci.* **26**, 786–789 (1969).
44. Sellers, W. D., A global climatic model based on the energy balance of the earth–atmosphere system, *J. Appl. Meteorology* **8**, 392–400 (1969).
45. Sellers, W. D., "Physical Climatology," The University of Chicago Press, Chicago, 1965.
46. Rasool, S. I., and Prabhakara, C., Heat budget of the Southern Hemisphere, "Problems of Atmospheric Circulation," R. V. Garcia and T. F. Malone, ed., Spartan Books, New York, 1966, pp. 76–92.
47. Kurihara, Y., A statistical-dynamical model of the general circulation of the atmosphere, *J. Atmos. Sci.* **27**, 847–870 (1970).
48. Saltzman, B., and Vernikar, A. D., An equilibrium solution for the axially symmetric component of the earth's macroclimate, *J. Geophys. Res.* **76**, 1498–1524 (1971).
49. Dickinson, R. E., Analytic model for zonal winds in the tropics, *Monthly Weather Review* **99**, 501–523 (1971).
50. Kasahara, A., and Washington, W. M., General circulation experiments with a six-layer NCAR model, including orography, cloudiness and surface temperature calculations, *J. Atmos. Sci.* **28**, 657–701 (1971).
51. Manabe, S., Smagorinsky, J., and Strickler, R. F., Simulated climatology of a general circulation model with a hydrological cycle, *Monthly Weather Review* **93**, 769–798 (1965).
52. Mintz, Y., Very long term global integration of the primitive equation of atmospheric motion, "WMO-IUGG Symposium on Research and Development of Long Range Forecasting, Geneva," WMO Technical Note 66, 1965, pp. 141–161.
53. Smagorinsky, J., General circulation experiments with primitive equations, 1, The basic experiment, *Monthly Weather Review* **93**, 265–276 (1963).

54. Martell, E. A., Residence times and other factors influencing pollution of the upper atmosphere, in "Man's Impact on the Climate," W. H. Matthews, W. W. Kellogg, and G. D. Robinson, eds., MIT Press, Cambridge, Mass., 1971, Chapter 35.

55. Junge, C. E., The nature and residence times of tropospheric aerosols, in "Man's Impact on the Climate," W. H. Matthews, W. W. Kellogg, and G. D. Robinson, eds., MIT Press, Cambridge, Mass, 1971, Chapter 23.

56. Gillette, D. A., and Blifford, I. H., Jr., Composition of tropospheric aerosols as a function of altitude, *J. Atmos. Sci.* **28**, 1199–1210 (1971).

57. Ludwig, J. H., Morgan, G. B., and McMullen, T. B., Trends in urban air quality, *EOS, Trans. Am. Geophys. Union* **51**, 468–475 (1970); reprinted as Chapter 25 in "Man's Impact on the Climate," W. H. Matthews, W. W. Kellogg, and G. D. Robinson, eds., MIT Press, Cambridge, Mass, 1971.

58. Ellis, H. T., and Pueschel, R. F., Solar radiation: absence of air pollution trends at Mauna Loa, *Science* **172**, 845–846 (1971).

59. Robinson, E., and Robbins, R. E., Emissions, concentrations, and rate of particulate atmospheric pollutants, Final Report, SRI Project SCC-8507, 1971.

60. Kellogg, W. W., Cadle, R. D., Allen, E. R., Lazrus, A. L., and Martell, E. A., Man's contributions are compared to natural sources of sulfur compounds in the atmosphere and oceans, *Science*, **175**, 587–596 (1972).

61. Went, F. W., Organic matter in the atmosphere, *Proc. Nat. Acad. Sci.* **46**, 212 (1960).

62. Cronin, J. F., Recent volcanism and the stratosphere, *Science* **172**, 847–849 (1971).

63. Lamb, H. H., Volcanic dust in the atmosphere: with a chronology and an assessment of its meteorological significance, *Phil. Trans. Royal Soc.* **266**, 425–533 (1970).

64. Junge, C. E., and Manson, J. E., Stratospheric aerosol studies, *J. Geophys. Res.* **66**, 2163–2182 (1961).

65. Cadle, R. D., Lazrus, A. L. Pollack, W. H., and Shedlovsky, J. P., Chemical composition of aerosol particles in the tropical stratosphere, in "Proc. Symp. Tropical Meteorology," Paper K-IV American Meteorological Society, Boston, 1970.

66. Cadle, R. D., Stratospheric particles, in "Man's Impact on the Climate," W. H. Matthews, W. W. Kellogg, and G. D. Robinson, eds., MIT Press, Cambridge, Mass., 1971, Chapter 37.

67. Lazrus, A. L., Gandrud, B., and Cadle, R. D., Chemical composition of air filtration samples of the stratospheric sulfate layer, *J. Geophys. Res.* **76**, 8083–8089 (1971).

68. Murcray, D. R., Kyle, T. G., Murcray, F. H., and Williams, W. J., Presence of HNO_3 in the upper atmosphere, *J. Opt. Soc. Am.* **59**, 1131 (1969).

69. Newell, R. E., Modification of stratospheric properties by trace constituent changes, *Nature* **227**, 697–699 (1970); reprinted as Chapter 38 in "Man's Impact on the Climate," W. H. Matthews, W. W. Kellogg, and G. D. Robinson, eds., MIT Press, Cambridge, Mass., 1971.

70. Park, J., The photochemical relation between water vapor and ozone in the stratosphere (abstract), *EOS, Trans. Am. Geophys. Union* **51**, 735 (1970).

71. Johnston, H., Reduction of stratospheric ozone by nitrogen oxide catalysts from SST exhaust, *Science* **173**, 517–522 (1971).

72. Crutzen, P. J., Ozone production rates in an oxygen–hydrogen–nitrogen oxide atmosphere, *J. Geophys. Res.* **76**, 7311–7327 (1971).

73. Nicolet, M., and Vergison, L., Nitrous oxide in the stratosphere, *Aeronomica Acta A* **1970**, No. 89 (Institut d'Aeronomie Spatiale de Belgique, Brussels).

74. Mastenbrook, J. H., The variability of water vapor in the stratosphere, *J. Atmos. Sci.* **28**, 1495–1501 (1971).

75. Komhyr, W. D., Barrett, E. W., Slocum, G., and Weickmann, H. K., Atmospheric total ozone increases during the 1960's, *Nature* **232**, 390–391 (1971).

Chapter 6

THE CARBON DIOXIDE CYCLE: RESERVOIR MODELS TO DEPICT THE EXCHANGE OF ATMOSPHERIC CARBON DIOXIDE WITH THE OCEANS AND LAND PLANTS

Charles D. Keeling

Scripps Institution of Oceanography
University of California at San Diego
La Jolla, California

PREFACE

The following survey of carbon dioxide in nature places major emphasis on describing how the injection of CO_2 into the atmosphere by man's industrial activity has perturbed the natural carbon cycle on a global scale. In a sense, this injection is a mammoth geochemical experiment. It permits us to observe the transient response of the air, the oceans, and the biosphere to a major disturbance taking place over the interval of only a few years. Our quantitative understanding of the carbon cycle is thus repeatedly challenged and refined.

Central to this process of refinement are geochemical models which organize our observations and logically test our deductions. Such models, if they could be perfected, would allow predictions of future changes in the carbon cycle—a matter of great importance as man's industry impinges ever more intensely on the global environment. Although these models today rest on insecure postulates, they are the best device we have to interpret complex and disorderly global geochemical data. Even their shortcomings are useful to indicate what further observations should be made.

The present chapter first investigates the largely mathematical problem of formulating and solving the governing equations of global CO_2 exchange. It next considers the chemical relations needed to put these equations to practical use and tests them out over a wide range in model parameters. Finally, an attempt is made to establish the set of model parameters which best explain the recent increase in atmospheric CO_2.

I have arranged the chapter in two parts with the discussion of the governing equations first and of the chemical relations second. The two parts are closely linked, however. Sections and equations are numbered in a single series and cross-references are given.

PART I

FORMULATION AND MATHEMATICAL SOLUTION OF THE MODEL EQUATIONS

1. INTRODUCTION

1.1. Introductory Remarks

The burning of coal, petroleum, and natural gas is bringing about a continuous rise in the concentration of atmospheric CO_2. The rate of increase in the past few years has been only about half that which would occur if all the CO_2 from combustion remained in the air, but this is no assurance that the fraction remaining in the air will always remain close to one-half. To predict correctly future atmospheric CO_2 concentrations it is essential first to establish from past variations the factors which determine the fraction of CO_2 from combustion (industrial CO_2) which remains airborne. This is the immediate goal of the present chapter.

Both oceans and land plants rapidly exchange carbon dioxide with the atmosphere, and are capable of storing large amounts of carbon. Both might be expected to absorb parts of the excess CO_2 from combustion.

In the case of the oceans, where CO_2 exchange depends on more or less obvious physical and chemical laws, a fairly realistic prediction of CO_2 uptake is possible on the basis of existing information. How land plants will respond to an increase in atmospheric CO_2 is not obvious, however. Plants require water, nutrients, and energy from the sun, and in many places these are in such short supply that additional growth cannot be induced by higher CO_2 concentrations in the ambient air. Even those plants which today may be growing more rapidly in response to the few percent increase in atmospheric CO_2 which has so far occurred are not capable of indefinite increase in growth if the CO_2 concentration in air rises to much higher levels. Predictions of future atmospheric CO_2 concentrations are thus unavoidably hindered by our lack of knowledge of the CO_2 response of plants. A major effort should therefore be made to determine the amount of CO_2 taken up by the oceans so that the activity of land plants can be estimated by difference. It is the general goal of the present chapter to assist this effort.

The mechanism of oceanic uptake is basically understood by marine chemists, but as yet no model for uptake has made full use of their knowledge. The paucity of oceanic chemical data to check these models has undoubtedly discouraged attempts to make them realistic, but this lack doesn't explain fully why all so far proposed models oversimplify the chemical relationships. It is clearly worthwhile to bring together all relevant existing chemical information and to determine, where possible, why previous models have led to conflicting predictions.

With this goal also in mind, the present chapter will review the capabilities of linear models which assume first-order transfer processes to govern chemical interaction between major domains of the air, sea, and land plants. In the framework of these so-called "reservoir" or "box" models, partitioning of a known CO_2 input between reservoirs can be determined by exact mathematical procedures. In particular, models with three, four, and five coupled reservoirs will be investigated. The chemistry of the oceans will be represented in considerable detail to reduce the chance of neglecting any chemical process which could critically affect CO_2 uptake over several hundred years. The land plants will be modeled as well as present information allows. Thus, the basis will be established for more realistic, but inevitably nonlinear, models which will be needed if accurate predictions are to be made of future increases of CO_2 in the atmosphere, oceans, and land biota.

1.2. Review of Previous Work

The simplest possible CO_2 exchange model neglects land plants altogether and assumes that the carbon in the ocean and atmosphere is so well mixed that the transfer of CO_2 in either direction at the air–sea boundary is proportional to the total amount of carbon in the upstream reservoir. Revelle and Suess [1] concluded from such a model that very little of the CO_2 artificially produced since the beginning of the industrial revolution has remained in the atmosphere.

Bolin and Eriksson [2] noted that a relatively thin layer of well-mixed water quickly readjusts to a change in atmospheric CO_2, but that the remaining ocean responds on average much more slowly. They accordingly divided the oceans into a surface layer of approximately 75 m depth corresponding to the water above the seasonal thermocline, and a sixty times greater "deep sea" layer beneath. The reservoirs were assumed to obey first-order exchange processes as though even the deep layer behaved as a well-mixed, or at least randomly-mixed reservoir. With this model, they predicted that well over half of the CO_2 from combustion has remained in the atmosphere. They speculated on the possible role of land plants in reducing this figure by as much as a third. Neither Revelle and Suess nor Bolin and Eriksson had access to direct atmospheric data to press them toward constructing a model with a predetermined partitioning of CO_2.

An intermediate layer between the surface layer and the deep sea was recognized by Broecker et al. [3] as accounting for a substantial part of the oceanic CO_2 uptake. This layer, which they called the "main thermocline" (the "troposphere" of DeFant [4]), lies between roughly 75 and 1000 m. It is imperfectly mixed and in poor exchange with the atmosphere, but nevertheless exchanges CO_2 considerably more rapidly than the cold, dense water beneath. Assuming diffusive uptake for that layer and first-order exchange for the other reservoirs, Broecker and his collaborators estimated that 25 % of the CO_2 from combustion had reached the intermediate layer, the surface layer taking up a mere 5 %. By assuming some direct exchange between the atmosphere and a deep sea "upcrop" in high latitudes, an additional uptake of up to 20 % could be attributed to the deep sea. Since the properties of the outcrop were uncertain, the amount directly taken up by the deep sea might, however, actually be negligible. If considerable exchange occurs through the outcrop, the recently observed increase in atmospheric CO_2 could be explained by oceanic uptake with no net growth of the land plants.

The high variability of these predictions of the CO_2 uptake arises mainly from the use of different assumptions about the behavior of the upper layers of ocean water. Bolin and Eriksson [2] evidently assumed the layer in direct exchange to be too small. Revelle and Suess [1] clearly assumed it to be too large. Of the three models, that of Broecker et al. [3] is the most reasonable, but is still too uncertain in the assignment of exchange parameters to settle whether the land biota plays a significant role in uptake or not. The model is not developed in sufficient detail to determine whether more reliable values of these parameters can be found. More seriously, the numerical results are given for a hypothetical single "spike" injection of CO_2 25 or 30 years ago rather than for the actual input which is approximately an exponentially increasing function of time. The authors do not show, nor can I discover, any way in which their numerical approach will lead to a correct solution of their model.

The more mathematically realistic model of Bolin and Eriksson is clearly a better basis to begin a more detailed investigation of the transient response of the oceans to an atmospheric chemical perturbation. They solved six coupled differential equations which govern, to a first-order approximation, the deviation from a steady state of inactive carbon (^{12}C plus ^{13}C) and radiocarbon (^{14}C) in a tandem three-reservoir system consisting of the atmosphere, ocean surface water, and the deep ocean water. They included in their equations for the air–sea exchange of CO_2 a correct expression for the change in the dissociative equilibrium of the carbonic acid system when a small amount of CO_2 is taken up by sea water.

Inclusion of radiocarbon as a chemical tracer of the departure from steady state added considerably to the power of their model (and also to that of Revelle and Suess) because the percent dilution of atmospheric radiocarbon by fossil fuel CO_2 (known widely as the "Suess effect") was approximately known from observed radiocarbon–inactive carbon ratios in tree rings, and the requirement that the model correctly predict this effect restricted the permissible range in the predicted increase of atmospheric CO_2. In fact, the striking discovery by Suess [5] that the dilution of atmospheric $^{14}CO_2$ in 1950 was only about 2 to 3 % at a time when the atmosphere had received a more than 10 % addition of ^{14}C-free fossil fuel CO_2 inspired the formulation of all models constructed before 1960.

The inclusion of radiocarbon by Bolin and Eriksson increased considerably the mathematical complexity of their model by adding a second set of first-order differential equations to the set for inactive carbon. The authors, in solving these equations by straightforward classical methods, did not attempt to circumvent the algebraic complexities arising when the constant coefficients of the equations are expressed in terms of measurable geochemical quantities. They also did not examine in full the consequences of having dropped a seemingly negligible term in the response function for each reservoir. As shown below, although the loss of this term had little effect on the predicted increase in atmospheric CO_2 for a given set of values of the model parameters, it produced a sizable enhancement in the predicted Suess effect through the coupling of errors between the expressions for inactive carbon and radiocarbon.

Bolin and Eriksson [2] briefly probed the problem of modeling the land plants and their response to an increase in CO_2. They wrote down a system of equations based on the assumption that all reservoir transfers were proportional to the mass of carbon in the reservoir from which the transfer emerged. They distinguished the living plants with a carbon mass one-half that of the atmosphere from a three times larger mass of dead carbon. Without solving this system of five differential equations, they produced a rough estimate of the atmospheric CO_2 increase which was evidently based on the reasoning that the living plants would take up an amount of CO_2 equal to roughly half the atmospheric increase.

Their method of depicting the land plants was copied from a paper by Craig [6], which in turn was based (without that author so stating) on a study by Eriksson and Welander [7]. These authors more correctly asserted, however, that plants assimilate CO_2 at a rate depending both on the concentration of atmospheric CO_2 and on the mass of the plants themselves. Craig made no computational use of this nonlinear relation, but Eriksson and Welander, by stepwise integration using an electronic computer, carried out numerical evaluations for hypothetical self-sustained oscillations triggered by a presumed departure from a preexisting dynamic equilibrium. Welander and Eriksson's treatment of the time response of the land plants is of special interest. For the living plants they introduced a generalized age-distribution function and found its functional form by supposing plant growth to be proportional to the assimilating mass while the probability of death increased linearly with plant age. With this pair of assumptions, they deduced a normal distribution of mass for plants at steady state with the atmosphere and established an equation for the time delay of the plant response to a perturbation in atmospheric CO_2. Subsequently, using a similar distribution function, Welander [8] treated the transient response of the oceans to hypothetical oscillatory variations in atmospheric CO_2. In his actual treatment of the oceans he confused the concepts of age and lifetime and produced some incorrect response curves, but his basic idea is useful and will almost surely aid in formulating more realistic nonlinear models needed to predict future changes in the CO_2 exchange system. A further discussion of nonlinear models, however, lies beyond the scope of the present chapter.

1.3. Preliminary Modeling Considerations

Over periods of interest to the industrial CO_2 problem, the ability of intermediate layers of ocean water to respond to an atmospheric transient more rapidly than deep water cannot be neglected without obtaining a distorted picture of the total transfer of CO_2 between the air and sea. Nevertheless, considerable advantage is gained if the first models to be investigated are not so mathematically complicated that one is tempted to simplify the representation of the atmospheric source or any of the chemical factors when no basis has been developed for testing the validity of the simplifying assumptions.

As the development below will indicate, the division of the ocean into even two layers leads to considerable complication in specifying the chemical processes involved in uptake of CO_2. To reduce the possibility of error in working out these chemical problems and to provide an enlightening comparison with earlier models, this paper will develop models having only two oceanic layers. Surprisingly, this does not altogether deny the possibility of modeling the role of the intermediate water since this layer can be included either by expanding the capacity of the surface layer or by assigning a more rapid response time to the subsurface exchange. Neither of these approaches permits the record of air–sea exchange to be used to gain insight into the true circulation of intermediate or deep ocean water, but, as will be demonstrated below, when adjusted to correctly represent events observed in the atmosphere over past decades, the two approaches give similar predictions of the future increase in atmospheric CO_2. This result suggests that the response of subsurface ocean water need not necessarily be modeled realistically to provide a correct prediction of events depending principally on processes occurring near the sea surface. Still, one can hardly be satisfied with models which give a distorted or oversimplified picture of the deep ocean circulation. Thus, once experience has been gained with two-layer models, attention should be given to better representing the subsurface oceanic exchange.

In the models to be developed in this chapter, the governing equations of Bolin and Eriksson will be modified to include isotopic fractionation factors. These factors are sufficiently close to unity that their inclusion produces only small changes in the calculations, but, since the factors are reasonably well known, no advantage is derived from leaving them out and producing several percent error in the final predictions.

The equations for transfer of carbon between the surface ocean water and deep layers will be further modified to take into account a gravitational particulate flux left out of the model of Bolin and Eriksson. Again, only small changes will occur in the final results.

In modeling the land plants, the immediate goal will be to depict their properties realistically enough to indicate their role in the uptake of atmospheric CO_2 relative to that of the oceans but simply enough to permit an exact solution of the linear differential equations for the response of all the reservoirs acting together. The resulting response functions can later be the point of departure for nonlinear models which more closely reflect the long-range plant behavior.

It adds very little useful detail to separate living and dead plant material, and I will combine their separate responses in a manner appropriate for small perturbations in atmospheric CO_2 concentration. Thus a reservoir called the "land biota" will here be defined to include the land plants and their detrital products. The inclusion of annual plants and the leaves of perennial plants in the same reservoir with long-lived woody material and its decomposition products in the soil, on the other hand, leads to a highly distorted estimate of response if this is assumed to depend on the ratio of total biomass to the total rate of assimilation of CO_2. Fully half of the CO_2 uptake by plants goes to producing short-lived organic materials which return most of their carbon to the atmosphere within two years, whereas the woody parts remain as organic carbon for decades or longer. Although no simple linear model will faithfully reproduce a system as complex as the land biota, the most serious errors are avoided if the biota is modeled by two reservoirs representing leaves and wood, respectively. The mathematically simpler case of a single biota reservoir will, however, also be considered below.

1.4. Physical Basis for the Models

It is worthwhile to consider briefly the physical justification for using linear reservoir models of the kind to be developed in this chapter, especially since the land biota will be treated in a somewhat unconventional manner. To begin the discussion, let us adopt the familiar "linear box model" assumption that the carbon transferred out of each reservoir is proportional to the total carbon mass of that reservoir, but provisionally add the restriction that each reservoir must be clearly distinguished by its physical properties, i.e., the whole atmosphere, all the land plants, and the entire world ocean. The governing equations to predict the redistribution of CO_2 added to the atmosphere at a rate $\gamma(t)$ are then of the form

$$\frac{dN_i}{dt} = \sum_{\substack{j=1 \\ j \neq i}}^{n} (k_{ji}N_j - k_{ij}N_i) + \gamma_i \qquad (1.1)$$

where N_i denotes the time-variable mass of carbon in reservoir i, t is the time, n (provisionally equal to 3) is the number of reservoirs, the k_{ij} are time-invariant "transfer coefficients" for transfer of carbon from reservoir i to j, and γ_i is nonzero only for the atmosphere, where it is denoted by $\gamma(t)$. As shown in Section 2, the set of equations (1.1) has the general solution

$$N_i = \sum_{k=1}^{n} C_{ki}(\lambda_k) \int_0^t e^{\lambda_k \tau} \gamma(t - \tau) \, d\tau + N_{i0} \qquad (1.2)$$

where λ_k are roots (eigenvalues) of the auxiliary equation of (1.1), the C_{ki} are ratios of polynomials in powers of the λ_k, with the polynomial coefficients made up of functions of the k_{ij}, the argument $(t - \tau)$ refers to time $t - \tau$, and τ is a "dummy" variable eliminated on integration. N_{i0} denotes the steady-state value of N_i.

For transfers of carbon *from* the lower atmosphere, the linear box model assumptions are realistic, at least for small perturbations from long-term mean conditions, because that reservoir is well mixed relative to the time scale of industrial CO_2 production, and outgoing transfers vary in direct response to changes in the CO_2 mixing ratio in air. Even if the stratosphere and lower atmosphere are regarded as a single well-mixed reservoir, only small errors occur in predicting the distribution of carbon in the lower atmosphere, as shown by Ekdahl and Keeling [9]. The linear box model assumptions are hardly appropriate, however, for the land plants and oceans, which are not well mixed. For these reservoirs, following the lead of Eriksson and Welander [3], we seek a function $f(\tau, t)$ which describes correctly the age distribution of carbon within the reservoir, i.e., the (possibly time-dependent) distribution of carbon mass with respect to the time interval, or, "age," τ, since that carbon was transferred from the atmosphere. If this function were known, the time-dependent masses N_i could be precisely computed relative to their steady-state values N_{i0} from convolution integrals of the general form

$$N_i = \int_0^t f_i(\tau, t) g_i(t - \tau) \, d\tau + N_{i0} \qquad (1.3)$$

where $g_i(t)$ denotes the fractional increase above steady state (at time t) in the carbon flux into reservoir i, this increase resulting from the atmospheric input, $\gamma(t)$. The variable τ, although still eliminated on integration, has now the definite meaning of an age.

As shown by Eriksson [10] for a reservoir i with a single boundary contacting reservoir j, if reservoir i is well mixed and the boundary transfer process is time invariant,

$$f(\tau) = f(\tau = 0)\, e^{-k_{ij}\tau} \qquad (1.4)$$

The reciprocal k_{ij}^{-1} of the distribution coefficient k_{ij} is the so-called "turn-over" or "residence" time of the reservoir, and k_{ij} itself can be regarded as a transfer coefficient similar to those employed in Eq. (1.1). From Eq. (1.4) it is evident that if a reservoir has only one boundary and if the carbon age distribution with respect to that boundary approaches the functional form of a declining exponential in age, the outgoing transfer of carbon can be approximately portrayed using a linear box model equation even if the reservoir is not well mixed.

Such an exponentially declining age distribution and single characteristic time k_{ij}^{-1} clearly do not apply to the land biota, however. As indicated above, about half of the photosynthetic uptake goes to producing annual and biennial plants and the leaves of long-lived plants, which altogether contain only a small fraction of the total biomass.

To allow for a flux preponderantly of "young" carbon but a storage preponderantly of "old" carbon, we require at least two characteristic time constants. This demand can be met in the simplest possible way if we subdivide the land biota into two spatially superimposed reservoirs representing pools of short- and long-lived carbon materials, where each reservoir is assumed to obey Eq. (1.4) so that formally we retain the linear box model representation. This approach, we emphasize, does not mean to imply that any part of the biota is well mixed.

We are confronted with an equally difficult problem in depicting the oceanic carbon age distribution. The oceans transport carbon both by water motion and by mixing between advecting layers. This circulation is so complicated that the time distribution over τ has never been deduced from existing hydrographic or chemical-tracer information. We will therefore submit to the common practice of subdividing the oceans into two layers of the linear box model type. As in the case of the biota, this subdivision does not imply that any part of the reservoir is well mixed although, as it happens, most or all of the surface layer is vertically mixed. For the subsurface waters the assumption of a single box model reservoir is a crude representation, and the relative sizes of the two ocean layers and the rate of exchange of carbon between them will therefore be made highly flexible to establish whether our predictions of atmospheric CO_2 increase depend critically on how the subsurface water is modeled.

Since we have no information to determine how an increase in CO_2 might indirectly influence the circulation of the oceans or the passage times of carbon atoms through the land biota, these will be regarded as time invariant. To avoid additional complexity, all boundary properties will be spatially averaged.

2. THREE-RESERVOIR ATMOSPHERE–OCEAN TANDEM MODEL

As a starting point, we will consider the exchange of carbon between the atmosphere and world ocean without regard for the influence of the land plants. This approach will permit the essential mathematical techniques to be introduced in a relatively simple form and will facilitate comparison with the treatment of Bolin and Eriksson [2]. The ocean will be divided into a deep water layer and a surface layer which includes the vertically well-mixed water above the thermocline and may also include some of the intermediate water beneath. The capacity of the surface layer plays a vital role in determining the partitioning of carbon and, as discussed above, the average depth of this layer will be considered as a variable parameter. The model is depicted in Fig. 1.

Each reservoir is characterized by the mass of inactive carbon, N_i, and radiocarbon, $*N_i$, which it contains. These masses will be treated as time-dependent variables with initial values N_{i0} and $*N_{i0}$. The latter represent conditions before the introduction of industrial carbon. The flux of carbon from one reservoir, i, to another, i', will be expressed as a product $k_{ii'}N_i$, where $k_{ii'}$ denotes a time-invariant exchange or "transfer" coefficient.

The three reservoirs will be identified by subscripts a, m, and d for the atmosphere, surface ("mixed") layer, and deep ocean, respectively. In writing the transfer equations the coefficients k_{am}, k_{ma}, etc. will be replaced by serially numbered factors k_j to reveal a mathematical regularity intrinsic to the tandem model. The numbering will begin with 3 (see Fig. 1), indices 1 and 2 being reserved for later use in connection with the land plants. The evaluation of the k_j will be considered in Sections 5–7. The k_j are assumed to be known when the model equations are solved for the transient case.

The carbon in the atmosphere is almost all in the form of CO_2, and the flux of carbon from the atmosphere to the ocean is directly proportional to the CO_2 mixing ratio in air. The oceans, on the other hand, contain numerous organic and inorganic chemical carbon compounds. The flux of carbon from the ocean surface to the atmosphere is not strictly proportional to the total concentration of carbon in the water but, rather, to the CO_2 partial pressure. At steady state the flux is thus properly expressed by a product $k'_{ma}P_{m0}$ ($\equiv k'_4P_{m0}$), where the transfer coefficient carries a prime to emphasize that it has different dimensions than the other transfer coefficients, and where P_{m0} denotes the steady-state CO_2 partial pressure averaged over

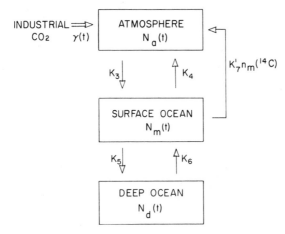

Fig. 1. Three-reservoir model of the CO_2 exchange between the atmosphere and ocean surface water, and between the latter and deep ocean water. The double-lined arrow indicates an external source to the atmosphere of inactive CO_2 from industrial processes, principally fossil fuel combustion. Single-lined straight arrows denote the directions of flow associated with CO_2 transfer coefficients k_j (shown) and $*k_j$ for radiocarbon. The bent arrow denotes the direction of readjustment of the radiocarbon distribution owing to the industrial CO_2 input $\gamma(t)$. This readjustment is expressed as a virtual source $k'_7 n_m$.

the ocean surface. For small temporal variations in surface water carbon content the change in the CO_2 evasion rate is nevertheless, to a first-order approximation, proportional to the change in total inorganic carbon. Considerable space will be given later on to elucidating this approximation. The case for large variations in carbon content has never before been worked out, but will be considered in Section 6, to estimate the limits over which the linear approximation is reasonable.

The transfer of carbon between the surface layer and deep ocean is by water transport and gravitational settling of particles derived from marine plants and animals [11]. As shown in Section 7, the flux by both modes of transport combined is very nearly proportional to the total carbon in each reservoir, but the time-dependent character of the transports differs for inactive carbon and radiocarbon.

For inactive carbon (^{12}C plus ^{13}C) the following conditions apply at steady state:

$$\left.\begin{array}{l} k_3 N_{a0} - k'_4 P_{m0} \qquad\qquad\qquad = 0 \\ -k_3 N_{a0} + k'_4 P_{m0} + k_5 N_{m0} - k_6 N_{d0} = 0 \\ -k_5 N_{m0} + k_6 N_{d0} \qquad\qquad\qquad = 0 \end{array}\right\} \qquad (2.1)$$

For radioactive carbon (^{14}C), a continual radioactive decay loss from each reservoir is balanced by an atmospheric source (originating in the stratosphere) which is equal to the sum of the decays. Denoting the ^{14}C decay constant by $*\lambda$:

$$
\left.
\begin{aligned}
*k_3*N_{a0} - *k_4'*P_{m0} &= *Q - *\lambda*N_{a0} \\
-*k_3*N_{a0} + *k_4'*P_{m0} + *k_5*N_{m0} - *k_6*N_{d0} &= -*\lambda*N_{m0} \\
-*k_5*N_{m0} + *k_6*N_{d0} &= -*\lambda*N_{d0}
\end{aligned}
\right\}
\quad (2.2)
$$

where the source, $*Q$, is equal to $*\lambda(*N_{a0} + *N_{m0} + *N_{d0})$.

Prior to the industrial era, the steady-state conditions (2.1) and (2.2) will be assumed to be appropriate. Deviations from steady state owing to the release of industrial CO_2 from combustion of fossil fuel and kilning of limestone [12] at a variable rate $\gamma(t)$ will, in general, be denoted by the symbol Δ, i.e., for reservoir i,

$$
\left.
\begin{aligned}
\Delta N_i &= N_i - N_{i0} \\
\Delta P_i &= P_i - P_{i0}
\end{aligned}
\right\}
\quad (2.2a)
$$

with corresponding expressions for radiocarbon. As already mentioned, if the deviation for the ocean surface layer (ΔN_m) is small, the perturbation in the ocean to atmosphere flux $k_4'\Delta P_m$ can, to a first approximation, be replaced by a term of the form $k_4\Delta N_m$ (see Section 6.3). The deviations ΔN_i, rewritten with the more compact symbol n_i, are then governed by the equations

$$
\left.
\begin{aligned}
(d/dt + k_3)n_a - k_4 n_m &= \gamma \\
-k_3 n_a + (d/dt + k_4 + k_5)n_m - k_6 n_d &= 0 \\
-k_5 n_m + (d/dt + k_6)n_d &= 0
\end{aligned}
\right\}
\quad (2.3)
$$

with the initial (preindustrial) conditions

$$
n_a = n_m = n_d = 0 \qquad \text{at } t = 0 \quad (2.4)
$$

Since fossil fuel and limestone have undergone long-term underground storage, their combustion releases only inactive CO_2. But since this release disturbs the dissociative equilibrium of the carbonic acid system in the ocean, ^{14}C is driven out of the ocean into the atmosphere. As shown by Bolin and Eriksson [2], this effect can be modeled mathematically by introducing into the surface layer a virtual sink for ^{14}C proportional to the increase in inactive carbon, n_m, in that layer. A corresponding virtual source appears in the atmosphere. We will postpone for the moment a derivation of the associated proportionality factor, which depends on thermodynamic relationships in the ocean, and simply denote this factor as k_7'. The coefficient k_7' carries a prime as a reminder that it has different dimensions and significance than the transfer coefficients k_3 to k_6. Somewhat analogous to the case of inactive carbon, the term $*k_4'\Delta*P_m$, to a first-order approximation, will be replaced (see Section 6.3) by the expression $*k_4*n_m + k_7'n_m$. We thus obtain the set of equations

$$
\left.
\begin{aligned}
(d/dt + *\lambda + *k_3)*n_a - *k_4*n_m \qquad\qquad &= k_7'n_m \\
-*k_3*n_a + (d/dt + *\lambda + *k_4 + *k_5)*n_m - *k_6*n_d &= -k_7'n_m \\
-*k_5*n_m + (d/dt + *\lambda + *k_6)*n_d \qquad\qquad &= 0
\end{aligned}
\right\} \quad (2.5)
$$

with the initial conditions

$$
*n_a = *n_m = *n_d = 0 \qquad \text{at } t = 0 \tag{2.6}
$$

To simplify the treatment, a small virtual source to represent the gravitational transfer from the surface layer to the deep ocean is omitted. This will, however, be included in the four-reservoir model discussed in Section 3.

The exact solution of (2.3) and (2.5) is conveniently obtained by the use of Laplace transforms. The inactive case will be considered first.

Each term in (2.3) is multiplied by e^{-st}, where s is a generalized frequency appearing in the transforms but not in the final solution. Integrating from $t = 0$ to ∞,

$$
\left.
\begin{aligned}
(s + k_3)\tilde{n}_a - k_4\tilde{n}_m \qquad\qquad &= \tilde{\gamma} \\
-k_3\tilde{n}_a + (s + k_4 + k_5)\tilde{n}_m - k_6\tilde{n}_d &= 0 \\
-k_5\tilde{n}_m + (s + k_6)\tilde{n}_d \qquad\qquad &= 0
\end{aligned}
\right\} \quad (2.7)
$$

where

$$
\left.
\begin{aligned}
\tilde{n}_i &= \int_0^\infty e^{-st} n_i \, dt \qquad i = a, m, d \\
\tilde{\gamma} &= \int_0^\infty e^{-st} \gamma \, dt \\
\int_0^\infty e^{-st}(dn_i/dt) \, dt &= -(n_i)_{t=0} + s\tilde{n}_i = s\tilde{n}_i
\end{aligned}
\right\} \quad (2.8)
$$

The coefficients of (2.7) form the array (matrix)

$$
\mathbf{D}(s) \equiv
\begin{pmatrix}
s + k_3 & -k_4 & 0 \\
-k_3 & s + k_4 + k_5 & -k_6 \\
0 & -k_5 & s + k_6
\end{pmatrix}
\tag{2.9}
$$

the determinant of which can be expanded to yield

$$
D(s) = s^3 + (k_3 + k_4 + k_5 + k_6)s^2 + (k_3k_5 + k_3k_6 + k_4k_6)s \tag{2.10}
$$

This polynomial contains no constant term because the k's in each column of $\mathbf{D}(s)$ sum to zero. This is a consequence of inactive carbon being conservative, i.e., no gains or losses in carbon occur in any reservoir except by transfers between reservoirs and owing to the external source $\gamma(t)$.

We can write (2.10) more compactly

$$D(s) = s(s^2 + As + B) \tag{2.11}$$

where

$$A = \sum_{j=3}^{6} k_j \tag{2.12}$$

$$B = \sum_{j=3}^{4} \sum_{j'=5}^{6} k_j k_{j'}, \qquad j' > j + 1 \tag{2.13}$$

and where $D(s)$ represents the determinant of (2.9) in polynomial form.

Solving the auxiliary equation $D(s) = 0$, we obtain the roots λ_k ($k = 1, 2, 3$). Specifically,

$$D_R(s) = (s - \lambda_1)(s - \lambda_2)(s - \lambda_3) \tag{2.14}$$

where

$$\left. \begin{array}{l} \lambda_1 = 0 \\[4pt] \lambda_2 = -A/2 - \sqrt{A^2/4 - B} \\[4pt] \lambda_3 = -A/2 + \sqrt{A^2/4 - B} \end{array} \right\} \tag{2.15}$$

and the subscript R indicates that $D(s)$ has been factored according to its roots. (If the reader wishes to verify that these roots are identical to those obtained when (2.3) is solved by standard methods, he may eliminate two of the n_i from (2.3) to obtain a third-order differential equation in the remaining n_i. Substituting e^{st} for this n_i then gives an auxiliary equation identical to (2.10), cf. Bolin and Eriksson [2], p. 155.)

The algebraic equations (2.7) are now solved to give

$$\left. \begin{array}{l} \tilde{n}_a = \dfrac{1}{D(s)} \begin{vmatrix} \tilde{\gamma} & -k_4 & 0 \\ 0 & s + k_4 + k_5 & -k_6 \\ 0 & -k_5 & s + k_6 \end{vmatrix} \\[28pt] \phantom{\tilde{n}_a} = \dfrac{\tilde{\gamma}[s^2 + (k_4 + k_5 + k_6)s + k_4 k_6]}{(s - \lambda_1)(s - \lambda_2)(s - \lambda_3)} \\[20pt] \text{Similarly,} \\[14pt] \tilde{n}_m = \dfrac{\tilde{\gamma} k_3(s + k_6)}{(s - \lambda_1)(s - \lambda_2)(s - \lambda_3)} \\[20pt] \tilde{n}_d = \dfrac{\tilde{\gamma} k_3 k_5}{(s - \lambda_1)(s - \lambda_2)(s - \lambda_3)} \end{array} \right\} \tag{2.16}$$

These transforms are of the general form

$$\tilde{n}_i = \frac{\tilde{\gamma} f_i(s)}{D(s)} \qquad i = a, m, d \tag{2.17}$$

where the f_i are first-row cofactors (signed minors) of the determinant $D(s)$. (If the source $\gamma(t)$ had occurred in the surface ocean layer or the deep ocean, the appropriate cofactors would have been obtained from the second or third row, respectively.)

The general solution of (2.3) is

$$n_i = C_{1i} e^{\lambda_1 t} + C_{2i} e^{\lambda_2 t} + C_{3i} e^{\lambda_3 t} + C_{Pi} \tag{2.18}$$

where the C_{ki} ($k = 1$ to 3) are integration constants and C_{Pi} are particular solutions depending on the time-dependent character of $\gamma(t)$. They will now be determined for two cases:

(1) $\gamma(t)$ is approximated by an exponential function. As shown by Bolin and Eriksson [2], the industrial CO_2 production can be represented to fair accuracy by

$$\gamma(t) = \gamma_0 e^{rt} \tag{2.19}$$

where

$$\left. \begin{array}{l} \gamma_0 = 4.96 \, N_a \times 10^{-4} \\ r = 0.029 \, \text{yr}^{-1} \end{array} \right\} \tag{2.20}$$

(2) $\gamma(t)$ is estimated from the set of annual values of the industrial production available in various compilations [12].

In the first case, the Laplace transform of $\gamma(t)$ is

$$\tilde{\gamma} = \frac{\gamma_0}{s - r} \tag{2.21}$$

whence

$$\tilde{n}_i = \frac{\gamma_0 f_i(s)}{(s - r)D(s)} = \frac{\gamma_0 f_i(s)}{(s - r)(s - \lambda_1)(s - \lambda_2)(s - \lambda_3)} \tag{2.22}$$

It is now possible to obtain expressions for the constants C_{ki} and C_{Pi} of (2.18) by making use of (2.22). Since e^{rt} is a solution of (2.3), we can write in place of (2.18)

$$n_i = C_{1i} e^{\lambda_1 t} + C_{2i} e^{\lambda_2 t} + C_{3i} e^{\lambda_3 t} + C_{4i} e^{rt} \tag{2.23}$$

If we transform (2.23) by multiplying by e^{-st} and then integrate from $t = 0$ to ∞, we obtain

$$\tilde{n}_i = \frac{C_{1i}}{s - \lambda_1} + \frac{C_{2i}}{s - \lambda_2} + \frac{C_{3i}}{s - \lambda_3} + \frac{C_{4i}}{s - r} \tag{2.24}$$

Since (2.22) can be expanded in partial fractions to yield an equation
with identical denominators, the problem reduces to matching coefficients
in $[1/(s - \lambda_k)]$. The algebra is considerably simplified if we apply the Lagrange
interpolation formula for partial fractions (Irving and Mullineux [13], p. 774)

$$\frac{f(s)}{g(s)} = \sum_{k=1}^{n} \frac{f(\lambda_k)}{g'(\lambda_k)(s - \lambda_k)} \tag{2.25}$$

where

$$g(s) = \prod_{k=1}^{n} (s - \lambda_k),\, g'(\lambda_k) \equiv \frac{dg}{ds} \text{ (evaluated at } \lambda_k) = \prod_{\substack{l=1 \\ l \neq k}}^{n} (\lambda_k - \lambda_l),$$

$f(s)$ is a polynomial of less than nth degree, and the λ_k are all different.

Expanding (2.22) by use of (2.25) with $n = 4$ and $\lambda_4 \equiv r$, and comparing
terms, we find

$$\left.\begin{aligned}
C_{1i} &= \frac{\gamma_0 f_i(\lambda_1)}{(\lambda_1 - r)(\lambda_1 - \lambda_2)(\lambda_1 - \lambda_3)} \\[2mm]
C_{2i} &= \frac{\gamma_0 f_i(\lambda_2)}{(\lambda_2 - r)(\lambda_2 - \lambda_1)(\lambda_2 - \lambda_3)} \\[2mm]
C_{3i} &= \frac{\gamma_0 f_i(\lambda_3)}{(\lambda_3 - r)(\lambda_3 - \lambda_1)(\lambda_3 - \lambda_2)} \\[2mm]
C_{4i} &= \frac{\gamma_0 f_i(r)}{(r - \lambda_1)(r - \lambda_2)(r - \lambda_3)}
\end{aligned}\right\} \tag{2.26}$$

where

$$\left.\begin{aligned}
f_a(\lambda_k) &= \lambda_k^2 + (k_4 + k_5 + k_6)\lambda_k + k_4 k_6 \\
f_m(\lambda_k) &= k_3(\lambda_k + k_6) \\
f_d(\lambda_k) &= k_3 k_5
\end{aligned}\right\} \tag{2.27}$$

(For the argument λ_4 ($\equiv r$) the f_i correspond to the factors S_{a0}, S_{m0}, and S_{d0}
of Bolin and Eriksson [2], p. 135.)

We may write the solution compactly in the form

$$n_i = \sum_{k=1}^{4} \frac{\gamma_0 f_i(\lambda_k)}{\displaystyle\prod_{\substack{l=1 \\ l \neq k}}^{4} (\lambda_k - \lambda_l)} e^{\lambda_k t} \tag{2.28}$$

where the $f_i(\lambda_k)$ are given by (2.27).

For the case where $\gamma(t)$ is estimated from a set of annual values, the
solution for $n_i(t)$ involves an integral of the product of $\gamma(t)$ and a kernel.
The latter can be determined by use of the convolution theorem for Laplace

transforms (Irving and Mullineux [13], p. 224), which states that if $\tilde{\gamma}(s)$ and $f_i(s)/D(s)$ are transforms of $\gamma(t)$ and $h_i(t)$, respectively, their product

$$\tilde{n}_i = \tilde{\gamma} \frac{f_i(s)}{D(s)} \tag{2.29}$$

is the transform of

$$n_i = \int_0^t h_i(t - u)\gamma(u)\, du = \int_0^t h_i(t)\gamma(t - u)\, du \tag{2.30}$$

where $h_i(t - u)$ and $\gamma_i(t - u)$ denote h_i and γ_i at time $(t - u)$.

To express h_i we first expand the quotient $f_i(s)/D(s)$, in partial fractions. This gives a result identical to expanding (2.22) except that the factor $(s - r)$ has been dropped. We find

$$\tilde{h}_i(s) \equiv \frac{f_i(s)}{D(s)} = \frac{C'_{1i}}{s - \lambda_1} + \frac{C'_{2i}}{s - \lambda_2} + \frac{C'_{3i}}{s - \lambda_3} \tag{2.31}$$

and

$$h_i(t) = C'_{1i}\, e^{\lambda_1 t} + C'_{2i}\, e^{\lambda_2 t} + C'_{3i}\, e^{\lambda_3 t} = \sum_{k=1}^{3} C'_{ki}\, e^{\lambda_k t} \tag{2.32}$$

where

$$\left.\begin{aligned}
C'_{1i} &= \frac{f_i(\lambda_1)}{(\lambda_1 - \lambda_2)(\lambda_1 - \lambda_3)} \\[2mm]
C'_{2i} &= \frac{f_i(\lambda_2)}{(\lambda_2 - \lambda_1)(\lambda_2 - \lambda_3)} \\[2mm]
C'_{3i} &= \frac{f_i(\lambda_3)}{(\lambda_3 - \lambda_1)(\lambda_3 - \lambda_2)}
\end{aligned}\right\} \tag{2.33}$$

or in general

$$C'_{ki} = \frac{f_i(\lambda_k)}{\displaystyle\prod_{\substack{l=1 \\ l \neq k}}^{3} (\lambda_k - \lambda_l)} \tag{2.34}$$

Thus we obtain as the solution for (2.3)

$$n_i = \sum_{k=1}^{3} C'_{ki} \int_0^t e^{\lambda_k(t - u)}\gamma(u)\, du \tag{2.35}$$

where the C'_{ki} are evaluated according to (2.34). [In (2.35) $(t - u)$ denotes a factor, not an argument.]

This result, indeed, applies generally to any expression of $\gamma(t)$ that can be integrated. For example, substitution of $\gamma_0\, e^{ru}$ for $\gamma(u)$ inside the integral of (2.35) leads to the earlier result (2.28). If annual values of $\gamma(t)$ are given, the integral can be accurately approximated by numerical integration without first replacing it with an analytical function. This approximation will be described in detail in Section 3.

The equations for radiocarbon will now be solved applying the same procedures. Transforming (2.5) we obtain a set of equations similar to (2.7):

$$
\left.
\begin{aligned}
(s + {}^*\lambda + {}^*k_3){}^*\tilde{n}_a - {}^*k_4{}^*\tilde{n}_m &= k_7'\tilde{n}_m \\
-{}^*k_3{}^*\tilde{n}_a + (s + {}^*\lambda + {}^*k_4 + {}^*k_5)\tilde{n}_m - {}^*k_6{}^*\tilde{n}_d &= -k_7'\tilde{n}_m \\
-{}^*k_5{}^*\tilde{n}_m + (s + {}^*\lambda + {}^*k_6){}^*\tilde{n}_d &= 0
\end{aligned}
\right\}
\qquad (2.36)
$$

The coefficients of (2.36) yield the array

$$
{}^*\mathbf{D}(s) = \begin{pmatrix}
s + {}^*\lambda + {}^*k_3 & -{}^*k_4 & 0 \\
-{}^*k_3 & s + {}^*\lambda + {}^*k_4 + {}^*k_5 & -{}^*k_6 \\
0 & -{}^*k_5 & s + {}^*\lambda + {}^*k_6
\end{pmatrix}
\qquad (2.37)
$$

Expanding the determinant of ${}^*\mathbf{D}(s)$, we obtain

$$
{}^*D(s) = (s + {}^*\lambda)[(s + {}^*\lambda)^2 + {}^*A(s + {}^*\lambda) + {}^*B]
\qquad (2.38)
$$

where

$$
\left.
\begin{aligned}
{}^*A &= \sum_{j=3}^{6} {}^*k_j \\
{}^*B &= \sum_{j=3}^{4} \sum_{j'=5}^{6} {}^*k_j{}^*k_{j'}, \qquad j' > j + 1
\end{aligned}
\right\}
\qquad (2.39)
$$

The roots of ${}^*D(s) = 0$ are

$$
\left.
\begin{aligned}
{}^*\lambda_1 &= -{}^*\lambda \\
{}^*\lambda_2 &= -{}^*A/2 - \sqrt{{}^*A^2/4 - {}^*B} - {}^*\lambda \\
{}^*\lambda_3 &= -{}^*A/2 - \sqrt{{}^*A^2/4 - {}^*B} - {}^*\lambda
\end{aligned}
\right\}
\qquad (2.40)
$$

As shown later, the roots ${}^*\lambda_2$ and ${}^*\lambda_3$ are of the order of 0.1 and 1 yr^{-1}, respectively. Therefore, the term ${}^*\lambda$ ($\simeq 10^{-4}\ yr^{-1}$) has very little influence on the value of ${}^*D(s)$ except as it affects ${}^*\lambda_1$. This latter root produces a term proportional to $e^{{}^*\lambda t}$ in the final expression for ${}^*n_i$, but, because no radioactive source is introduced when industrial CO_2 is added to the atmosphere, the coefficient of this term will turn out to be zero. The constant ${}^*\lambda$ also contributes very nearly negligibly to the numerators of the expression for the transforms ${}^*\tilde{n}_i$ (and hence to ${}^*n_i$).

Solving algebraically for the $*\tilde{n}_i$:

$$*\tilde{n}_a = \frac{1}{*D(s)} \begin{vmatrix} k'_7\tilde{n}_m & -*k_4 & 0 \\ -k'_7\tilde{n}_m & s + *\lambda + *k_4 + *k_5 & -*k_6 \\ 0 & -*k_5 & s + *\lambda + *k_6 \end{vmatrix} \quad (2.41)$$

$$= k'_7\tilde{n}_m(s + *\lambda) \begin{vmatrix} 1 & -*k_4 & 0 \\ 0 & 1 & 1 \\ 0 & -*k_5 & s + *\lambda + *k_6 \end{vmatrix}$$

$$= \frac{k'_7\tilde{n}_m(s + *\lambda)(s + *\lambda + *k_5 + *k_6)}{(s + *\lambda)(s - *\lambda_2)(s - *\lambda_3)}$$

$$= \frac{k'_7\tilde{n}_m(s + *\lambda + *k_5 + *k_6)}{(s - *\lambda_2)(s - *\lambda_3)}$$

In the second form of (2.41), the determinant has been simplified by replacing the second row by the sum of all three rows and then factoring out $k'_7 n_m(s + *\lambda)$. In the fourth expression, $(s + *\lambda)$ has been cancelled between the numerator and the denominator. Finally, replacing $\tilde{n}_m$ by its specific form according to (2.16) and then proceeding in the same way for the other $*n_i$,

$$
\left.
\begin{aligned}
*\tilde{n}_a &= \frac{\tilde{\gamma}(s + *\lambda + *k_5 + *k_6)k_3(s + k_6)k'_7}{(s - \lambda_1)(s - \lambda_2)(s - \lambda_3)(s - *\lambda_2)(s - *\lambda_3)} \\
*\tilde{n}_m &= \frac{-\tilde{\gamma}(s + *\lambda + *k_6)k_3(s + k_6)k'_7}{(s - \lambda_1)(s - \lambda_2)(s - \lambda_3)(s - *\lambda_2)(s - *\lambda_3)} \\
*\tilde{n}_d &= \frac{-\tilde{\gamma}*k_5 k_3(s + k_6)k'_7}{(s - \lambda_1)(s - \lambda_2)(s - \lambda_3)(s - *\lambda_2)(s - *\lambda_3)}
\end{aligned}
\right\} \quad (2.42)
$$

If we write λ_4 and λ_5 for $*\lambda_2$ and $*\lambda_3$, the general solution of (2.5) is

$$*n_i = \sum_{k=1}^{5} *C_{ki} \int_0^t e^{\lambda_k(t-u)}\gamma(u)\,du \quad (2.43)$$

The coefficients $*C_{ki}$ are evaluated by expanding (2.42) in partial fractions using the Lagrange interpolation formula, i.e.,

$$*C_{ki} = \frac{*f_i(\lambda_k)f_m(\lambda_k)k'_7}{g'(\lambda_k)} \quad (2.44)$$

where

$$*f_a(\lambda_k) = (\lambda_k + *\lambda + *k_5 + k_6)$$
$$*f_m(\lambda_k) = -(\lambda_k + *\lambda + *k_6)$$
$$*f_d(\lambda_k) = -*k_5 \qquad\qquad\qquad (2.45)$$
$$f_m(\lambda_k) = k_3(\lambda_k + k_6)$$
$$g'(\lambda_k) = \prod_{\substack{l=1 \\ l \neq k}}^{5} (\lambda_k - \lambda_l)$$

3. FOUR-RESERVOIR TANDEM MODEL WITH LAND BIOTA

The four reservoirs (see Fig. 2) will be denoted by subscripts b (land biota), a (atmosphere), m (ocean surface mixed layer), and d (deep ocean). As in Section 2, the transfer coefficients will be numbered in ascending order to take advantage of a mathematical regularity intrinsic to the tandem model. The coefficients k_1 and k_2 will represent the transfers involving the biota (the land plants and their detritus), and k_3 through k_6 will have the same meaning as in Section 2. For inactive carbon, the following conditions apply

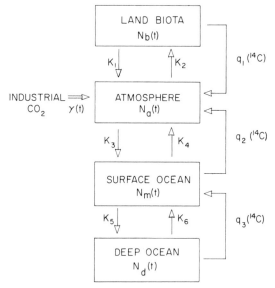

Fig. 2. Four-reservoir model of the CO_2 exchange between the atmosphere, land biota (biosphere), and oceans. The model extends the three-reservoir model of Fig. 1 by including the biota and two additional virtual radiocarbon sources q_1 and q_3. The source q_3 reflects a readjustment of the radiocarbon distribution within the oceans owing to a gravitational flux from the surface layer to the deep ocean; q_1 reflects a readjustment owing to industrial CO_2 entering the biota.

at steady state:

$$
\left.
\begin{aligned}
k_1 N_{b0} - k_2 N_{a0} \qquad\qquad\qquad &= 0 \\
-k_1 N_{b0} + (k_2 + k_3) N_{a0} - k_4' P_{m0} \qquad &= 0 \\
-k_3 N_{b0} + k_4' P_{m0} + k_5 N_{m0} - k_6 N_{d0} &= 0 \\
-k_5 N_{m0} + k_6 N_{d0} \qquad\qquad\qquad &= 0
\end{aligned}
\right\}
\tag{3.1}
$$

For radiocarbon, again denoting the decay constant by $*\lambda$,

$$
\left.
\begin{aligned}
*k_1 *N_{b0} - *k_2 *N_{a0} \qquad\qquad\qquad\qquad &= -*\lambda *N_{b0} \\
-*k_1 *N_{b0} + (*k_2 + *k_3) *N_{a0} - *k_4' *P_{m0} \qquad &= *Q - *\lambda *N_{a0} \\
-*k_3 *N_{a0} + *k_4' *P_{m0} + *k_5 *N_{m0} - *k_6 *N_{d0} &= -*\lambda *N_{m0} \\
-*k_5 *N_{m0} + *k_6 *N_{d0} \qquad\qquad\qquad\qquad &= -*\lambda *N_{d0}
\end{aligned}
\right\}
\tag{3.2}
$$

where

$$
*Q = *\lambda \sum_i *N_{i0}
\tag{3.3}
$$

For inactive carbon, deviations from steady state are governed by the equations

$$
\left.
\begin{aligned}
(d/dt + k_1) n_b - k_2 n_a \qquad\qquad &= 0 \\
-k_1 n_b + (d/dt + k_2 + k_3) n_a - k_4 n_m &= \gamma \\
-k_3 n_a + (d/dt + k_4 + k_5) n_m - k_6 n_d &= 0 \\
-k_5 n_m + (d/dt + k_6) n_d \qquad\qquad &= 0
\end{aligned}
\right\}
\tag{3.4}
$$

with the initial conditions

$$
n_b = n_a = n_m = n_d = 0 \qquad \text{at } t = 0
\tag{3.5}
$$

Again, the $n_i(t)$ denote the deviation of the N_i from steady state owing to the release into the atmosphere of industrial CO_2 at a rate $\gamma(t)$, and the flux $k_4' P_{m0}$ has been replaced, to a first-order approximation, by $k_4 n_m$.

The equations for transient response of radiocarbon to industrial CO_2 injection must reflect indirect readjustments in the radiocarbon content of the ocean and biota owing to an increase in the exchange of inactive carbon between these reservoirs and the atmosphere. This readjustment is, to a first approximation (see Section 5), linearly related to the changes in the inactive content of both the biota and the atmosphere and can be expressed as a virtual sink·

$$
-q_1 = k_9' n_b - k_{10}' n_a
\tag{3.6}
$$

for the biota, with a corresponding source q_1 for the atmosphere. For realistic (always nonnegative) values of k_9' and k_{10}', the term involving n_a will be larger than that involving n_b; hence the designation of (3.6) as a sink.

For the ocean surface layer, the adjustment owing to exchange of inactive carbon with the atmosphere is written as before:

$$-q_2 = -k'_7 n_m \tag{3.7}$$

with a corresponding source q_2 for the atmosphere. The transfer of radiocarbon between the ocean surface layer and deep ocean by gravitational settling of particles introduces still another adjustment, which can be written

$$q_3 = k'_8 n_m \tag{3.8}$$

for the ocean surface layer with a corresponding sink $-q_3$ for the deep-ocean reservoir.

Since no actual source of radiocarbon is provided by industrial CO_2, the sum of the virtual sources is zero.

We thus arrive at the following governing equations for deviations from steady state for radiocarbon:

$$
\left.
\begin{aligned}
(d/dt + {}^*\lambda + {}^*k_1){}^*n_b - {}^*k_2{}^*n_a &= k'_9 n_b - k'_{10} n_a = -q_1, \\
- {}^*k_1{}^*n_b + (d/dt + {}^*\lambda + {}^*k_2 + {}^*k_3){}^*n_a - {}^*k_4{}^*n_m & \\
= -k'_9 n_b + k'_{10} n_a + k'_7 n_m &= q_1 + q_2, \\
- {}^*k_3{}^*n_a + (d/dt + {}^*\lambda + {}^*k_4 + {}^*k_5){}^*n_m - {}^*k_6{}^*n_d & \\
= -k'_7 n_m + k'_8 n_m &= -q_2 + q_3, \\
- {}^*k_5{}^*n_m + (d/dt + {}^*\lambda + {}^*k_6){}^*n_d = -k'_8 n_m &= -q_3
\end{aligned}
\right\}
\tag{3.9}
$$

with the initial conditions

$$^*n_b = {}^*n_a = {}^*n_m = {}^*n_d = 0 \qquad \text{at } t = 0 \tag{3.10}$$

The solution given below of the sets of equations (3.4) and (3.9) with the initial conditions (3.5) and (3.10) makes use of the Laplace transform method as explained in detail for the three-reservoir model in Section 2.

Multiplying both sets of equations, (3.4) and (3.9), by e^{-st} and integrating from time 0 to ∞ results in sets of algebraic equations for the transforms $\tilde{n}_i$ and $^*\tilde{n}_i$. The coefficients of these sets form the array

$$
\mathbf{D}(U) =
\begin{pmatrix}
U + k_1 & -k_2 & 0 & 0 \\
-k_1 & U + k_2 + k_3 & -k_4 & 0 \\
0 & -k_3 & U + k_4 + k_5 & -k_6 \\
0 & 0 & -k_5 & U + k_6
\end{pmatrix}
\tag{3.11}
$$

where the determinant for inactive carbon, $D(s)$, is evaluated by setting $U = s$, and that for radiocarbon, $^*D(s)$, is evaluated by setting $U = s + {}^*\lambda$ and replacing each k_j by $^*k_j$.

The transforms are taken in the order $\{\tilde{n}_b, \tilde{n}_a, \tilde{n}_m, \tilde{n}_d\}$ and $\{*\tilde{n}_b, *\tilde{n}_a, *\tilde{n}_m, *\tilde{n}_d\}$. The transformed sources are $\{0, \tilde{\gamma}, 0, 0\}$ for inactive carbon and $\{-\tilde{q}_1, \tilde{q}_1 + \tilde{q}_2, -\tilde{q}_2 + \tilde{q}_3, -\tilde{q}_3\}$ for radiocarbon.

Expanding $D(s)$ and $*D(s)$, we obtain two quartic equations in s:

$$D(s) = s(s^3 + As^2 + Bs + C) \tag{3.12}$$

$$*D(s) = (s + *\lambda)[(s + *\lambda)^3 + *A(s + *\lambda)^2 + *B(s + *\lambda) + *C] \tag{3.13}$$

where

$$
\left.
\begin{aligned}
A &= \sum_{j=1}^{6} k_j \\[2mm]
B &= \sum_{j=1}^{4} \sum_{j'=3}^{6} k_j k_{j'}, \qquad j' > j + 1 \\[2mm]
C &= \sum_{j=1}^{2} \sum_{j'=3}^{4} \sum_{j''=5}^{6} k_j k_{j'} k_{j''}, \qquad j' > j + 1, j'' > j' + 1
\end{aligned}
\right\} \tag{3.14}
$$

with identical expressions for $*A$, $*B$, and $*C$ in terms of the $*k_j$.

The polynomials (3.12) and (3.13) are set equal to zero and solved for s to obtain the roots λ_1 to λ_4 and $*\lambda_1$ to $*\lambda_4$. The roots will be numbered in order of increasing absolute magnitude of λ_k and $*\lambda_k$. As in the three-reservoir model, $\lambda_1 = 0$ and $*\lambda_1 = \lambda$. The latter root will again produce in the $*n_i$ a term with a zero coefficient.

Solving for the $\tilde{n}_i$ and $*\tilde{n}_i$ algebraically, we obtain

$$\tilde{n}_i = \frac{\tilde{\gamma} f_i(s)}{\prod\limits_{k=1}^{4} (s - \lambda_k)} \tag{3.15}$$

$$*\tilde{n}_i = \sum_{n=1}^{3} \frac{\tilde{q}_n \, *f_{ni}(s)}{\prod\limits_{k=2}^{4} (s - *\lambda_k)} \tag{3.16}$$

The factors $\tilde{q}_n$ are obtained by transforming the q_n of (3.6)–(3.8). For example,

$$\tilde{q}_1 = -k'_9 \tilde{n}_b + k'_{10} \tilde{n}_a \tag{3.17}$$

Substituting for the $\tilde{n}_i$ by (3.15)

$$\tilde{q}_1 = \frac{\tilde{\gamma}[-k'_9 f_b(s) + k'_{10} f_a(s)]}{\prod\limits_{k=1}^{4} (s - \lambda_k)} \tag{3.18}$$

Similarly

$$\tilde{q}_2 = \frac{\tilde{\gamma} k_7' f_m(s)}{\prod_{k=1}^{4} (s - \lambda_k)} \qquad (3.19)$$

$$\tilde{q}_3 = \frac{\tilde{\gamma} k_8' f_m(s)}{\prod_{k=1}^{4} (s - \lambda_k)} \qquad (3.20)$$

Introducing the composite factor

$$*f_i(s) = [-k_9' \, f_b(s) + k_{10}' \, f_a(s)]*f_{1i}(s) + [k_7'*f_{2i}(s) + k_8'*f_{3i}(s)] \, f_m(s) \qquad (3.21)$$

(3.16) can be rewritten in the compact form

$$*\tilde{n}_i = \frac{\tilde{\gamma} \, *f_i(s)}{\prod_{k=1}^{7} (s - \lambda_k)} \qquad (3.22)$$

where

$$\left.\begin{array}{l} \lambda_5 = *\lambda_2 \\ \lambda_6 = *\lambda_3 \\ \lambda_7 = *\lambda_4 \end{array}\right\} \qquad (3.23)$$

The specific factors $f_i(s)$ and $*f_{ni}(s)$ are listed in Appendix A.

The solutions of (3.4) and (3.9) for the n_i and $*n_i$ are the inverse transforms of (3.15) and (3.22). These are readily deduced if we note, as in Section 2, that the coefficients of the solutions of (3.4) and (3.9) are given when the factors which multiply the transformed source, $\tilde{\gamma}$, are expanded as partial fractions $C_{ki}/(s - \lambda_k)$. Finally the convolution theorem is used to transform each product $[C_{ki}/(s - \lambda_k)]\tilde{\gamma}$.

Proceeding as in Section 2 for the three-reservoir model,

$$n_i = \sum_{k=1}^{4} \frac{f_i(\lambda_k)}{\prod_{\substack{l=1 \\ l \neq k}}^{4} (\lambda_k - \lambda_l)} \int_0^t e^{\lambda_k(t-u)} \gamma(u) \, du \qquad (3.24)$$

$$*n_i = \sum_{k=1}^{7} \frac{*f_i(\lambda_k)}{\prod_{\substack{l=1 \\ l \neq k}}^{7} (\lambda_k - \lambda_l)} \int_0^t e^{\lambda_k(t-u)} \gamma(u) \, du \qquad (3.25)$$

where $f_i(\lambda_k)$ and $*f_i(\lambda_k)$ are evaluated for each of the roots by substituting λ_k for s in (3.21) and (A.1)–(A.4).

To solve numerically the integral

$$J(t) = \int_0^t e^{\lambda_k(t-u)}\gamma(u)\,du \qquad (3.26)$$

we approximate $\gamma(t)$ by a step function based on successive increments of industrial production. Let γ_j refer to the average of $\gamma(t)$ between times t_j and t_{j+1}. (The symbol j here is not related to its earlier use to subscript a transfer coefficient.) Then, for the integration from t_0 ($=0$) to t_n,

$$
\left.
\begin{aligned}
J(t_n) &= \sum_{j=0}^{n-1} \int_{t_j}^{t_{j+1}} e^{\lambda_k(t_n-u)}\gamma_j\,du \\
&\simeq \sum_{j=0}^{n-1} \frac{\gamma_j}{-\lambda_k}\left(e^{\lambda_k(t_n-t_{j+1})} - e^{\lambda_k(t_n-t_j)}\right) \\
&= \sum_{j=0}^{n-1} \frac{\gamma_j}{-\lambda_k}\left[e^{\lambda_k(t_n-t_{j+1})}\left(1 - e^{\lambda_k(t_{j+1}-t_j)}\right)\right]
\end{aligned}
\right\} \qquad (3.27)
$$

If we evaluate for constant time intervals ε,

$$
\left.
\begin{aligned}
t_{j+1} - t_j &= \varepsilon \\
t_n - t_{j+1} &= (n-j-1)\varepsilon
\end{aligned}
\right\} \qquad (3.28)
$$

$$
\left.
\begin{aligned}
J_\varepsilon(n) &= \frac{1-e^{\varepsilon\lambda_k}}{-\lambda_k}\sum_{j=0}^{n-1} e^{(n-j-1)\varepsilon\lambda_k}\gamma_j \\
&= \frac{1-e^{\varepsilon\lambda_k}}{-\lambda_k}\underset{n}{[}\ \underset{n-1}{[}\ \cdots\underset{2}{[}\ \underset{1}{[}\gamma_0]e^{\varepsilon\lambda_k} + \gamma_1\underset{1}{]}\cdots\ \underset{2}{]}\ e^{\varepsilon\lambda_k} + \gamma_{n-1}\underset{n}{]}
\end{aligned}
\right\} \qquad (3.29)
$$

The second form of (3.29) suggests a convenient stepwise method to evaluate $J_\varepsilon(n)$:

$$J_\varepsilon(1) = \frac{1-e^{\varepsilon\lambda_k}}{-\lambda_k}\gamma_0$$

$$J_\varepsilon(2) = J_\varepsilon(1)e^{\varepsilon\lambda_k} + \frac{1-e^{\varepsilon\lambda_k}}{-\lambda_k}\gamma_1 \qquad (3.30)$$

$$J_\varepsilon(n) = J_\varepsilon(n-1)e^{\varepsilon\lambda_k} + \frac{1-e^{\varepsilon\lambda_k}}{-\lambda_k}\gamma_{n-1}$$

4. FIVE-RESERVOIR BRANCHED MODEL WITH DIVIDED LAND BIOTA

As discussed in Section 1, the response of the land biota to CO_2 uptake can be better represented by postulating two biota reservoirs rather than one. In this way, short- and long-cycling carbon can be distinguished by

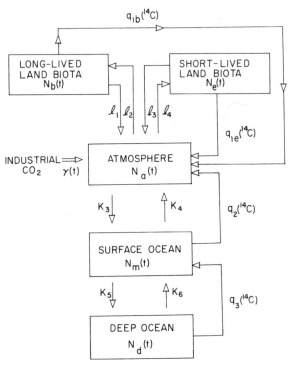

Fig. 3. Five-reservoir model of CO_2 exchange equivalent to the four-reservoir model of Fig. 2 except that the land biota has been split into two subreservoirs. The symbols are explained in the text.

means of independently chosen transfer coefficients. A model with this provision (see Fig. 3) will now be developed.

So that the governing equations conform as closely as possible with the four-reservoir model already discussed, the former transfer coefficients are retained for the atmosphere–ocean exchanges, and those for exchange with the land biota are replaced by pairs: l_1 and l_3 for k_1, and l_2 and l_4 for k_2. The symbol N_b is retained to denote the long-lived biota and N_e is introduced for the short-lived biota. (The latter reservoir I have informally dubbed the "ephemesphere" from $\varepsilon\phi\eta\mu\varepsilon\rho\iota\sigma$, meaning "short-lived.") The first virtual radiocarbon source q_1 of the earlier model is split into two virtual sources denoted q_{1b} and q_{1e}, and the virtual-source coefficients k_9' and k_{10}' are replaced by pairs l_7', l_9' and l_8', l_{10}', respectively. These modifications alter the general properties of the four-reservoir model so slightly that it is hardly required to describe again the full setup of the governing equations and their method of solution. The derivation of the analytic functions is more tedious owing to their arising from fourth- rather than third-order determinantal cofactors, but no basically new problems arise. The reader is referred to Appendix B for the details.

PART II

CHEMICAL SPECIFICATION
AND NUMERICAL RESULTS

PREFACE

In Part I of this two-part survey of the carbon dioxide cycle in nature, mathematical models were developed to represent the exchange of CO_2 between the atmosphere, oceans, and land plants. The governing model equations were linearized in terms of transfer coefficients which multiplied the average mass of carbon in each reservoir. The significance of these coefficients was touched upon, but their relations to observable geochemical quantities were not developed for practical calculations. Such relations will now be worked out explicitly. Afterwards, in the final two sections below, summaries of carbon data will be made the basis for numerical solutions of the three- and five-reservoir models.

5. DERIVATION OF TRANSFER COEFFICIENTS FOR THE LAND BIOTA

Because of the mathematical advantage in representing the transfer of carbon between reservoirs as being proportional to the first power of a single concentration or mass variable, it is worthwhile to establish at the outset the limitation on prediction which this linearization of the problem poses to modeling the response of the biota to an increase in atmospheric CO_2.

Let F_{ij0} denote the steady-state (preindustrial) flux of carbon from reservoir i to reservoir j, where one of the reservoirs is the atmosphere and the other is a portion of the land biota. If the amounts of carbon in the emitting and receiving reservoirs change from steady-state values N_{i0} and N_{j0}, respectively, to new values N_i and N_j owing to addition of industrial CO_2, the carbon flux F_{ij0} will change to a new value, say $F_{ij0} + \Delta F_{ij}$. This change can be expressed, at least for small variations from steady state, by writing for the steady-state flux the function

$$F_{ij0} = k_{ij} N_{i0}^{\beta_i} N_{j0}^{\beta_j} \qquad (5.1)$$

and for the general (time-dependent) flux

$$F_{ij} = F_{ij0} + \Delta F_{ij} = k_{ij} N_i^{\beta_i} N_j^{\beta_j} = F_{ij0}(N_i/N_{i0})^{\beta_i}(N_j/N_{j0})^{\beta_j} \qquad (5.2)$$

where k_{ij}, β_i, and β_j are constants. Expanding in a Taylor series (Sokolnikoff and Redheffer [14], p. 36) with $N_i - N_{i0} = n_i$,

$$\Delta F_{ij} = F_{ij0}\left\{ \beta_i \frac{n_i}{N_{i0}} + \beta_j \frac{n_j}{N_{j0}} + \frac{\beta_i(\beta_i - 1)}{2}\left(\frac{n_i}{N_{i0}}\right)^2 \right.$$

$$\left. + \frac{\beta_j(\beta_j - 1)}{2}\left(\frac{n_j}{N_{j0}}\right)^2 + \frac{\beta_i \beta_j n_i n_j}{N_{i0} N_{j0}} + \cdots \right\} \qquad (5.3)$$

According to studies on individual plants grown in air with near normal CO_2 concentrations, the uptake of CO_2 by photosynthesis roughly obeys the relation

$$F_{ab} = F_{b0} + \Delta F_{ab} = F_{b0}(N_a/N_{a0})^{1/2}(N_b/N_{b0}) \qquad (5.4)$$

where a and b refer to the atmosphere and biota reservoirs, respectively, and F_{b0} is written in place of F_{ab0} to simplify the symbolism. The return flux to the atmosphere, F_{ba}, is so nearly proportional to the first power of the mass of the biota that it need not be approximated by a Taylor expansion. Direct atmospheric measurements indicate that in 1970 the accumulated increase in atmospheric CO_2 owing to a total industrial CO_2 input of about 18% was roughly half of the input [15], i.e.,

$$\frac{n_a}{N_{a0}} \simeq 0.09 \qquad (5.5)$$

A substantial part of the industrial CO_2 has without doubt been absorbed by the oceans, so that the relative growth of the biota as a whole can hardly have exceeded

$$\frac{n_b}{N_{b0}} = 0.5\frac{n_a}{N_{a0}} \qquad (5.6)$$

and may be considerably less. A similar figure is also likely to be appropriate to any major portion of the biota. From these considerations it follows that the ratio of the sum of the second-order terms to the sum of the first-order terms in (5.3) (with $i = a$ and $j = b$) cannot be greater than about 2%, while the third-order terms are entirely negligible. Our present knowledge of the global average photosynthetic flux is less precise than 2%. Accounting for second-order terms could not greatly improve the calculation of past changes in the CO_2 system. Clearly, however, if the assimilation of the biota approximately obeys equation (5.1) or some other nonlinear function which agrees with (5.4) only to first order, we cannot expect to produce realistic predictions for very many years into the future on the basis of a simple linear approximation. For example, according to several studies in glass houses [16-19], the assimilation by plants at very high CO_2 concentrations is roughly proportional to the logarithm of the ambient CO_2 concentration. Thus we might replace (5.2), as it applies to the flux from the atmosphere to the land biota, by an expression of the form

$$F_{ab} = F_{b0}[1 + \beta_a \ln(N_a/N_{a0})](N_b/N_{b0})^{\beta_b} \qquad (5.7)$$

where the last factor allows, as before, for a rough proportionality with the size of the biota N_b. Since the first-order terms in the Taylor expansion of (5.7) are the same as for (5.2), the first-order expansion of (5.2) is still appropriate for studying past variations, but again is not a good approximation for long-range predictions.

Actually, no direct evidence exists that the biota as a whole has increased in size since the industrial era or that it can grow indefinitely in size according to a relation similar to (5.2) or (5.7). For at least some plant communities, growth is limited by nutrient and water supply or unfavorable temperature or light conditions. Assimilation responds very slightly, if at all, to increases in atmospheric CO_2. Owing to agriculture, considerable amounts of CO_2 have probably been released from former virgin lands recently developed to harvest crops.

In any case, because direct observational data are few and scattered, the global average response of the biota can hardly be established reliably by any means independent of the existing record of the atmospheric CO_2 coupled with estimates of ocean exchange such as the present paper considers. We are thus practically obliged to consider the rate of increase of the biota as an unknown and to use the results of the model to set limits on possible past changes in biota size.

With regard to radiocarbon, we are principally concerned with the overall storage capacity of the biota and the rate at which carbon passes through the reservoir. Small changes in size, however uncertain, will not seriously influence the model calculations of the Suess effect.

With these considerations in mind, expressions will now be derived for the transfer of carbon between the atmosphere and land biota in terms of observable fluxes and masses. The discussion will consider only one biota reservoir (designated by subscript b), but the derived mathematical relationships can be readily extended later to a two-reservoir model which distinguishes short- and long-lived components (see Section 4).

Ideally, observations used to define the initial conditions of the model would have been made at steady state. Actually, man's agricultural activities have disturbed the natural biota in limited areas for millenia, while in recent years profound changes in plant communities have occurred over most of the world. Today the earth's remaining virgin lands are being more rapidly brought under exploitation than at any previous time in history. The majority of observations of land plant fluxes and masses have been made on these recently exploited communities, so that the relatively few existing observations of virgin lands must be awarded disproportionately large significance when world totals are assembled. Within these limitations, I will accept as known quantities the steady-state flux F_{b0}, of CO_2 assimilated by land plants and the mass of carbon, N_{b0}, associated with these plants and their dead remains. The amount of carbon in the form of atmospheric CO_2, denoted by N_{a0}, is roughly known for the late 19th century (see Section 8).

The isotopic fractionation factors for assimilation, α_{ab}, and respiration, α_{ba}, of radiocarbon are also known. These factors are close to unity and can be evaluated from laboratory and field measurements of the related rare stable isotope ^{13}C with more than the accuracy required for modeling ^{14}C. The ^{14}C factors are almost exactly the square of those for ^{13}C.

Seasonal variations in flux will not be considered because their time scale is short compared to the several decades of industrial CO_2 input.

In general, then, uptake and release of inactive carbon will be assumed to be given by the expressions

$$F_{ab} = F_{b0}(N_a/N_{a0})^{\beta_a}(N_b/N_{b0})^{\beta'_b} \qquad (5.8)$$

$$F_{ba} = F_{b0}(N_b/N_{b0})^{\beta_b} \qquad (5.9)$$

(A prime has been placed on the second exponent of (5.8) to distinguish it from the exponent which appears in (5.9).)

Since both assimilation and respiration depend more or less directly on the size of the biota, both β_b and β'_b are approximately equal to unity. The exponent β_a is highly uncertain, however, and will be considered as an adjustable parameter. For the sake of generality we shall carry along all three exponents in the development below.

For radiocarbon, the transfer rates vary as the isotopic ratio of the rare isotope varies in the source material. Their magnitudes are further modified owing to isotopic fractionation. Hence, in general,

$$\text{uptake:} \quad *F_{ab} = \alpha_{ab}(*N_a/N_a)F_{ab} \qquad (5.10)$$

$$\text{release:} \quad *F_{ba} = \alpha_{ba}(*N_b/N_b)F_{ba} \qquad (5.11)$$

At steady state, for inactive carbon, the flux in either direction is F_{b0}, while for radiocarbon,

$$\underset{\text{uptake}}{\alpha_{ab}(*N_{a0}/N_{a0})F_{b0}} = \underset{\text{release}}{\alpha_{ba}(*N_{b0}/N_{b0})F_{b0}} + \underset{\text{decay}}{*\lambda*N_{b0}} \qquad (5.12)$$

where the term $*\lambda*N_{b0}$ accounts for radioactive decay within the biota.

Solving (5.12) for $*N_{b0}/N_{b0}$, and writing R_{i0} for the ^{14}C/inactive C ratio $*N_{i0}/N_{i0}$, we obtain

$$R_{b0} = \frac{\alpha_{ab}}{\alpha_{ba}} \cdot \frac{R_{a0}}{1 + *\lambda N_{b0}/(\alpha_{ba}F_{b0})} \qquad (5.13)$$

Reliable data on R_{b0}, the average preindustrial radioisotopic ratio of carbon in land plants, are unavailable and R_{b0} cannot be estimated now because of the very large unnatural increase in radiocarbon from nuclear weapons testing since 1953. We are thus obliged to use (5.13) to obtain a value for R_{b0} instead of being able to use observational data to solve for F_{b0} or check on the validity of the transfer equations (5.8) and (5.9).

For the transient case, the general equations are expanded according to (5.3). Retaining only the first-order terms, we obtain for inactive carbon

$$\Delta F_{ab} = F_{b0}\left\{\frac{\beta_a n_a}{N_{a0}} + \frac{\beta'_b n_b}{N_{b0}}\right\} \qquad (5.14)$$

$$\Delta F_{ba} = F_{b0}\left\{\frac{\beta_b n_b}{N_{b0}}\right\} \qquad (5.15)$$

For radiocarbon, we first substitute explicitly for F_{ab} and F_{ba} in (5.10) and (5.11) according to (5.8) and (5.9). Then expanding about the steady-state values, retaining only first-order terms,

$$\Delta^* F_{ab} = \alpha_{ab} R_{a0} F_{b0} \left\{ \frac{^*n_a}{^*N_{a0}} + (\beta_a - 1)\frac{n_a}{N_{a0}} + \beta'_b \frac{n_b}{N_{b0}} \right\} \tag{5.16}$$

$$\Delta^* F_{ba} = \alpha_{ba} R_{b0} F_{b0} \left\{ \frac{^*n_b}{^*N_{b0}} + (\beta_b - 1)\frac{n_b}{N_{b0}} \right\} \tag{5.17}$$

For inactive carbon, the difference between biota uptake and release is equal to the rate of growth in biota size, i.e.,

$$dn_b/dt = \Delta F_{ab} - \Delta F_{ba} \tag{5.18}$$

Substituting from (5.14) and (5.15) and collecting terms in n_a and n_b, we obtain

$$\frac{dn_b}{dt} = \frac{F_{b0}}{N_{a0}} \beta_a n_a - \frac{F_{b0}}{N_{b0}} (\beta_b - \beta'_b) n_b \tag{5.19}$$

The factors multiplying n_a and n_b can now be identified with the transfer coefficients in the corresponding reservoir equation in (3.4)

$$k_1 = \frac{F_{b0}}{N_{b0}} (\beta_b - \beta'_b) \tag{5.20}$$

$$k_2 = \frac{F_{b0}}{N_{a0}} \beta_a \tag{5.21}$$

Similarly for radiocarbon,

$$(d/dt + {}^*\lambda)^* n_b = \Delta^* F_{ab} - \Delta^* F_{ba} \tag{5.22}$$

whence

$$(d/dt + {}^*\lambda)^* n_b = \frac{\alpha_{ab} F_{b0}}{N_{a0}} {}^*n_a + \frac{\alpha_{ab} R_{a0} F_{b0}}{N_{a0}} (\beta_a - 1) n_a - \frac{\alpha_{ba} F_{b0}}{N_{b0}} {}^*n_b$$

$$- \frac{\alpha_{ba} R_{b0} F_{b0}}{N_{b0}} (\beta_b - 1) n_b + \frac{\alpha_{ab} R_{a0} F_{b0}}{N_{b0}} \beta'_b n_b \tag{5.23}$$

The corresponding equation [cf. (3.9)] in terms of the k_i's is

$$(d/dt + {}^*\lambda)^* n_b = -{}^*k_1 {}^*n_b + {}^*k_2 {}^*n_a + k'_9 n_b - k'_{10} n_a \tag{5.24}$$

whence

$$^*k_1 = \frac{\alpha_{ba} F_{b0}}{N_{b0}} \tag{5.25}$$

$$^*k_2 = \frac{\alpha_{ab} F_{b0}}{N_{a0}} \tag{5.26}$$

$$k'_9 = \frac{\alpha_{ab} R_{a0} F_{b0}}{N_{b0}} \left\{ \beta'_b + \frac{1 - \beta_b}{1 + {}^*\lambda N_{b0}/(\alpha_{ba} F_{b0})} \right\} \tag{5.27}$$

$$k'_{10} = \frac{\alpha_{ab} R_{a0} F_{b0}}{N_{a0}} (1 - \beta_a) \tag{5.28}$$

In the expression for the virtual source coefficient k'_9, R_{b0} is replaced according to (5.13). The isotopic ratio R_{a0}, which occurs as a factor in (5.27) and (5.28), will not actually need to be evaluated because the final solution of the reservoir equations will be expressed in terms of dimensionless ratios n_i/N_{a0} and ${}^*n_i/{}^*N_{a0}$ ($= {}^*n_i/R_{a0} N_{a0}$). Thus, excepting the β's, only observable quantities appear in the final expressions for the reservoir transfer coefficients.

6. DERIVATION OF TRANSFER COEFFICIENTS FOR AIR–SEA EXCHANGE

6.1. Influence of the Buffer Factor

Of the carbon-bearing gases which occur in the atmosphere, only CO_2 is in high enough concentration to affect the total transfer of carbon between the air and the sea significantly. Because the atmosphere is relatively well mixed and the transfer of CO_2 across the air–sea boundary is proportional to CO_2 partial pressure, it follows that transfer of CO_2 from the atmosphere to the oceans is very nearly proportional to the total mass of carbon in the air as prescribed by the reservoir model via Eqs. (2.1), (2.3), (3.1), and (3.4). This is not the case for the oceans, however, because these contain, besides CO_2, other carbon-bearing compounds which strongly influence the CO_2 partial pressure and thus the CO_2 transfer. Before the reservoir models can be solved, it is therefore necessary to express, for surface ocean water, the relation between CO_2 partial pressure, P_m, and total mass of *inorganic* carbon in surface sea water, $N_{\tilde{m}}$. This is equivalent to evaluating an instantaneous sea surface buffer factor ζ defined by

$$\frac{\delta P_m}{P_{m0}} = \zeta \frac{\delta N_{\tilde{m}}}{N_{\tilde{m}0}}$$

$$= \zeta \frac{\Sigma C}{\Sigma C_0} \tag{6.1}$$

where δ denotes the infinitesimal change in pressure or concentration owing to an infinitesimal transfer of CO_2 across the air–sea boundary, and in the second form of (6.1) ΣC denotes the concentration of all inorganic carbon in the water. Preindustrial values are denoted by P_{m0}, $N_{\tilde{m}0}$, and ΣC_0. The factor ζ for a given sea water is derivable from laboratory experiments. It varies with ΣC, so that a nonlinear term occurs in the governing reservoir model equations for air–sea exchange. Since this term produces mathematical complications, we now will investigate the errors produced in model calculations if the buffer factor is assumed to be constant.

To determine ζ for sea water of any given initial composition, we need specifically to consider dissolved CO_2 gas and its principal dissociation products. These are related by the chemical reaction equations

$$CO_2 + H_2O \rightleftarrows H^+ + HCO_3^- \tag{6.2}$$

$$HCO_3^- \rightleftarrows H^+ + CO_3^= \tag{6.3}$$

If CO_2 is added or removed from sea water, the amount in solution almost instantaneously readjusts with the bicarbonate, HCO_3^-, and carbonate, $CO_3^=$, ions to preserve constant the ratios

$$K_1 = \frac{[H^+][HCO_3^-]}{[CO_2]} \tag{6.4}$$

$$K_2 = \frac{[H^+][CO_3^=]}{[HCO_3^-]} \tag{6.5}$$

where brackets denote concentrations in any convenient units, say, moles per liter of sea water; $[H^+]$ stands for the concentration of the hydrogen ion.

In addition, the partial pressure, denoted in general by p_{CO_2}, remains proportional to the concentration of dissolved CO_2 so that

$$K_0 = \frac{[CO_2]}{p_{CO_2}} \tag{6.6}$$

where K_0 is the gas solubility coefficient.

The quantities K_0, K_1, and K_2 are *stoichiometric* quotients which are invariant to CO_2 exchange at the sea surface, but which vary with temperature, T, and the total salt content (salinity) of sea water, S. Because the equations involving these factors will be put to use only in connection with CO_2 exchange over a moderate range in values of $[H^+]$, it will not be necessary to consider other dissolved inorganic species such as H_2CO_3, or ion pairs of the dissociation products of H_2CO_3 with sea salts such as $NaCO_3^-$ or $MgHCO_3^+$ ([20], p. 167). To avoid specifying these and yet allow for their influence on the conservation of mass, $[CO_2]$ will be understood to represent the concentration of H_2CO_3 as well as CO_2, and $[HCO_3^-]$ and $[CO_3^=]$ will include the contribution of all ion pairs of HCO_3^- and $CO_3^=$, respectively. The total concentration of inorganic carbon in sea water is then the sum

$$\Sigma C = [CO_2] + [HCO_3^-] + [CO_3^=] \tag{6.7}$$

Using (6.4), (6.5), and (6.6) to eliminate $[HCO_3^-]$ and $[CO_3^=]$ from (6.7), we obtain the relation

$$p_{CO_2} = \Phi(\Sigma C) \tag{6.8}$$

where

$$\Phi = \left(K_0 + \frac{K_0 K_1}{[H^+]} + \frac{K_0 K_1 K_2}{[H^+]^2} \right)^{-1} \tag{6.9}$$

Before (6.8) can be used to evaluate the buffer factor ζ, the hydrogen ion concentration $[H^+]$ must be determined. The latter is not an independent variable, because the sum of the electrical charges of the ions of sea water must be zero regardless of what changes occur in individual ion concentrations. The detailed equation which describes this conservation of charge involves numerous ions, such as Na^+ and Cl^-, whose concentrations in surface sea water will not vary owing to changes in carbon species. Not even all of the chemical species which react with hydrogen ion need be taken into account. Several are in too low concentration to be significant (e.g., phosphate). Several others (e.g., silicate, sulfate, and fluoride) remain almost completely dissociated from or associated with hydrogen ion over the range in $[H^+]$ which we need consider. The borate ion, however, cannot be neglected, since its affinity for hydrogen ion is close to that of the carbonate ion, and its concentration not greatly less. Thus, the conservation of charge can be written

$$A + [H^+] = [HCO_3{}^-] + 2[CO_3{}^=] + [H_2BO_3{}^-] + [OH^-] \tag{6.10}$$

where A stands for the net charge of all the ions (both positive and negative) which do not vary significantly with changes in ΣC or p_{CO_2} in the range of interest. Because the quantity A reduces to zero if acid is added to sea water just sufficient to neutralize all the ions on the right side of (6.10), it is often called the "titration alkalinity." It can be experimentally determined to a high accuracy and thus permits an indirect evaluation of $[H^+]$.

The last two terms in (6.10) take part in equilibria governed by the stoichiometric quotients

$$K_B = \frac{[H^+][H_2BO_3{}^-]}{[H_3BO_3]} \tag{6.11}$$

$$K_W = \frac{[H^+][OH^-]}{[H_2O]} \quad \text{where by convention } [H_2O] = 1 \tag{6.12}$$

With the aid of (6.4), (6.5), (6.11), (6.12), and an expression for the sum of the dissociated and undissociated borate,

$$\Sigma B = [H_2BO_3{}^-] + [H_3BO_3] \tag{6.13}$$

we obtain

$$A = \left(\frac{K_0 K_1}{[H^+]} + \frac{2K_0 K_1 K_2}{[H^+]^2} \right) p_{CO_2} + \frac{(\Sigma B) K_B}{[H^+] + K_B} + \frac{K_W}{[H^+]} - [H^+] \tag{6.14}$$

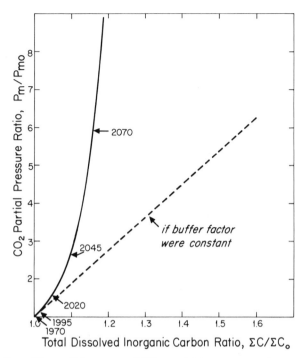

Fig. 4. Variation of the partial pressure of CO_2 exerted by average surface ocean water, P_m, with the total dissolved inorganic carbon, ΣC, relative to initial (preindustrial) values (denoted by P_{m0} and ΣC_0). The relation was derived using the constants of Buch as described in Section 8.1. The curve near 1970 (denoted by an arrow) is shown in greater detail in Fig. 5. The time arrows are suggested by the model of Section 9.3.

In this expression $[H^+]$ is the only unknown quantity for a given value of p_{CO_2}. Solving (6.14) for $[H^+]$ in terms of p_{CO_2}, we can afterwards solve (6.8) for ΣC and hence find the factor ζ. A plot of p_{CO_2} *versus* ΣC (Fig. 4) has been prepared with initial values appropriate for average surface sea water before the industrial era, as estimated in Section 8.1. For relative changes in p_{CO_2} up to 7% (approximately the increase from 1700 to 1970), the curve deviates only slightly from a straight line with a constant slope near 9 (Fig. 5). This average factor agrees closely with the related factor $d \ln p_{CO_2}/d \ln \Sigma C$ evaluated by Broecker *et al.* [3] because up to 1970 P_m and ΣC differ only slightly from P_{m0} and ΣC_0. Bolin and Eriksson [2] obtained a higher factor of 12.5 because they neglected the influence of borate. Neither group of authors discussed the variability of ζ with changing ΣC or referred their factor to preindustrial conditions. Uncertainty in the computations of ζ owing to imprecision in K_1 and K_2 is discussed in Sections 8.1 and 9.3. The only influence of organic carbon on ζ is to increase the total carbon relative to the fractional increase, i.e., $\delta N_m/N_m$ is slightly smaller than $\delta N_{\tilde{m}}/N_{\tilde{m}}$. This effect

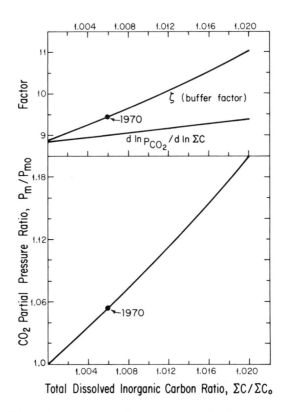

Total Dissolved Inorganic Carbon Ratio, $\Sigma C / \Sigma C_o$

Fig. 5. Buffer conditions of average ocean surface water appropriate between 1700 and A.D. 2000. All curves are plotted *versus* the relative increase in total dissolved inorganic carbon $\Sigma C/\Sigma C_0$. *Upper curve*: Plot of ζ, the instantaneous increase in CO_2 partial pressure per unit increase in ΣC relative to initial (preindustrial) values. This curve illustrates the gradually decreasing capability of the oceans to absorb additional CO_2. *Middle curve*: Same ratio as upper curve except relative to contemporary values. This latter ratio was used by previous investigators in place of ζ. *Lower curve*: Plot of Fig. 4 on an expanded scale.

is allowed for in the derivation of the transfer coefficients in Section 6.3 below.

6.2. Steady-State Exchange

The physical state of the sea surface is regarded by most investigators as principally controlling the rates of CO_2 transfer at the air–sea boundary. Some laboratory studies suggest a relation to turbulence (see, e.g., Riley and Skirrow [20], p. 314), but detailed knowledge of the mechanism of transfer is lacking, and we cannot rule out the possibility that above a (conceivably even modest) threshold of turbulence, the transfer rate approaches a limit owing to the onset of predominantly molecular scale factors. In any case, major changes in sea state with resultant changes in the global average transfer rate are not likely to have occurred during the industrial era, because, on a global scale, factors such as weather and sea temperature have not changed very much during this period. We therefore invoke here the probably justifiable assumption that the CO_2 transfer rate has varied only because of changes in partial pressure in the air and sea.

Let F_{am} and F_{ma} denote the time-dependent flux of inactive CO_2 from the air to the sea and the reverse. Consistent with the foregoing discussion,

$$F_{am} = k_{am} N_a \tag{6.15}$$

$$F_{ma} = k'_{ma} P_m \tag{6.16}$$

where k_{am} and k'_{ma} are time-independent transfer coefficients and P_m denotes the time-dependent partial pressure of CO_2 in the surface layer.

The transfers of radiocarbon will obey similar expressions except that the coefficients will, owing to isotopic fractionation, differ by a few percent from those of inactive carbon. Using asterisks to denote radiocarbon,

$$*F_{am} = \alpha_{am} k_{am} *N_a \tag{6.17}$$

$$*F_{ma} = \alpha'_{ma} k'_{ma} *P_m \tag{6.18}$$

where α_{am} and α'_{ma} denote the fractional departures in transfer rates from the values for inactive carbon. At steady state, for inactive carbon

$$k_{am} N_{a0} = k'_{ma} P_{m0} \tag{6.19}$$

while for radiocarbon

$$\alpha_{am} k_{am} *N_{a0} = \alpha'_{ma} k'_{ma} *P_{m0} + *\lambda(*N_{m0} + *N_{d0}) \tag{6.20}$$

where the term $*\lambda(*N_{m0} + *N_{d0})$ accounts for radiocarbon which, after transfer to the oceans, decays radioactively and therefore does not return to the atmosphere. To offset this decay, a net transfer of radiocarbon to the ocean occurs. The partial pressure of $^{14}CO_2$ at steady state is thus lower in the ocean water phase, even though the partial pressures of inactive CO_2 are equal in both phases. The observational data available to estimate the transfer rates are ^{14}C/inactive C ratios

$$R_i = *N_i/N_i \tag{6.21}$$

and the total amount of inactive carbon in the sea ($N_m + N_d$). These are actually known only after considerable industrial CO_2 had entered the oceans. The steady-state values R_{i0} and ($N_{m0} + N_{d0}$) thus can be obtained only approximately, by anticipating the results of reservoir model calculations. The values so obtained may, however, be checked afterwards for consistency. With the help of the fractionation factors as determined in the laboratory, and having specified the relative sizes of the separate reservoirs and hence the relative mass of carbon assigned to N_{m0}, we can in principle determine the forward transfer coefficient k_{am}. The reverse coefficient is a dependent variable which afterwards can also be calculated.

The chemical relationships derived in Section 6.1 are needed now to reexpress the transfer out of the ocean surface layer in terms of the total inorganic CO_2, distinguished from total carbon (including organic), as in Section 6.1, by the symbol N_i (initial value N_{i0}, ^{14}C/inactive C ratio R_i, etc.).

To proceed, we rearrange (6.20) with the decay term on the left side, thus:

$$\alpha_{am} k_{am} {}^*N_{a0} - {}^*\lambda({}^*N_{m0} + {}^*N_{d0}) = \alpha'_{ma} k'_{ma} {}^*P_{m0} \tag{6.22}$$

Dividing (6.22) by (6.19) and using (6.21),

$$\alpha_{am} R_{a0} - \frac{{}^*\lambda}{k_{am}} \left(R_{m0} \frac{N_{m0}}{N_{a0}} + R_{d0} \frac{N_{d0}}{N_{a0}} \right) = \alpha'_{ma} \frac{{}^*P_{m0}}{P_{m0}} \tag{6.23}$$

Total dissolved radiocarbon $\Sigma^{14}C$ varies with the partial pressure of dissolved $^{14}CO_2$ gas according to a relation similar to (6.8), which we can write

$$p_{14_{CO_2}} = {}^*\Phi\, (\Sigma^{14}C) \tag{6.24}$$

where

$$^*\Phi = \left({}^*K_0 + \frac{{}^*K_0 {}^*K_1}{[H^+]} + \frac{{}^*K_0 {}^*K_1 {}^*K_2}{[H^+]^2} \right)^{-1} \tag{6.25}$$

The apparent constants $^*K_0$, $^*K_1$, and $^*K_2$ differ from their equivalents for inactive carbon by no more than a few percent.

Writing the stoichiometric relations specifically for the ocean surface layer at steady state, we have

$$P_{m0} = \Phi_0 N_{\tilde{m}0}/W_m \tag{6.26}$$

$$^*P_{m0} = {}^*\Phi_0 {}^*N_{\tilde{m}0}/W_m \tag{6.27}$$

where W_m denotes the volume of the ocean surface layer.

Next, the factor

$$\alpha_{ma} = ({}^*\Phi_0/\Phi_0)\alpha'_{ma} \tag{6.28}$$

is introduced to denote the overall radiosotopic fractionation between CO_2 in the air and total dissolved CO_2 in surface water at steady state. This factor includes the boundary kinetic factor α'_{ma} in series with an equilibrium factor $^*\Phi_0/\Phi_0$.

Eliminating α'_{ma} between (6.23) and (6.28), we obtain

$$\alpha_{am} R_{a0} - \frac{{}^*\lambda}{k_{am}} \left(R_{m0} \frac{N_{m0}}{N_{a0}} + R_{d0} \frac{N_{d0}}{N_{a0}} \right) = \alpha_{ma} \frac{\Phi_0}{{}^*\Phi_0} \frac{{}^*P_{m0}}{P_{m0}} \tag{6.29}$$

$$= \alpha_{ma} R_{m0} \tag{6.30}$$

where the second equality makes use of (6.21), (6.26), and (6.27), and where the negligible difference between $R_{\tilde{m}0}$ and R_{m0} is disregarded.

If the preindustrial ^{14}C/inactive C ratios R_{m0} and R_{d0} were known, it would be possible, from this expression, to calculate k_{am} without it being required to specify exactly how the ocean is to be divided into surface and subsurface reservoirs. This is because only the integrated radiocarbon $(R_{m0}N_{m0} + R_{d0}N_{d0})$ and the isotopic gradient at the boundary (corrected for fractionation) are critical to the evaluation. The preindustrial ratio R_{m0} was never measured, however, and we proceed more logically is we use trial values of k_{am}, first to obtain R_{m0} via (6.30) and then to obtain the time-dependent value R_m from our model. The latter can then be compared with the few direct observations which preceded nuclear weapons testing in 1954. If this approach is adopted, the choice of size of the ocean surface layer is, however, more important, since the predicted variations in R_m depend strongly on the assumed volume of the surface layer. Nevertheless, for models which predict a reasonable value for the atmospheric Suess effect, the range in values of R_m will turn out to be less than the uncertainty in the measured isotopic ratio.

The radiocarbon content of deep water is not yet appreciably influenced by industrial CO_2 so that R_{d0} could be estimated by assuming equality with recent observations of R_d. It will be more convenient, however, to specify k_{dm} and compare calculated values of R_d with the observations. To proceed, we first derive from (6.30) the expression

$$\frac{R_{m0}}{R_{a0}} = \frac{\alpha_{am}}{\alpha_{ma} + (*\lambda/k_{am})[N_{m0}/N_{a0} + (R_{d0}/R_{m0})(N_{d0}/N_{a0})]} \tag{6.31}$$

To evaluate the ratio R_{d0}/R_{m0}, we anticipate from (7·12) in Section 7 the relation

$$\frac{R_{d0}}{R_{m0}} = \frac{k_{dm}\alpha_{dm}}{*\lambda + k_{dm}} \tag{6.32}$$

Excepting k_{am} and k_{dm}, all the quantities on the right side of these two expressions can be estimated from observations. Hence R_{m0}/R_{a0} and R_{d0}/R_{a0} can be evaluated. As pointed out in Section 5 after Eq. (5.27), it is not necessary to evaluate R_{a0} separately: all relevant equations involve ratios of the form R_{l0}/R_{u0}

6.3. Transient Exchange

The equations for variations from steady state can be derived directly from (6.15)–(6.20).

For inactive carbon, with Δ to denote departures of a flux or partial pressure from steady state,

$$\Delta F_{am} = k_{am}n_a \tag{6.33}$$

$$\Delta F_{ma} = k'_{ma} \Delta P_m$$

$$= k_{am} \frac{N_{a0}}{P_{m0}} \Delta P_m \tag{6.34}$$

For radiocarbon,

$$\Delta^* F_{am} = \alpha_{am} k_{am} {}^* n_a \tag{6.35}$$

$$\Delta^* F_{ma} = \alpha'_{ma} k'_{ma} \Delta^* P_m$$

$$= \alpha_{ma} \frac{\Phi_0}{{}^*\Phi_0} k_{am} \frac{N_{a0}}{P_{m0}} \Delta^* P_m \tag{6.36}$$

where the last expression makes use of (6.23) and (6.28).

To replace the variable quotients $\Delta P_m/P_{m0}$ and $\Delta^* P_m/P_{m0}$ with expressions involving total inorganic carbon will require that time-dependent factors be introduced. These must be approximated by constants before the linear reservoir model equations (3.4) and (3.9) can be solved.

The general time-dependent expression corresponding to the steady-state equation (6.26) is

$$P_m = \Phi N_{\tilde{m}}/W_m \tag{6.37}$$

or, relative to initial values,

$$\frac{P_m}{P_{m0}} = \frac{\Phi}{\Phi_0} \frac{N_{\tilde{m}}}{N_{\tilde{m}0}} \tag{6.38}$$

For finite variations in P_m relative to initial values,

$$\frac{\Delta P_m}{P_{m0}} = \frac{\Phi - \Phi_0}{\Phi_0} + \frac{\Phi}{\Phi_0} \frac{n_{\tilde{m}}}{N_{\tilde{m}0}} \tag{6.39}$$

Similarly, for radiocarbon [cf. (6.27)],

$$\frac{\Delta^* P_m}{{}^*P_{m0}} = \frac{{}^*\Phi - {}^*\Phi_0}{{}^*\Phi_0} + \frac{{}^*\Phi}{{}^*\Phi_0} \frac{{}^*n_{\tilde{m}}}{{}^*N_{\tilde{m}0}} \tag{6.40}$$

The functions Φ and ${}^*\Phi$ depend implicitly on $N_{\tilde{m}}$ because they are functions of $[H^+]$, which depends on ΣC, and hence on $N_{\tilde{m}}$. No other dependence need be considered since the alkalinity, temperature, and salinity of surface water are expected to remain constant.

We may better understand the significance of (6.39) and (6.40) by first considering infinitesimal variations in P_m. If we differentiate (6.38) with due regard for the dependency of Φ on $N_{\tilde{m}}$, we find that δP_m is related to infinitesimal variations in $N_{\tilde{m}}$ by the general expression

$$\frac{\delta P_m}{P_{m0}} = \frac{\Phi}{\Phi_0} \left[\left(\frac{\partial \ln \Phi}{\partial \ln N_{\tilde{m}}} \right)_A + 1 \right] \frac{\delta N_{\tilde{m}}}{N_{\tilde{m}0}} \tag{6.41}$$

where the subscript A is a reminder that the alkalinity is to be held constant as Φ varies with $N_{\tilde{m}}$. The combined factor before $\delta N_{\tilde{m}}/N_{\tilde{m}0}$ is equivalent to the buffer factor ζ of (6.1). Since the preindustrial value of ζ for surface ocean water had an average value near 9, P_m varies much faster than $N_{\tilde{m}}$ as CO_2 is taken up at the ocean surface (Fig. 4). Thus ζ rises rapidly with increasing P_m as shown in Fig. 5. Since $d \ln P_m / d \ln \Sigma C$ itself increases slowly with P_m (Fig. 5), the increase in ζ with P_m is even faster than direct proportionality. As a result, ocean surface water, according to the transfer equations just developed, is much less able to absorb large additional amounts of CO_2 than is predicted by previously published models, all of which assume that the buffer factor remain indefinitely constant, with a value as estimated for the preindustrial era.

For calculations up to 1970, a linear relation between P_m and $N_{\tilde{m}}$ is, nevertheless, not a serious contradiction to the correct relation, as can be shown by drawing a chord between the origin and the value of ζ for 1970 in the lowest curve of Fig. 5. To adopt this straight-line relation to represent P_m versus $N_{\tilde{m}}$ is equivalent to employing (6.41) in finite difference form with exact agreement for one specified year, i.e.,

$$\frac{\Delta P_m}{P_{m0}} = \zeta \frac{n_{\tilde{m}}}{N_{\tilde{m}0}} \tag{6.42}$$

which [cf. (6.39)] leads to

$$\zeta = \frac{\overline{\Phi} - \Phi_0}{\Phi_0} \frac{N_{\tilde{m}0}}{\overline{n}_{\tilde{m}}} + \frac{\overline{\Phi}}{\Phi_0} \tag{6.43}$$

where ζ, $\overline{\Phi}$, and $\overline{n}_{\tilde{m}}$ denote values of ζ, Φ, and $n_{\tilde{m}}$ for the year specified. To evaluate ζ and $\overline{\Phi}$, we first find $N_{\tilde{m}}$ and Φ as functions of P_m as described in Section 6.1. Then an estimate of $\overline{n}_{\tilde{m}}$ is obtained for the models in question by solving the model equations with the preindustrial buffer factor ζ_0 in place of ζ. Finally, by reiteration, a consistent set of values can be established for ζ, $\overline{\Phi}$, and $\overline{n}_{\tilde{m}}$.

Substituting (6.42) into (6.34),

$$\Delta F_{ma} = k_{am} \frac{N_{a0}}{N_{\tilde{m}0}} \zeta n_m \tag{6.44}$$

where n_m is written in place of $n_{\tilde{m}}$ because the organic fraction of inactive carbon in ocean water has been declared to be invariant (see Section 6.1). This last equation is identical in form with that of Bolin and Eriksson [2].

For radiocarbon, we proceed by multiplying (6.40) by (6.27), dividing the product equation by (6.26), and finally multiplying each side of the quotient equation by $\Phi_0 / ^*\Phi_0$. This leads to the expression

$$\frac{\Phi_0}{^*\Phi_0} \frac{\Delta ^*P_m}{P_{m0}} = R_{\tilde{m}0} \frac{^*\Phi - ^*\Phi_0}{^*\Phi_0} + \frac{^*\Phi}{^*\Phi_0} \frac{^*n_{\tilde{m}}}{N_{\tilde{m}0}} \tag{6.45}$$

As was the case for inactive carbon up to 1970, a linear approximation is no serious contradiction of the true relationship as far as our interest extends (in this case only to 1954).

Defining an average radioactive buffer factor $*\bar{\zeta}$ analogous to $\bar{\zeta}$,

$$*\bar{\zeta} = \frac{*\bar{\Phi} - *\Phi_0}{*\Phi_0} \cdot \frac{N_{\tilde{m}0}}{\bar{n}_{\tilde{m}}} + \frac{*\bar{\Phi}}{*\Phi_0} \tag{6.46}$$

we obtain in place of (6.45)

$$\frac{\Phi_0}{*\Phi_0} \frac{\Delta *P_m}{P_{m0}} = R_{\tilde{m}0} \left(*\bar{\zeta} - \frac{*\bar{\Phi}}{*\Phi_0} \right) \frac{n_{\tilde{m}}}{N_{\tilde{m}0}} + \frac{*\bar{\Phi}}{*\Phi_0} \frac{*n_{\tilde{m}}}{N_{\tilde{m}0}} \tag{6.47}$$

where, as in the case of (6.42), the expression is essentially exact for a single specified year. Assuming any reasonable value for $\bar{n}_{\tilde{m}}$, one finds that $*\bar{\Phi}/*\Phi_0$ is equal to $\bar{\Phi}/\Phi_0$ within one part in 10^5 and $*\bar{\zeta}$ is equal to $\bar{\zeta}$ within one part in 10^4. Hence (6.47) can be evaluated using the factors for inactive carbon.

Since radiocarbon redistributes between the inorganic and organic fractions with relatively little time delay,

$$\frac{*n_{\tilde{m}}}{N_{\tilde{m}0}} \simeq \frac{*n_m}{N_{m0}} \tag{6.48}$$

With this approximation, and with $n_m = n_{\tilde{m}}$ and $R_{m0} \simeq R_{\tilde{m}0}$ as discussed earlier, we obtain in place of (6.36)

$$\Delta *F_{ma} = \alpha_{ma} k_{am} \frac{N_{a0}}{N_{\tilde{m}0}} \left[R_{m0} \left(*\bar{\zeta} - \frac{*\bar{\Phi}}{*\Phi_0} \right) n_m + \frac{*\bar{\Phi}}{*\Phi_0} \frac{N_{\tilde{m}0}}{N_{m0}} *n_m \right] \tag{6.49}$$

The equivalent expression of Bolin and Eriksson [2], (p. 137) written in our notation is

$$\Delta *F_{ma} = k_{am} \frac{N_{a0}}{N_{\tilde{m}0}} [R_{a0}(\zeta - 1)n_m + *n_m] \tag{6.50}$$

Using a value of $\bar{n}_m$ appropriate to 1954 (approximately $0.005 \times N_{m0}$), the factors $(*\bar{\zeta} - *\bar{\Phi}/*\Phi_0)/(\zeta - 1)$ and $*\bar{\Phi}/*\Phi_0$ exceed unity by 3% and 5%, respectively, while the product $\alpha_{ma} R_{m0}$ is about 8% less and $N_{\tilde{m}0}/N_{m0}$ 4% less than unity. Thus Bolin and Eriksson's equation is as precise as is justified when neglecting isotopic fractionation and organic carbon.

It remains to identify these results with the governing equations of the reservoir model. For the atmospheric reservoir with a CO_2 source $\gamma(t)$, the instantaneous rate of change in inactive carbon is

$$dn_a/dt = -\Delta F_{ab} + \Delta F_{ba} - \Delta F_{am} + \Delta F_{ma} + \gamma \tag{6.51}$$

Adding the corresponding equation for the land biota and transferring γ to the other side of the equality sign, we obtain

$$d/dt\,(n_a + n_b) - \gamma = -\Delta F_{am} + \Delta F_{ma} \tag{6.52}$$

Similarly combining and rearranging the first two equations of (3.4):

$$d/dt\,(n_a + n_b) - \gamma = -k_3 n_a + k_4 n_m \qquad (6.53)$$

Matching terms between (6.52) and (6.53) with the help of (6.33) and (6.44), we find

$$k_3 = k_{am} \qquad (6.54)$$

$$k_4 = k_{am} \frac{N_{a0}}{N_{\tilde{m}0}} \zeta \qquad (6.55)$$

For radiocarbon the corresponding rate equations are

$$(d/dt + {}^*\lambda)({}^*n_a + {}^*n_b) = -\Delta^* F_{am} + \Delta^* F_{ma} \qquad (6.56)$$

$$= -{}^*k_3 {}^*n_a + {}^*k_4 {}^*n_m + k_7' n_m \qquad (6.57)$$

With the help of (6.35) and (6.49) we find

$$^*k_3 = \alpha_{am} k_{am} \qquad (6.58)$$

$$^*k_4 = \alpha_{ma} k_{am} \frac{N_{a0}}{N_{m0}} \frac{\overline{\Phi}}{\Phi_0} \qquad (6.59)$$

$$k_7' = \alpha_{ma} k_{am} \frac{N_{a0}}{N_{\tilde{m}0}} R_{m0} \left(\zeta - \frac{\overline{\Phi}}{\Phi_0} \right) \qquad (6.60)$$

where in (6.59) and (6.60) the radioactive quantities ${}^*\overline{\zeta}$ and ${}^*\Phi/{}^*\Phi_0$ are replaced by those for inactive carbon in view of the observations made after (6.47).

7. DERIVATION OF TRANSFER COEFFICIENTS FOR EXCHANGE WITHIN THE OCEAN

7.1. Steady-State Exchange

The transport of carbon between surface and deep ocean water proceeds by both water motion (advection and eddy diffusion) and by gravitational settling of organic particles. Although the former mechanism is clearly dominant, some 10% of the carbon arrives in deep water via the second mechanism (Keeling and Bolin [11], p. 50). The existence of this gravitational flux can have only a minor influence on the transient adjustment of inactive carbon between the atmosphere and oceans as a whole because the flux does not respond appreciably to small changes in CO_2 partial pressure. For radiocarbon, however, the gravitational transport responds to changes in ^{14}C/inactive C ratio in the ocean surface layer and therefore enters more directly into the relationships for transient adjustment resulting from CO_2 exchange at the air–sea boundary.

To express these ideas in the framework of a reservoir model, let F_g denote the time-dependent gravitational flux of inactive carbon with steady-state value F_{g0}, and $k_{md}N_m$ the corresponding flux by water transport, where the latter is defined as proportional to the mass of carbon in the ocean surface layer. If F_{md} and F_{dm} denote the time-dependent total downward and upward flux of inactive carbon respectively,

$$F_{md} = k_{md}N_m + F_g \qquad (7.1)$$

$$F_{dm} = k_{dm}N_d \qquad (7.2)$$

where k_{md} and k_{dm} are time-invariant transfer coefficients. Equation (7.2) has no term for gravitational flux since gravity carries carbon only downward as provided for in (7.1).

Next let W_m and W_d denote the mass of water in the ocean surface layer and deep ocean, respectively. The products $k_{md}W_m$ and $k_{dm}W_d$ then express the downward and upward flux of water, respectively, across the boundary between the ocean surface layer and deep ocean. To preserve continuity of water within the ocean,

$$k_{md}W_m = k_{dm}W_d \qquad (7.3)$$

By analogy, the time-dependent radiocarbon fluxes are given by the expressions

$$*F_{md} = k_{md}*N_m + \alpha_{mg}(*N_{\tilde{m}}/N_{\tilde{m}})F_g \qquad (7.4)$$

$$*F_{dm} = k_{dm}*N_d \qquad (7.5)$$

where the factor α_{mg} allows for radioisotopic fractionation associated with the gravitational flux. The carbon in this flux is derived from inorganic carbon by the photosynthesis of marine plants which preferentially assimilate inactive carbon. Thus α_{mg} is less than unity, as discussed in Section 8 below. No radioisotopic fractionation is to be expected in the transport of carbon by water motion since turbulence in this motion prevents the influence of molecular scale forces.

At steady state, for inactive carbon

$$k_{md}N_{m0} + F_{g0} = k_{dm}N_{d0} \qquad (7.6)$$

while for radiocarbon

$$k_{md}*N_{m0} + \alpha_{mg}(*N_{\tilde{m}0}/N_{\tilde{m}0})F_{g0} = k_{dm}*N_{d0} + *\lambda*N_{d0} \qquad (7.7)$$

where the term in $*\lambda$ provides for radioactive decay in the deep ocean.

Consistent with (7.3), (7.6), and (7.7), the magnitudes of the flux constants and F_{g0} can be expressed in terms of concentrations N_{m0}/W_m, N_{d0}/W_d, and the ^{14}C/inactive C ratios

$$\left. \begin{aligned} R_{m0} &\simeq R_{\tilde{m}0} = *N_{\tilde{m}0}/N_{\tilde{m}0} \\ R_{d0} &= *N_{d0}/N_{d0} \end{aligned} \right\} \qquad (7.8)$$

Solving (7.6) for F_{g0}, we find

$$F_{g0} = k_{dm} W_d \left(\frac{N_{d0}}{W_d} - \frac{N_{m0}}{W_m} \right) \tag{7.9}$$

where the factor in parentheses is obtained by making use of (7.3). This equation shows that the gravitational flux is balanced by a net transport proportional to the product of the water flux and the concentration gradient between reservoirs.

Similarly, for radiocarbon,

$$\alpha_{mg} R_{m0} F_{g0} = k_{dm} W_d \left(\frac{*N_{d0}}{W_d} - \frac{*N_{m0}}{W_m} \right) + *\lambda *N_{d0} \tag{7.10}$$

where due regard is given to isotopic fractionation and radioactive decay.

The flux coefficient k_{dm} can be solved for by eliminating F_{g0} between (7.9) and (7.10). The result may be written in convenient reciprocal form:

$$k_{dm}^{-1} = *\lambda^{-1} \frac{W_d}{*N_{d0}} \left\{ \alpha_{mg} R_{m0} \left(\frac{N_{d0}}{W_d} - \frac{N_{m0}}{W_m} \right) - \left(\frac{*N_{d0}}{W_d} - \frac{*N_{m0}}{W_m} \right) \right\} \tag{7.11}$$

$$= *\lambda^{-1} \left\{ \frac{R_{m0}}{R_{d0}} \alpha_{dm} - 1 \right\} \tag{7.12}$$

where we introduce the pseudo-fractionation factor

$$\alpha_{dm} = \alpha_{mg} + \frac{N_{m0}}{W_m} \frac{W_d}{N_{d0}} (1 - \alpha_{mg}) \tag{7.13}$$

On the basis of the observable quantities as evaluated in Section 8, the evaluation of k_{dm}^{-1} by (7.12) is 2.6 % less than by an approximation in which radioisotopic fractionation is neglected ($\alpha_{mg} = 1$).

Since identically

$$k_{dm} N_{d0} = \left(\frac{N_{d0}}{N_{m0}} k_{dm} \right) N_{m0} \tag{7.14}$$

the steady-state downward flux would be correctly evaluated relative to k_{dm} if the steady-state gravitational flux F_{g0} were neglected and we wrote for the reservoir constant in the downward direction

$$k_{md} = \frac{N_{d0}}{N_{m0}} k_{dm} \tag{7.15}$$

as was done by Bolin and Eriksson [2] (p. 134).

Equation (7.15) is, however, an unsatisfying basis for establishing transient adjustments between reservoirs because it leads to an overestimation of the transient flux downward to the extent that F_g fails to increase as rapidly as $N_m k_{md}$.

7.2. Transient Exchange

Actually, because plant nutrients are in more limited supply than bicarbonate ions in ocean surface water, the rate of assimilation of carbon by marine plants is probably independent of the small changes in total carbon which result from the injection of industrial CO_2 into the atmosphere. Thus F_g is probably best regarded as a constant, equal to F_{g0}. The transient exchange of inactive carbon is then given by (cf. 7.1, 7.2, 7.3)

$$\Delta F_{md} = k_{md} n_m \tag{7.16}$$

$$= \frac{k_{dm} W_d}{W_m} n_m \tag{7.17}$$

$$\Delta F_{dm} = k_{dm} n_d \tag{7.18}$$

while for radiocarbon (cf. 7.4, 7.5)

$$\Delta *F_{md} = \frac{k_{dm} W_d}{W_m} *n_m + \alpha_{mg} F_{g0} \left(\frac{*N_{\tilde{m}}}{N_{\tilde{m}}} - \frac{*N_{\tilde{m}0}}{N_{\tilde{m}0}} \right) \tag{7.19}$$

$$\Delta *F_{dm} = k_{dm} *n_d \tag{7.20}$$

When the model is solved numerically for the time period of interest (up to 1954), the variations $*n_{\tilde{m}}/*N_{\tilde{m}0}$ and $n_{\tilde{m}}/N_{\tilde{m}0}$ turn out to be of the same order as $*n_a/*N_{a0}$, i.e., about 30% of the relative increase in atmospheric CO_2, n_a/N_{a0}. Since the second term in (7.19) contributes only about 10% to $\Delta *F_{md}$, the ^{14}C/inactive C ratio, $*N_{\tilde{m}}/N_{\tilde{m}}$, is sufficiently well approximated by the first-order terms of a Taylor series expansion [cf. (5.3)]. Thus we obtain (with the minor additional assumption $*N_{\tilde{m}}/N_{\tilde{m}} \simeq *N_m/N_m$)

$$\Delta *F_{md} = \left(\frac{k_{dm} W_d}{W_m} + \frac{\alpha_{mg} F_{g0}}{N_{m0}} \right) *n_m - \left(\frac{\alpha_{mg} R_{m0} F_{g0}}{N_{m0}} \right) n_m \tag{7.21}$$

$$= \alpha_{dm} k_{dm} \frac{N_{d0}}{N_{m0}} *n_m - \alpha_{mg} R_{m0} k_{dm} \left(\frac{N_{d0}}{N_{m0}} - \frac{W_d}{W_m} \right) n_m \tag{7.22}$$

where the second expression, (7.22), makes use of (7.9) and (7.13).

Comparing these equations term for term with the reservoir equations (3.4) and (3.9), we obtain

$$k_5 = \frac{W_d}{W_m} k_{dm} \tag{7.23}$$

$$k_6 = k_{dm} \tag{7.24}$$

$$*k_5 = \alpha_{dm} k_{dm} (N_{d0}/N_{m0}) \tag{7.25}$$

$$*k_6 = k_{dm} \tag{7.26}$$

TABLE 1. Data for Surface Ocean Water

Property	Notes	40°N to 70°N (NCSW)	40°N to 40°S (WSW)	40°S to 70°S (SCSW)	Areal average
Area ($\times 10^{18}$ cm^2)	1	0.40	2.40	0.73	—
T, temperature (°C)	2	8.37	25.27	6.89	19.59
S, salinity (g kg^{-1})	2	32.92	35.20	34.30	34.76
pH, $-\log_{10}$ [H$^+$]	2	8.10	8.20	8.08	—
A, titration alkalinity (meq liter^{-1})	2	2.293	2.405	2.342	—
DOC, dissolved organic carbon (mg liter^{-1})	3	—	—	—	1.0
$\Sigma B/Cl$, borate-chlorinity ratio [mmol liter^{-1}/(g kg^{-1})]	4	—	—	—	0.02126
$(p_{CO_2})_0$, assigned values of CO$_2$ partial pressure for 1700 A.D. (ppmv)	5	280	300	280	290

1. Sverdrup et al. [45] (p. 15).
2. Keeling and Bolin [11] (pp. 31–34) based on Russian oceanographic data of the International Geophysical Year near 180° longitude.
3. Williams [46]. This figure is highly approximate.
4. Greenhalgh and Riley, unpublished results of samples representative of all the major ocean basins as quoted by Riley and Chester [47] (p. 81). The original dimensionless ratio of 0.00023 was converted using an atomic weight of 10.81 and disregarding difference between 1 kg and 1 liter for sea water.
5. See text.

TABLE 2. Calculated Parameters for Surface Ocean Water

Property	Notes	NCSW Buch	NCSW Lyman	WSW Buch	WSW Lyman	SCSW Buch	SCSW Lyman	Areal average Buch	Areal average Lyman
Source of K_1, K_2, K_B		Buch	Lyman	Buch	Lyman	Buch	Lyman	Buch	Lyman
Stoichiometric factors									
K_0 ($\times 10^2$)	1	4.780		2.869		4.994		3.347	
K_1 ($\times 10^7$) $\Big\}$ CO_2 system	2	7.607	7.817	10.632	10.844	7.482	7.666	9.747	9.952
K_2 ($\times 10^{10}$)	2	6.079	5.000	9.706	8.103	5.995	4.975	8.501	7.076
K_B ($\times 10^9$), borate	2	—	1.393	—	2.124	—	1.376	—	1.881
K_W ($\times 10^{15}$), water	3	2.480		10.090		2.167		6.463	
Time-invariant chemical concentrations									
A (meq liter^{-1})	4	2.368	2.361	2.443	2.447	2.428	2.421	2.435	2.434
ΣB (mmol liter^{-1})	5	0.0387		0.0414		0.0404		0.0409	
Cl (g kg^{-1})	6	18.22		19.49		18.99		19.24	
Initial chemical concentrations									
pH	7	8.264	8.265	8.271	8.280	8.264	8.266	8.271	8.278
ΣC, dissolved inorganic C (mmol liter^{-1})	8	2.093	2.117	2.024	2.060	2.149	2.171	2.057	2.089
Buffer parameters*									
ζ_0, initial buffer factor	9	10.36	11.18	8.43	8.83	10.39	11.16	8.88	9.36
$d \ln P_m / d \ln \Sigma C$ in 1954	9	10.56	11.44	8.53	8.95	10.59	11.42	9.00	9.51
ζ, buffer factor in 1954	9	11.08	12.05	8.85	9.30	11.11	12.02	9.36	9.92
$\bar{\zeta}$, average buffer factor (for 1954)	9	10.71	11.61	8.64	9.06	10.74	11.59	9.11	9.64
Φ_0 ratio of p_{CO_2} to ΣC for inactive and	10	1.338	1.323	1.482	1.456	1.303	1.290	1.4098	1.3882
$^*\Phi_0$	10							1.3850	1.3638
$\bar{\Phi}$ radiocarbon, initial and 1954 values	10	1.402	1.392	1.539	1.515	1.366	1.357	1.4667	1.4479
$^*\bar{\Phi}$	10							1.4409	1.4224

Values for 1934 assume $n_m / N_{m0} = 0.003$

1. The following interpolation equation was used, based on the data of Murray and Riley [25] (p. 539).

$$\log_{10} K_0 = +2622.38/T - 15.5873 + 0.0178471T - Cl(0.0117950 + 2.77676 \times 10^{-5} T)$$

where p_{CO_2} is in atmospheres and $[CO_2]$ in moles liter^{-1}.

2. Values based on the experimental work of Lyman [21] were calculated from the equations of Edmond and Gieskes [22] (pp. 1271–1272) as discussed in the text. From the experimental work of Buch et al. [42] and Buch [49,50], the following equations were derived with due regard for the differences in definitions and pH scales as discussed by Edmond and Gieskes [22] (pp. 1265–1266).

$$\log_{10} K_1 = -3523.46/T + 15.6500 - 0.034153T + 0.074709\sqrt{Cl} + 0.0023483\,Cl$$

$$\log_{10} K_2 = -2902.39/T + 6.4980 - 0.02379T + 0.45322\sqrt{Cl} - 0.035226\,Cl$$

In the original definitions of the K's [(6.4), (6.5), (6.11)] the ratios $K_i/[H^+]$ are dimensionless, and thus independent of the concentration units chosen for A, ΣC, and ΣB. The units of $[H^+]$ must, however, be specified. By a widely accepted convention $[H^+]$ is replaced in the defining equations by a thermodynamic activity. For the range of conditions of interest here, this activity is lower than $[H^+]$ in moles liter^{-1} by a constant factor of about 0.9 (Culberson et al. [51], p. 16). Only ratios $K_i/[H^+]$ appear in the equations needed in this paper, except for the last term in (6.14) which is practically negligible relative to the sum of terms. Thus, no adjustment has been attempted to convert $[H^+]$ from the activity scale to moles liter^{-1}.

3. Obtained by linear interpolation from the table of Harvey [52] (p. 160). Variation with salinity was neglected.

4. Zonal values are copied from Table 5. Areal averages were calculated from areal averaged ΣC and $p_{CO_2} = 290$ ppmv via (6.8) and (6.14) as described in the text.

5. Calculated from the value of $\Sigma B/Cl$ in Table 1 using values of Cl listed directly below.

6. Calculated from salinity S via the relation of Knudsen (Riley and Chester [47], p. 14): $S = 1.805\,Cl + 0.030$.

7. Calculated as an intermediate step in preparing Table 5 (see text).

8. Zonal values are copied from Table 5. Areal averages are zonal values weighted according to the areas listed in Table 1.

9. Calculated from values of p_{CO_2} and ΣC as described in the text.

10. Calculated from values of $[H^+]$ as described in the text.

$$k'_8 = \alpha_{mg} R_{m0} k_{dm} \left(\frac{N_{d0}}{N_{m0}} - \frac{W_d}{W_m} \right) \qquad (7.27)$$

where α_{dm} is defined by (7.13).

8. OBSERVATIONAL DATA

8.1. Global Summaries

The observed and derived quantities used to solve the above global models are listed in Tables 1–4 together with information on the source of the data. Values of some related parameters which help explain the calculations are also listed. For ocean surface water, to obtain insight into the problem of spatial variability, I have separately considered three zones of the Pacific Ocean for which data were assembled by Keeling and Bolin [11]. Although hardly equivalent to world averages, these data are probably adequate input values for models having no areal resolution.

The chemical constants K_1 and K_2 were critically measured at one temperature (20°C) more than 30 years ago by Buch and coworkers and independently by Lyman [21]. For K_1 the two sets of results, for varying salinity, are nearly concordant, as shown by Edmond and Gieskes [22] (p. 1270), but for K_2, which is more important to a determination of the buffer factor ζ, the discrepancy between the investigators is over 10%. Recently (after model calculations for this chapter were completed) new determinations by Hansson [23] from 5 to 30°C have become available. These agree closely with those of Lyman at 20°C and support reasonably well the temperature variation adopted below. To establish how sensitive the models are to the uncertainty in K_1 and K_2, some model calculations in Section 9 and the data of Table 2 are worked out twice, first using values of Buch and then using those of Lyman. Calculations not quoted twice are based on Buch's constants. For borate dissociation, a highly accurate value is not required in calculating the buffer factor. Lyman's determination of K_B at 20°C will be accepted. To express the temperature variation of K_1 and K_2 in sea water, I have accepted the temperature coefficients adopted by Edmond and Gieskes [22] based on extrapolation to infinite dilution of measurements in NaCl solution. For boric acid, use is made of the temperature coefficients of Owen and King [24], as expressed by Edmond and Gieskes [22] (p. 1272).

The solubility of CO_2 gas in sea water, expressed by K_0, has for the first time recently been determined over a range of temperature and salinity by Murray and Riley [25]. The new results agree well with earlier predicted values based on data for sodium chloride solution (see Edmond and Gieskes [22], p. 1266), and are clearly reliable enough for use here.

To obtain values of alkalinity and total inorganic carbon for preindustrial times the following procedure was used. First, the CO_2 partial pressure p_{CO_2}

TABLE 3. Data for Preindustrial Land Biota, Atmosphere, and Deep Ocean Water

		Notes	Value
Land Biota			
N_{e0},	mass of short-lived organic carbon	1	7.5×10^{16} g
N_{b0},	mass of long-lived organic carbon	1	1.56×10^{18} g
F_{e0},	production rate for N_{e0}	1	3.0×10^{16} g yr^{-1}
F_{b0}.	production rate for N_{b0}	1	2.6×10^{16} g yr^{-1}
Atmosphere			
	mass of dry air	2	5.119×10^{21} g
	mixing ratio of CO_2 in dry air	3	290 ppmv
N_{a0},	mass of carbon in atmospheric CO_2	4	6.156×10^{17} g
Oceans			
ΣC,	dissolved inorganic carbon in sea water below 1000 m	5	28.0 mg liter^{-1}
DOC,	dissolved organic carbon in sea water below 1000 m	6	0.4 mg liter^{-1}
N_{s0},	mass of carbon in world oceans	7	3.89×10^{19} g
N_{s0}/N_{a0}		8	63.2
$W_m + W_d$,	volume of world oceans	9	1.37×10^{21} liter

1. See Table 7. The figures differ only slightly from estimates by SCEP [32] (p. 162).
2. Verniani [53] (p. 390).
3. Value rounded from 293 ppmv as found by Bray [29] (p. 226) for the mean of observations from 1857 to 1906. Bray's value (his Table 5, lines 11, 12) was based on selected series of observations in which the values within a series differed by no more than $\pm 30\%$ from their mean; thus approximately in the range 240–350 ppmv. The uncertainty in the 293 ppmv value probably exceeds ± 10 ppmv. The increase in atmospheric CO_2 owing to industrial CO_2, less than 3 ppmv before 1900, was not taken into account.
4. Obtained by the computation $N_a = 290 \times (12.011/28.964) \times 5.119 \times 10^{21}$. The ratio in parentheses is the atomic weight of carbon divided by the molecular weight of dry air as given by Verniani [53] (p. 390).
5. Eriksson [54] (p. 4). Original value is 2.33 mmol liter^{-1}.
6. Menzel and Ryther [55] (p. 334).
7. The estimates assume that the above concentrations of ΣC and DOC apply to the entire ocean.
8. Calculated from values quoted above.
9. Sverdrup et al. [45] (p. 12).

was calculated from the data of Table 1 using (6.14). The values obtained were then employed in (6.8) to calculate ΣC.

In this calculation the values of p_{CO_2}, as shown in Table 5, come out unreasonably high. Direct measurements of p_{CO_2} near 180° longitude, although too few to establish reliable averages (see Keeling [26] and Gordon et al. [27]), suggest that the average values of Table 1 must lie within about 20 ppmv of the atmospheric values. (Strictly speaking, the unit ppmv expresses a mixing ratio in parts per million by volume, but it can here also refer to partial pressure in atmospheres $\times 10^{-6}$, since we are dealing with time-averaged air masses for which the barometric pressure at sea level is

TABLE 4. Additional Quantities Used in the Reservoir Models

		Notes	Value
Carbon 14/12 fractionation factors			
α_{ab}	for production of organic carbon in the biosphere	1	0.964
α_{ba}	for respiration of plants and soil	2	1.000
α_{am}	for oceanic uptake of CO_2 at the air–sea boundary	3	0.972
α_{ma}	for oceanic release of CO_2 at the air–sea boundary	4	0.955
α_{mg}	for production of particulate carbon in the oceans	5	0.966
$^*K_0/K_0$	associated with CO_2 gas solubility	6	1.000
$^*K_1/K_1$	associated with first dissociation of carbonic acid	7	1.018
$^*K_2/K_2$	associated with second dissociation of carbonic acid	7	1.000
Miscellaneous			
$^*\lambda^{-1}$	reciprocal of carbon-14 decay constant	8	8267 yr^{-1}

1. Steady-state average values of $\delta^{13}C$ are: for atmospheric CO_2, -7%, for living matter on land, -25% (Degens [56], p. 322); hence $^{13}\alpha_{ab} = 0.982$. The fractionation factors for radiocarbon are the square of the corresponding carbon-13 factors (Riley and Skirrow [20], p. 288).
2. Respiration of plants and soils converts organic carbon to CO_2 with little or no fractionation (Park and Epstein [57], p. 137).
3. Based on a laboratory study of Craig [58].
4. Steady-state equilibrium factor for carbon-13 (equal to $^{13}\alpha_{am}/^{13}\alpha_{ma}$) is approximately 1.009 (Craig [40], p. 693; Mook [59], p. 177). Hence $\alpha_{ma} = 0.972/(1.009)^2 = 0.955$.
5. From direct measurements of $\delta^{13}C$ and ΣC distributions in ocean water, Craig [40] (p. 693) estimates $\delta^{13}C$ values: for the particulate flux, -15% for ΣC in surface water, $+2\%$; hence $^{13}\alpha_{mg} = 0.983$. α_{mg} is obtained as the square (cf. Ref. 1).
6. The fractionation factor found by Vogel [60] (p. 218) is 0.99964. A value of unity will be assumed.
7. The overall fractionation, CO_2 gas to total dissolved inorganic carbon, is as given in Ref. 5. Since the fractionation factor for CO_2 gas to solid calcium carbonate has almost the same value (Vogel [60], p. 218) $^*K_2/K_2$ must be near unity, although no direct measurements exist to check this hypothesis. Since $^*K_0/K_0$ is also nearly unity, to a close approximation $^*\Phi/\Phi = ^*K_1/K_1$.
8. Value adopted by the Twelfth Nobel Symposium [61].

very nearly one atmosphere.) A relatively narrow range in p_{CO_2} values is also suggested by a steady-state model of Keeling and Bolin [11] (p. 52, Table 12), which predicts differences:

$$\left.\begin{aligned}
P_{scsw} - P_a &= -9 \text{ ppmv} \\
P_{wsw} - P_a &= 10 \text{ to } 18 \text{ ppmv}
\end{aligned}\right\} \tag{8.1}$$

where the three- and four-letter subscripts refer to the designated zones and P_a refers to atmospheric CO_2.

The measurements of pH and alkalinity listed in Table 1 were made between 1957 and 1959, after significant industrial CO_2 had disturbed the preindustrial CO_2 balance between air and sea. As the model calculations in Section 9 reveal, the average p_{CO_2} of water in 1958 probably lagged some 5 ppmv below equilibrium with atmospheric CO_2, which at that time had a

TABLE 5. Comparison of Ocean Surface Water Values of CO_2 System for Different Values of the Stoichiometric Quotients

Source for K_1	Source for K_2	Alkalinity			p_{CO_2}			Total inorganic carbon		
(L = Lyman, B = Buch)		NCSW	WSW	SCSW	NCSW	WSW	SCSW	NCSW	WSW	SCSW
Preliminary calculation of p_{CO_2}										
L	L	2.293	2.405	2.405	422	371	443	2.130	2.075	2.185
B	B	2.293	2.405	2.405	423	364	423	2.107	2.039	2.163
B	L	2.293	2.405	2.405	434	378	434	2.131	2.075	2.185
L	B	2.293	2.405	2.405	412	357	412	2.106	2.039	2.162
Adjustment of alkalinity to give corrected p_{CO_2}										
L	L	2.361	2.447	2.421	300	320	300		matrix	
B	B	2.368	2.443	2.428	300	320	300		unchanged	
B	L	2.367	2.452	2.426	300	320	300			
L	B	2.361	2.437	2.422	300	320	300			
Adjustment of total inorganic carbon to give preindustrial values										
L	L		matrix		280	300	280	2.117	2.060	2.171
B	B		unchanged		280	300	280	2.093	2.024	2.149
B	L				280	300	280	2.118	2.060	2.172
L	B				280	300	280	2.093	2.024	2.148

mixing ratio of 313 ppmv (Pales and Keeling [28], p. 6066). Thus we might expect for the period 1958 to 1960 [using for $(P_{wsw} - P_a)$ the lower figure in Eq. (8.1)]

$$P_{ncsw} = P_{scsw} = 300 \text{ ppmv}$$
$$P_{wsw} = 320 \text{ ppmv} \tag{8.2}$$

with an uncertainty of about ± 10 ppmv.

The most likely explanation for the high value of p_{CO_2} as derived from the oceanographic data is that slightly acid material was present in the sample containers before sampling. According to a private communication from one of the original investigators, all the alkalinity values are in doubt to some extent on this account and because of calibrating difficulties. The derived ΣC values are thus more likely to be correct than the observed alkalinities. Consequently, as the next step, new alkalinity values were calculated to correspond to the p_{CO_2} values of (8.2) by solving (6.8) for $[H^+]$ using the calculated values of ΣC, and then applying the new $[H^+]$ values in (6.14). Four sets of values of the adjusted alkalinity are shown in Table 5 to illustrate the influence of K_1 and K_2. The adjusted alkalinity values apply to the pre-industrial period as well as 1957–1959 because the titration alkalinity is not influenced by gain or loss of CO_2 from sea water.

As the third step, to obtain preindustrial values for ΣC, I have accepted Bray's [29] value of 293 ppmv (see Table 2) for preindustrial p_{CO_2}. From (8.1) it then follows that before appreciable industrial CO_2 entered the oceans,

$$P_{scsw} = P_{ncsw} = 280 \text{ ppmv}$$
$$P_{wsw} = 300 \text{ ppmv} \tag{8.3}$$

with an uncertainty of about ± 10 ppmv. Using (8.3) and adjusted alkalinity values, (6.14) was solved for $[H^+]$ and the values used in (6.8) to estimate preindustrial ΣC.

Finally, to obtain global averages, the zonal estimates for preindustrial ΣC were weight-averaged according to the areas listed in Table 1, p_{CO_2} was set equal to 290 ppmv (a rounded value used in all the model calculations), and, for consistency, a new alkalinity value was calculated to correspond to the global values of p_{CO_2} and ΣC at the weighted average temperature and chlorinity listed in Tables 1 and 2, respectively. The buffer factor ζ and its average, $\bar{\zeta}$, as defined by (6.43) for the typical case $\bar{n}_m/N_{m0} = 0.5\%$ were then calculated as described in Sections 6.1 and 6.3. Associated values of Φ and Φ_0 and their radiocarbon equivalents were also calculated. The cross pairing of K_1 and K_2 values by different investigators was not carried beyond the calculations shown in Table 5 because the results seemed less reasonable than using values of one investigator for both constants.

The basis for obtaining the atmospheric and deep ocean data listed in Tables 3 and 4 is explained in the tables. The choice of data for the land biota is discussed in Section 8.2.

8.2. Carbon Fluxes and Masses of the Land Biota

Land plants have produced a reservoir of organic carbon which each year exchanges with the atmosphere more than ten times as much CO_2 as is currently produced by the combustion of fossil fuels. A substantial part of the annual uptake of CO_2 by plants involves the production of leaves, litter, short-lived animals, and algae, which retain carbon for a few years or less. The rest of the uptake and some carbon from the above materials goes to wood, large roots, long-lived animals, and upper-soil humus, where it remains in organic phases for substantially longer periods. Most of it, nevertheless, returns to the atmosphere within 100 years. A small fraction of the uptake, however, is incorporated in deep-soil humus, very old plants, and incipient geologic deposits including peat, lignite, and coal.

Although the uptake of CO_2 by the land biota during a growing season produces about equal amounts of short- and long-lived materials, the latter far exceed the former in total abundance by virtue of their much longer storage or turnover times. Likewise, the amount of very old material, and especially geological deposits, far exceeds the amount of all younger materials combined.

To analyze the response of the land biota to changes in concentration of atmospheric CO_2 and radiocarbon, we need to know how the flux of CO_2 exchanged with plants and animals is divided between carbon-bearing materials of varying storage time, and the relative abundance of these materials. Published estimates of photosynthesis and respiration rates and inventories of the amounts of carbon in various reservoirs do not give us this information directly. Materials are usually distinguished by their most readily observed physical, chemical, or biological properties, e.g., coal *versus* peat or living organisms *versus* dead organic remains. Transfer rates are often tabulated with respect to specific plant groups and plant parts and then integrated to obtain overall rates. Indeed, classical techniques such as harvesting and gas exchange measurements can hardly be expected reliably to separate short- and long-lived components. Special tracer studies, e.g., using radiocarbons, are needed, but as yet too little use has been made of these to formulate accurate world-wide averages.

The atmospheric fluctuations we are concerned with here have characteristic times ranging from a few years to a few decades. Thus the distinction between "short-lived" as, say, 1–10 years and "long-lived" as 10–100 years is a permissible first approximation to the distant goal of establishing a continuous relationship between elements of the CO_2 flux entering and leaving the land biota on the one hand, and associated elements of the uptaken CO_2 ranked according to storage time within the organic phases on the other. But even this first approximation can be only crudely established owing to the paucity of observational data. The approach I have used below is to assemble provisional estimates of turnover times for short- and long-lived materials within broad ecosystem areas and the most probable partitioning of net seasonal uptake of CO_2 by these materials. These estimates are hardly better than educated guesses, but they are probably preferable to making no estimates at all, because an assumption that each ecosystem is a single reservoir which turns over carbon at a rate summing cycles of all storage times produces a much too rapid calculated response of the land biota as a whole.

I define the turnover time as the mass of carbon in some part of the biota (such as leaves or wood or a combination of such categories) divided by the flux of carbon entering that reservoir via photosynthesis. If the reservoir is not changing in biomass, an equal amount of organic carbon is withdrawn by respiration or some other form of transfer and ultimate oxidation.

Of interest is only that flux which causes carbon to pass from CO_2 to an organic phase and back to CO_2; transfers from leaves to roots or from living to nonliving phases need not be specifically taken into account. A further simplification will be made by accepting values of net primary production (NPP) from the literature as representing the amount of carbon entering the biota. Carbon respired by the plants from day to day is thus not considered at all, which seems justified because it remains such a short time in the organic phase. No clear distinction exists between long- and short-cycled carbon, but it seems reasonable to define the short-cycled carbon as that carbon remaining in the biota for times comparable to the cycling time for carbon in leaves. This time is longer going from tropical to temperate to boreal regions. Annual and biennial plants are considered to cycle carbon principally on a short-term basis, so that long cycling times apply only to the woody parts of shrubs and trees and their organic decay products, and deeper humus underlying short-lived plants.

With the help of several biologists I decided (or guessed) what part of the net primary production is used by the plants to produce leaves and labile carbon and which returns to the atmosphere, mostly via decay of litter, without having been incorporated in the long-living parts of the plants. Using the categories of Olson [30], I propose the division of flux and turnover times shown in Table 6.

The proportions of NPP assigned in Table 6 may seem arbitrary but are actually based on an earlier survey of existing data to deduce them from turnover times and estimates of the amount of short-cycled carbon. The high proportion for tropical forests and low proportion for boreal forests reflect this study. The turnover times are approximately as agreed on in discussions with the biologists.

For forest and woodland the pool of short-cycled forest carbon can be calculated with the numbers of Table 6, and subtracted from the figures of Olson for the total carbon biomass. When this is done, the turnover time of the long-cycled carbon in forests comes out to be about 20 years for most categories. This time is certainly too short and reflects Olson's omission of

TABLE 6. Characteristics of Short-Cycled Carbon in the Land Biota*

Ecosystem	Proportion of total net primary production	Turnover time (yr)
Woodland or Forest		
Temperate cold-deciduous	1/2	3
Conifer: boreal and mixed	1/3	5
Rainforest: temperate	1/2	3
Dry woodland	1/2	3
Rainforest: tropical and subtropical	2/3	2
Non-forest		
Tundra-like	1/2	6
Grassland	1/2	2
Agricultural	2/3	1
Desert and semidesert	1/2	3
Wetlands	1/2	1

* Ecosystems are named by the scheme of Olson [61] (p. 234), except for the last two categories, which are taken from the SCEP [32] (p. 151–162).

long-cycled litter and humus. A more reasonable average turnover time is 40 years, which leads to the conclusion that the nonliving carbon pool of forests is about the same size as the live carbon biomass. This is consistent, for example, with an estimate of Lieth [31] that total nonliving "younger organic matter" is about 710×10^{15} g of carbon; Olson's figure for the live carbon biomass is 562×10^{15} g. I reject, however, Leith's very low figure of 124×10^{15} g for the live carbon biomass. Using 40 years for the turnover time of the long-cycle carbon in forests leads to too little carbon in tropical rain forest and dry wood land (perhaps because we have estimated the proportion of long-cycle flux incorrectly). I have adjusted the assumed times in these categories to preserve reasonable estimates of the nonliving carbon biomass.

For nonforest ecosystems, most of the living material and litter is short-lived and, if above ground, is soon oxidized after death. Much of the humus, on the other hand, is turned over slowly. How much of the uptake of CO_2 goes to long-cycled carbon depends on how much goes to the deeper humus layers, and few data exist to guide us. For agricultural lands a figure of 2/3 was chosen for the proportion of short-lived net primary production to reflect harvesting of crops. For the other categories 1/2 was chosen after noting that this fraction agrees approximately with that for forest uptake. Actually, the estimate cannot be far off using an equal partition between short- and long-cycled fractions provided the corresponding turnover times can be established. For living materials and litter, turnover times can be estimated from field observations. The same values serve well enough for short-cycled humus. The turnover times of the deeper humus are not well known, however.

TABLE 7. Subdivision of World Inventory of Terrestrial Organic Carbon into Material with Short and Long Turnover Times

Reservoir	Carbon inventory (10^{15} g)					Net primary production (10^{15} g yr^{-1})			Turnover time (yr)	
	Short lifetime	Long lifetime	Total	Living*	Dead	Short lifetime	Long lifetime	Total*	Short lifetime	Long lifetime
1	2	3	4	5	6	7	8	9	10	11
(1) Woodland or forest										
(a) Temperate and boreal										
deciduous	12.0	160.0	172.0	80.0	92.0	4.0	4.0	8.0	3	40
coniferous	15.0	240.0	255.0	120.0	135.0	3.0	6.0	9.0	5	40
rain forest	1.8	24.0	25.8	12.0	13.8	0.6	0.6	1.2	3	40
dry woodland	4.2	140.0	144.2	70.0	74.2	1.4	1.4	2.8	3	100
(b) Tropical and subtropical										
rain forest	20.0	400.0	420.0	200.0	220.0	10.0	5.0	15.0	2	80
Total	53.0	964.0	1017.0	482.0	535.0	19.0	17.0	36.0		
(2) Nonforest										
tundra-like	3.6	60.0	63.6	7.2	56.4	0.6	0.6	1.2	6	100
grassland	7.8	234.0	241.8	18.2	223.6	3.9	3.9	7.8	2	60
agricultural	4.0	160.0	164.0	15.0	149.0	4.0	2.0	6.0	1	80
desert and semidesert	4.8	64.0	68.8	19.2	49.6	1.6	1.6	3.2	3	40
wetlands	1.0	40.0	41.0	4.0†	37.0	1.0	1.0	2.0†	1	40
Total	21.2	558.0	579.2	63.6	515.6	11.1	9.1	20.2		
									mean	mean
Grand total	74.2	1522.0	1596.2	545.6	1050.6	30.1	26.1	56.2	2.5	58
Rounded	75	1560	1635	550	1085	30	26	56	2.5	60

* All values except for wetlands from Olson [66] (p. 234).

The resulting inventories, production figures, and turnover times are summarized in Table 7. The figures for agricultural land, grassland, tundra, and wetlands reflect an opinion of biologists that for these ecosystems the amount of nonliving carbon is of the order of ten times the living carbon, while for deserts and semideserts three times is reasonable. The totals agree closely with a briefer table prepared for SCEP [32] (p. 162) and would agree almost identically if the turnover times of short-cycled tropical and subtropical rain forest were raised to 3 years.

Although the subdivision by storage time cannot be checked against other published data, one can compare totals of living and dead material and net primary production. These have been recently reviewed by Bowen [33] (pp. 46–53) and summarized by Whittaker [34]. The former author states that such world averages are uncertain to at least ±50%. In view of the differences between estimates of the most recent investigators, this assertion seems entirely reasonable. One may hope that new investigations now in progress will soon lead to more precise data.

9. NUMERICAL RESULTS AND DISCUSSION

9.1. Additional Observational Data

Essentially three independent observations are available to check the predictions of the foregoing models:

(1) The fraction of industrial CO_2 currently remaining airborne.
(2) The relative dilution of atmospheric radiocarbon dioxide by industrial CO_2 (the Suess effect) since the beginning of the industrial era.
(3) The ^{14}C/inactive carbon ratio of ocean water relative to that of atmospheric CO_2.

As for the first observation, the fraction remaining airborne has been estimated by Keeling et al. [35] for the period 1959 to 1969 on the basis of closely agreeing observations of atmospheric CO_2 at the South Pole and Hawaii. If the amount of industrial CO_2 produced during this period is as estimated by Keeling [12], the airborne fraction is 49% ± 12%, where the uncertainty is expressed at the 95% confidence level, and includes the uncertainty in estimating industrial CO_2 production.

To estimate the cumulative airborne fraction for 1700–1954, model calculations were carried out for the entire period 1700–1970. These indicated that the cumulative fraction to 1954 was about 45%, or 5.0% of the preindustrial atmospheric concentration.

The dilution of radiocarbon by industrial CO_2 between the late nineteenth century and mid-twentieth century can be estimated from measurements of the radiocarbon content of dendrochronologically dated tree rings of which the most accurate and extensive data are reported by Suess [36],

Houtermans et al. [37], and Lerman et al. [38]. These data, which indicate a dilution of -2.5% in 1954, should be corrected for the dilution which had already occurred before the late nineteenth century. Model calculations indicate that this correction is about -0.3%. Thus the full Suess effect in 1954 (as defined above) was evidently -2.8%. The uncertainty as estimated by the scatter in the individual available ^{14}C measurements is about 0.3%. The analytical error, about 0.2% for an individual analysis, accounts for much of the scatter. The largest additional uncertainty probably arises from variations in the ^{14}C content of the tree rings owing to uptake of CO_2 produced near the trees by plant respiration or combustion of fossil fuel.

The average ^{14}C/inactive C ratios of surface and deep ocean water for 1954 can be estimated from fractionation-corrected data compiled by Broecker [39]. These data, as published, were normalized to eliminate the influence of isotopic fractionation by means of an adjustment based on the departure of the $^{13}C/^{12}C$ ratio from a standard value. It is not possible from these corrected data to deduce exactly the radioisotopic ratios of the original samples, but reasonably correct adjustments can be made on the basis of independent observations of $^{13}C/^{12}C$ in water, such as reported by Craig [40]. I have deduced that

$$R_m/R_{a0} = 0.96 \pm 0.02$$

$$R_d/R_{a0} = 0.83 \pm 0.04$$

in 1954 for surface and deep ocean water, respectively. The quoted uncertainties are as stated by Broecker et al. [3]. The distinction between the preindustrial ratio R_{a0} and the isotopic standard, of importance in establishing the Suess effect, is here too small to matter.

9.2. Comparison of Three- and Five-Reservoir Models

Before seeking a model which realistically reproduces the response of the carbon cycle to industrial CO_2 emissions, the several models of Sections 3 and 4 will be examined to gain insight into how the predicted distributions of inactive carbon and radiocarbon change as details are progressively added to the system of governing equations.

The results of Bolin and Eriksson [2] are a suitable starting point for discussion. These authors solved the governing equations of the three-reservoir model for a range of values of the transfer times $\tau_{am} (\equiv k_{am}^{-1})$ and $\tau_{dm} (\equiv k_{dm}^{-1})$ in the vicinity of reasonable estimates of 5 and 500 years, and they represented the industrial CO_2 input $\gamma(t)$ with an exponential function $\gamma_0 e^{rt}$ [see (2.19) and (2.20)], where γ_0 and r were obtained to best fit a table of decadal values of Revelle and Suess [1]. They neglected, as second order, the influence of radioactive decay and isotopic fractionation, and they simplified the solution of the governing equations by setting to zero the coefficients C_{3i} and $*C_{3i}$ for the terms in $e^{\lambda_3 t}$ and $e^{*\lambda_3 t}$ [cf. (2.23)] before evaluating the other

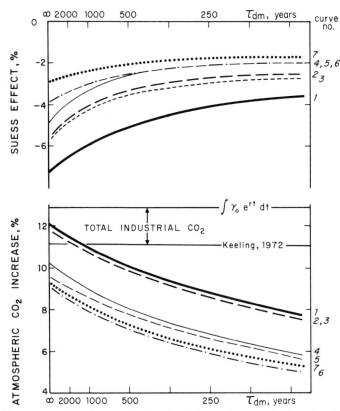

Fig. 6. *Upper curves*: The Suess effect (fractional change in atmospheric radiocarbon owing to dilution by industrial CO_2) in 1954 as predicted by various reservoir models, all with $\tau_{am} = 5$ yr. *Lower curves*: Predicted increase in inactive atmospheric CO_2 from 1700 to 1954 for the same models. Seven cases are shown, as follows:

Curve	Equation solving	Isotopic fractionation factors	CO_2 production	N_{m0}/N_{a0}*	Buffer factor	Biota and gravity flux
1	approximate	$= 1$	$\gamma_0 e^{rt}$	1.2	12.5	neither
2	exact	$= 1$	$\gamma_0 e^{rt}$	1.2	12.5	neither
3	exact	exact	$\gamma_0 e^{rt}$	1.2	12.5	neither
4	exact	exact	annual values	1.2	12.5	neither
5	exact	exact	annual values	2.0	12.5	neither
6	exact	exact	annual values	2.0	9.1	neither
7	exact	exact	annual values	2.0	9.1	both

* Ratio of the preindustrial mass of inactive carbon in the ocean surface layer to that in the atmosphere.

coefficients via the initial conditions. This last procedure is not the same as dropping a small term relative to larger terms, because it ignores one of the initial conditions and yields approximate values for all of the coefficients C_i and $^*C_i$.

Their values of the Suess effect and of n_a for 1954, recomputed for $\tau_{am} = 5$ years, are plotted as the upper and lower curves 1 of Fig. 6. If the Suess effect in 1954 was about -2.8% and the atmospheric CO_2 increase n_a/N_{a0} about 5% as indicated in Section 9.1, their results are clearly too high for n_a/N_{a0} and too negative for the Suess effect even if the deep ocean to surface layer transfer time τ_{dm} is unrealistically assigned the short time of 250 years. If the air to sea transfer time τ_{am} is reduced to 2 years, the results are only negligibly improved, as they point out in their paper.

If the same values of all model parameters are retained, but their system of governing equations is solved exactly according to (2.35) and (2.43), the values of n_a/N_{a0} are very slightly reduced (Fig. 6, lower curve 2). The Suess effect, however, is shifted so considerably (Fig. 6, upper curve 2) that the predicted values approach the correct result. Allowing for isotopic fractionation (Fig. 6, upper and lower curves 3) slightly offsets the improved prediction of the Suess effect, but does not change the calculation for n_a.

Bolin and Eriksson's method of calculating the Suess effect has two drawbacks. First, although dropping the terms in $e^{\lambda_3 t}$ for inactive carbon has only a slight effect on n_a, it causes n_m to shift by about 20%. Because transients in radiocarbon depend directly on n_m through the virtual source $k_7' n_m$ [cf. (2.5)], this considerable error is directly transmitted to the determination of the Suess effect. Second, in the radiocarbon equations it is less justified to drop the terms in $e^{*\lambda_3 t}$ because $^*\lambda_3$ is only about 10 times larger than $^*\lambda_2$, whereas for inactive carbon the corresponding ratio of λ's is about 200. (This relative difference in root ratios arises because sea water buffering directly influences only inactive carbon. For example, if the buffer factor is of the order of 10, the ratio of roots for inactive carbon is about 10 times that for radiocarbon.)

The predicted value of the Suess effect is further reduced if the source function $\gamma(t)$ is represented by better estimates of the production of CO_2 from fossil fuel. Bolin and Eriksson overestimated production in two ways. First, they accepted production data based on unreasonably high values for the carbon fraction of solid fuels. Revelle [41] later reduced these earlier figures by over 10%. Even Revelle's later figures may be over 10% too high, because he overestimated the carbon content of lignite and did not allow for incomplete combustion [12].

Second, Bolin and Eriksson, in order to provide their model with a single smooth exponential function over a wide time interval, sought a close fit for CO_2 production for the beginning of Revelle's record (1880) and for a forecast period (1950–2010) which was hard to overlook because the forecast itself assumed pure exponential growth. World fuel production actually rose

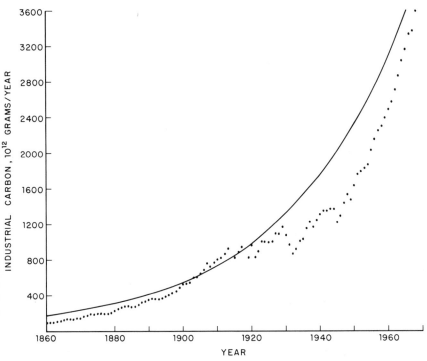

Fig. 7. Rate of production of "industrial" CO_2 from combustion of fossil fuels and kilning of limestone plotted in units of 10^{12} grams of carbon per year. *Dots*: annual values based on fuel production data of the United Nations as interpreted by Keeling [12]. *Curve*: exponential relation of Bolin and Eriksson [2].

very slowly between 1930 and 1950, and Bolin and Eriksson's analytic function overestimates the production for just the period of greatest importance in producing correct estimates for 1954 (Fig. 7). For 1940–1960, their values of γ are perhaps as much as 50% too high [12].

When the three-reservoir model is evaluated exactly [via (2.35) and (2.43)] with the revised annual production values of Keeling [12] introduced directly in the convolution integral [see (3.30)], the predicted Suess effect for 1954 is close to the observed value (Fig. 6, upper curve 4). On the other hand, the predicted relative increase in inactive CO_2, n_a/N_{a0}, although considerably lowered (Fig. 6, lower curve 4), is not yet close to the value of about 5% based on direct observations. Furthermore, the fraction of CO_2 remaining airborne is not significantly changed; it remains near 70% as in the cases using the function $\gamma_0 e^{rt}$ of Bolin and Eriksson.

To improve the agreement between prediction and observation of n_a, two additional oceanic parameters will now be examined before considering more complicated models. First, in regard to the proper size to assign to the surface ocean layer, Bolin and Eriksson chose $N_{m0}/N_{a0} = 1.2$, based on a suggestion of Craig [6] that water above the oceanic thermocline should be treated apart from deep water in modeling steady-state CO_2 exchange. The thermocline depth of 75 m which Craig adopted was appropriate to temperate and low latitudes, but is not realistic for polar regions. Here mixing occurs to well below 100 m, and, at least in winter and in restricted areas, involves the entire water column. For the time period of industrial CO_2 production, a single annual overturn is practically as effective as constant mixing. The deep polar surface layer should thus be included in N_{m0} or otherwise allowed for, as Craig [42,43] himself later pointed out. No detailed study as yet has been made to determine from the physical properties of sea water a proper global average depth of that part of the near-surface ocean column which is vertically homogeneous during some part of the year. A ratio N_{m0}/N_{a0} of 2, or even of 3, does not seriously violate existing knowledge. Even ratios greater than about 3, although no longer physically realistic, are still of interest as parametric representations of the underlying intermediate water. As seen in Fig. 6, upper and lower curves 5, increasing N_{m0}/N_{a0} to 2 shifts the Suess effect and n_a, if not conspicuously, at least measurably for values of the deep ocean to surface layer transfer time greater than 1000 years.

A more important adjustment in n_a involves the buffer factor ζ, for which a value of 12.5 was computed by Bolin and Eriksson. The most appropriate value of ζ for the year 1954 is probably between 9.1 and 9.8, according to the calculations discussed in Section 8. The calculated Suess effect is found to be practically independent of ζ, but n_a is considerably reduced (Fig. 6, lower curve 6). The lack of dependence of the Suess effect on the buffer factor has not, to my knowledge, been previously pointed out, although often tacitly assumed (see e.g., Baxter and Walton [44]). This lack is expressed by the similarity of equations (6.39) and (6.40), and the near identity of the ratios Φ/Φ_0 and $*\Phi/*\Phi_0$ (see Table 2).

The Suess effect is further reduced when account is taken of the exchange capacity of the land biota as provided by the five-reservoir model of Section 5. The land biota contains more carbon than the atmosphere and exchanges it rapidly enough with the air to modify considerably the radiocarbon distribution. Bolin and Eriksson examined the limiting case of infinitely rapid mixing between the atmosphere, land biota, and ocean surface layer, assuming no radiocarbon exchange with the deep sea. Their calculation did not depend on a mathematical solution of the governing model equations, and their estimated Suess effect of -2.0 to -2.5% is indeed close to the five-reservoir model prediction using reasonable parameterization. With the input values of Section 8, and with N_{m0}/N_{a0} equal to 2, the five-reservoir model predicts a Suess effect of about -2% for reasonable transfer times (Fig. 6, upper curve 7). This is actually somewhat less than the observed value.

The predicted Suess effect in 1954, as discussed in the next section, is practically independent of the assumed value of the biota growth factor β_a because up to 1954 such growth cannot be enough to influence appreciably the capacity of the biota to exchange carbon. Thus our analysis of the Suess

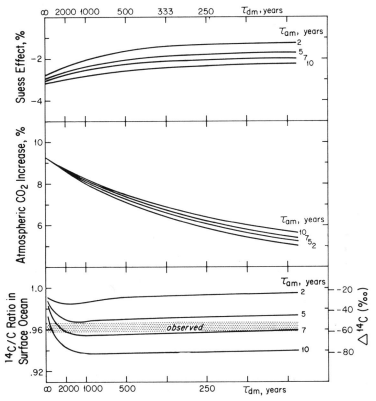

Fig. 8. *Upper curves*: Suess effect in 1954 predicted by five-reservoir model for various values of the air to sea transfer time τ_{am}, and deep ocean to surface layer transfer time τ_{dm}. The carbon pool of the land biota is assumed constant and $N_{m0}/N_{a0} = 2$. *Middle and lower curves*: Increase in inactive atmospheric CO_2 from 1700 to 1954 and ^{14}C/inactive C ratio of inorganic carbon in ocean surface water in 1954 (relative to preindustrial ^{14}C/inactive C ratio of atmospheric CO_2) as predicted using the same model parameters as for the upper curves. From highest to lowest curve in each set of four, the upper and lower set are for $\tau_{am} = 2, 5, 7$, and 10 yr; the middle set is for $\tau_{am} = 10, 7, 5$, and 2 yr. The upper and middle curves for $\tau_{am} = 5$ are the same as curves 7 of Fig. 6. Shaded line with lower set of curves denotes approximately the observed value of R_m/R_{a0} near 1954. The scale on the right side of the lower set of curves expresses R_m as the per mil variation from a conventional standard after correction for fractionation as quoted by Broecker [39].

effect is now essentially complete. In the next section the predicted Suess effect will be used to restrict the range in values of some of the model parameters. As for the predicted values of n_a/N_{a0}, these are unsatisfactory for any reasonable choice of model parameters so far considered, and this problem will receive detailed consideration in the next section. In Fig. 8 and Table 8, the Suess effect and n_a/N_{a0} predicted by the five-reservoir model are reproduced over the range in τ_{am} and τ_{dm} selected by Bolin and Eriksson.

To narrow the choice of transfer times, it is worthwhile to take into account the isotopic ratios of the ocean surface layer and deep ocean implied

TABLE 8. Solution of the Five-Reservoir Midel with Constant Land Biomass ($\beta_a = 0$)

Sources of K_1 and K_2* B = Buch L = Lyman	Transfer times		Relative increase in carbon masses in 1954			Suess effect atmosphere (%)	^{14}C/inactive C ratio		
	Air to sea τ_{am} (yr)	Deep ocean to surface ocean τ_{dm} (yr)	Atmosphere n_a/N_{ao} (%)	Surface ocean n_m/N_{ao} (%)	Deep ocean n_d/N_{ao} (%)	(%)	Surface ocean 1700 R_{mO}	Surface ocean 1954 R_m	Deep ocean 1700 R_{dO}
B	2	1500	8.41	1.71	1.02	-2.24	1.004	0.986	0.847
L			8.53	1.63	0.98				
B	5	1500	8.47	1.67	1.00	-2.54	0.984†	0.968†	0.831†
L			8.58	1.61	0.96				
B	7	500	7.35	1.38	2.41	-2.32	0.967	0.957	0.909
L			7.48	1.33	2.33				
B	7	1000	8.16	1.57	1.40	-2.58	0.969	0.956	0.862
L			8.28	1.51	1.34				
B	7	1500	8.50	1.65	0.98	-2.71	0.972†	0.957†	0.820†
L			8.61	1.59	0.94				
B	7	2000	8.68	1.70	0.76	-2.78	0.974	0.958	0.782
L			8.78	1.63	0.73				
B	10	1500	8.55	1.62	0.96	-2.93	0.953	0.940	0.804
L			8.65	1.56	0.92				

* Average buffer factor $\bar{\xi}$ is approximately 9.3 using Buch's constants, 9.8 using Lyman's constants. $N_{mO}/N_{aO} = 2$ for all cases.
† Values are applicable to Table 9.

by the choice of τ_{am} and τ_{dm}. As discussed in Section 9.1, these isotopic ratios, R_m and R_d, are approximately known for 1954, i.e., the same time as the most recent measurements of the atmospheric Suess effect. The five-reservoir model not only predicts the steady values R_{m0} and R_{d0} via Eqs. (6.31), (6.32), and (7.13), but also, as shown in Table 8, these ratios corrected for the disturbance produced by industrial CO_2. For the deep sea, a correction to R_{d0} is too small to be worth tabulating, but for R_m the correction is significant and depends to a considerable degree on the choice of model parameters.

For $N_{m0}/N_{a0} = 2$, the transfer times most nearly matching the observed values $R_m = 0.96$, $R_d = 0.83$ are $\tau_{am} = 7$ years, and $\tau_{dm} = 1500$ years. The Suess effect for this choice of parameters is -2.7%, in good agreement with the observed value. This result is independent of whether Buch's or Lyman's dissociation quotients are used to compute the buffer factor. As in all other cases so far considered, the value of n_a/N_{a0} is too large (8.5 % if Buch's quotients are used, 8.6 % if Lyman's are used).

Before closing this analysis of model dependence on parameters, several additional checks on the model will be mentioned.

(1) If the gravitational flux F_g is ignored by declaring the deep ocean concentration to be equal to the observed surface concentration, the n_i are not changed at all. (Even the Suess effect is but negligibly altered.) If, however, W_d/W_m is directly replaced by N_{d0}/N_{m0} in (7.23) as was done by Bolin and Eriksson [2] and in all examples using the three-reservoir model, n_a is underestimated as shown by Fig. 6, lower curve 6 as compared with lower curve 7.

(2) If the exponential representation of $\gamma(t)$ [(2.19), (2.20)] is transformed into annual increments [cf. text after (3.26)]

$$\gamma_j = \gamma_0 \int_{t_j}^{t_{j+1}} e^{ru}\, du \qquad (9.1)$$

and these are then employed in the convolution integrals (2.35), the resulting values of all the n_i agree with the analytical representation (2.28) to the order of one part in 10^4. The same order of agreement is found for the radiocarbon values $*n_i$. The convolution integral method of treating the annual industrial CO_2 data thus is seen to introduce no significant error into the calculations.

(3) Neglect of radioactive decay in solving the governing equations (3.9) produces insignificant errors in the Suess effect, as is expected owing to the long time constant for decay, $*\lambda^{-1}$, compared to the period of industrial activity and by the cancellation of the dominant terms in $*\lambda$ when deriving (3.16).

9.3. Search for a Best Fit with Observed Atmospheric Variations

All combinations of model parameters so far considered produce predictions of the increase in atmospheric CO_2 concentration which are

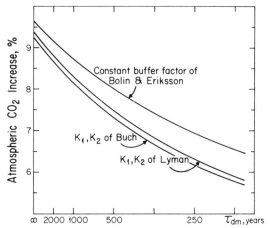

Fig. 9. *Upper curve*: Increase in inactive atmospheric CO_2 from 1700 to 1954 predicted by five-reservoir model for various values of the deep ocean to surface layer transfer time τ_{dm}, where an air to sea transfer time of 6 years, a constant land biota carbon pool, and constant buffer factor of 12.5 are assumed. *Middle and lower curves*: Same models except that the buffer factor is calculated according to the dissociation quotients of Buch and Lyman ($\bar{\zeta} = 9.1$ and 9.6, respectively).

higher than the observed value. For reasonable values of the transfer coefficients the discrepancy is considerable.

The five-reservoir model approaches compatibility with the observations if:

(1) the average buffer factor $\bar{\zeta}$ is reduced,
(2) the transfer rate from the deep ocean to the surface layer is decreased,
(3) the assumed volume W_m of the surface ocean layer is increased, or
(4) the land biota is assumed to grow in total mass by assigning a positive value to the growth factor β_a defined by (5.7).

The sensitivity of the five-reservoir model to the value assigned to the average buffer factor $\bar{\zeta}$ is as shown in Fig. 9. The cases plotted are for a constant land biomass, and are therefore equivalent to the three-reservoir model of Section 3 except for the inclusion of an oceanic gravitational flux. A shift from Bolin and Eriksson's value of $\bar{\zeta}$ to that based on Buch causes an appreciable drop in the predicted values of n_a, but differences in n_a on substituting Lyman's constants for Buch's are not striking. Recent determinations of K_1 and K_2 by Hansson [23] tend closely to confirm Lyman's measurements. It thus seems unlikely, on the basis of present evidence, that more accurate values of the dissociation quotients will lower the calculated value of $\bar{\zeta}$ significantly.

A source of uncertainty is the use of a global average buffer factor when the factor varies considerably from warm to cold water (see Table 2). Considerable evidence suggests that polar waters play a more important role in CO_2 transfer than warm waters, so that if any error has been made it is likely to be in the direction of underestimating the effective global average buffer factor, and thereby n_a.

If the transfer time from deep ocean water to the surface layer τ_{dm} is decreased, the predicted value of n_a is reduced. As the results plotted in Fig. 6 have already revealed, highly unrealistic values of τ_{dm} are required, however, to approach a correct value for n_a. Alternatively, if the assumed volume of the ocean surface layer is made large enough, the predicted n_a again approaches the correct value. Variations in either parameter also affect the predicted Suess effect. This occurs such that the correlation between predicted atmospheric variations in inactive carbon and radiocarbon is very nearly the same whichever of the two parameters is varied (Fig. 10), as though the only feature of subsurface water critical to predicting atmospheric changes up to 1954 is that it provides a certain capability for diluting industrial CO_2 regardless of the mode of subsurface transport.

Since a two-layer oceanic reservoir model has no direct provision for portraying the moderately rapidly circulating intermediate layer, such a model is certain to underestimate the capacity of the oceans to dilute industrial CO_2 if the deep ocean to surface layer transfer time τ_{dm} is chosen to give the correct average radiocarbon isotopic ratio of all subsurface water, and at the same time the surface layer is modeled to exclude the intermediate layer. Some form of adjustment of the two-layer model is clearly justified to provide for exchange of the intermediate water.

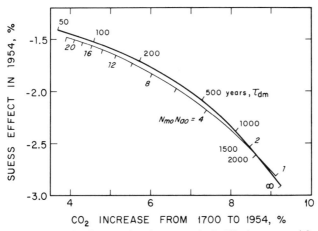

Fig. 10. *Upper curve*: Relation between inactive atmospheric CO_2 increase and Suess effect in 1954 for various deep ocean to surface layer transfer times τ_{dm}, assuming a constant ocean surface layer. *Lower curve*: Same relation for various ocean surface layer volumes as expressed by the ratio of the preindustrial mass of carbon in the surface layer to that in the atmosphere, N_{mo}/N_{ao}, assuming τ_{dm} constant. Upper curve is for N_{mo}/N_{ao} of 2, lower curve for τ_{dm} of 1500 years. Both curves are for an air to sea transfer time τ_{am} of 5 years and are for a five-reservoir model assuming a constant land biota carbon biomass.

The relations in Fig. 10 were derived for two extreme cases. A more realistic model with specific provision for the intermediate layer would almost surely yield predictions of n_a between these extremes. Thus if the Suess effect is specified, any oceanic model predicting it might also be expected to yield a nearly unique prediction of n_a irrespective of how the exchange of surface and subsurface water was parameterized.

Testing this hypothesis directly, although a desirable undertaking, cannot improve model predictability greatly unless more information can be found to prescribe the additional oceanic transfer parameters which must be introduced. The experience of Keeling and Bolin [11] with a three-reservoir oceanic model indicates that if additional chemical tracers are introduced as controls on the model parameters, the steady-state relations rapidly become highly complicated as reservoirs are added. The multiple reservoir steady-state models of Broecker [39], based principally on radiocarbon data, may be more realistic, but available oceanographic data do not sufficiently restrict the choice of transfer parameters to decide which (if any) of these models is to be preferred. We probably must wait until more accurate chemical data are gathered for the world oceans before settling the question of the role of intermediate water and the general problem of the response of the subsurface oceans to changes in atmospheric CO_2. Furthermore, the prediction of n_a given by oceanic models of any degree of complexity will agree with fact only if the land biota is correctly modeled, as discussed below.

Returning to a consideration of the two-layer model, we have further to determine the influence of the air to sea transfer time τ_{am}. If the existing prenuclear era isotopic values of surface water can be accepted, τ_{am} lies between 5 and 7 years (Fig. 8). A change in τ_{am} produces more change in the Suess effect than in n_a so that a redetermination of the relation of Fig. 10 using τ_{am} raised from 5 to 7 essentially shifts the curves downward. For pairs of values of N_{m0}/N_{a0} and k_{dm} which correctly predict n_a/N_{a0} in 1954, the predicted Suess effect is $-1.84\% \pm 0.02\%$ for $\tau_{am} = 7$, $-1.60\% \pm 0.03\%$ for $\tau_{am} = 5$, where the uncertainty reflects the alternate choices of a large surface layer or rapid subsurface transfer time.

These values lie well below the estimates of the Suess effect based on tree ring studies (-2.8%). Unless the tree ring or atmospheric CO_2 data are substantially wrong, it seems impossible that any oceanic model can fully explain the rise in n_a.

The five-reservoir model allows us only one additional option: to assume an increase in the carbon biomass of the land biota in response to increasing atmosphere CO_2. The rate of increase required to bring model and observations into agreement is small enough that the Suess effect is but slightly affected. For an assumed Suess effect (at $\beta_a = 0$) between -2.0 and -2.5% and assumed air to sea transfer times $\tau_{am} = 5$ and 7, the relation between biota growth and atmospheric CO_2 increase is as shown in Fig. 11. It is clear that the values of β_a cannot be established unless both the Suess

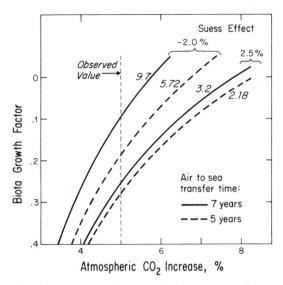

Fig. 11. Increase in inactive atmospheric CO_2 from 1700 to 1954 as predicted by a five-reservoir model with expanding land biota carbon pool ($\beta_a > 0$) or diminishing pool ($\beta_a < 0$). The size of the ocean surface layer was chosen to yield a Suess effect of -2.0% or -2.5% for a stationary land biota ($\beta_a = 0$). The values of N_{m0}/N_{a0} used in calculating each curve are shown as italicized numerals (9.7, 5.72, etc.). The deep ocean to surface layer transfer time τ_{dm} is assumed to be 1500 years. Solid curves are for an air to sea transfer time τ_{am} of 7 years, dashed curves for 5 years. The biota growth factor β_a denotes the fractional increase in rate of assimilation of CO_2 by plants relative to increase in atmospheric CO_2.

effect and τ_{am} are well determined. The results of covarying β_a and τ_{am} for the case where $N_{m0}/N_{a0} = 2.0$ are summarized in Table 9. The best fit to a Suess effect of -2.8% and $R_m = 0.96$ (for 1954) is given approximately by a model with $\tau_{am} = 7$ years, $\tau_{dm} = 1500$ years, and $\beta_a = 0.30$. Such a model predicts a growth in the land biota of 1.6% up to 1954.

This model is close to the extreme case where the intermediate water is totally ignored (except as included trivially in the deep ocean transport). This unreasonable result suggests that the Suess effect for 1954 was in actual fact less than -2.8%. As for the possibility that the biota increased in mass by 1 or 2% up to 1954 (an implied increase of up to 4% by 1972), the present means of measuring the global land biomass do not approach the precision required to answer this question. Thus we are led to conclude (as anticipated in Section 1.1) that measurements of atmospheric CO_2 coupled with model calculations delimit the possible changes in biomass more precisely than direct measurements of biomass. On the other hand, model calculations have not yet produced unequivocal predictions for the carbon cycle in the atmosphere and oceans. Further efforts in both modeling and data gathering are needed if we are to gain a satisfactory quantitative understanding of the exchange of atmospheric carbon dioxide with the oceans and land biota.

TABLE 9. Solution of the Five-Reservoir Model with Increasing Land Biomass*

Source of K_1 and K_2 B = Buch L = Lyman	Air to sea transfer time τ_{am} (yr)	Growth factor β_a	Relative increase in carbon masses in 1954				Suess effect, atmosphere (%)	Average buffer factor ζ
			Land biota $(n_b + n_e)/N_{a0}$ (%)	Atmosphere n_a/N_{a0} (%)	Surface ocean n_m/N_{a0} (%)	Deep ocean n_d/N_{a0} (%)		
B	5	0.1	2.09	6.84	1.37	0.84	−2.52	9.2
L	5	0.1	2.11	6.92	1.31	0.80		9.8
B	5	0.2	3.58	5.70	1.14	0.72	−2.51	9.2
L	5	0.2	3.61	5.75	1.09	0.69		9.7
B	5	0.3	4.67	4.86	0.98	0.63	−2.50	9.1
L	5	0.3	4.71	4.89	0.94	0.60		9.6
B	7	0.1	2.10	6.87	1.35	0.83	−2.69	9.2
L	7	0.1	2.12	6.94	1.29	0.79		9.8
B	7	0.2	3.59	5.71	1.13	0.71	−2.68	9.2
L	7	0.2	3.62	5.76	1.08	0.68		9.7
B	7	0.3	4.69	4.87	0.97	0.62	−2.67	9.1
L	7	0.3	4.72	4.90	0.92	0.59		9.6

* Deep ocean to surface ocean transfer time τ_{dm} is 1500 years and $N_{m0}/N_{a0} = 2$ for all cases. The ^{14}C/inactive carbon ratios for given τ_{am} and τ_{dm} are the same as for $\beta_a = 0$ (values marked with † in Table 8).

Appendix A

FACTORS TO SOLVE FOUR-RESERVOIR MODEL

The specific formulas for the $f_i(s)$ and $*f_i(s)$ of equations (3.15) and (3.16) are as follows:

$$
\left.
\begin{aligned}
f_b(s) &= k_2\{s^2 + (k_4 + k_5 + k_6)s + k_4 k_6\} \\[4pt]
f_a(s) &= (s + k_1)\{s^2 + (k_4 + k_5 + k_6)s + k_4 k_6\} \\[10pt]
f_m(s) &= (s + k_1)\{k_3 s + k_3 k_6\} \\[4pt]
f_d(s) &= (s + k_1)\{k_3 k_5\}
\end{aligned}
\right\} \tag{A.1}
$$

$$
\left.
\begin{aligned}
*f_{1b} &= -\{(s + *\lambda)^2 + (*k_3 + *k_4 + *k_5 + *k_6)(s + *\lambda) + *k_3*k_5 \\
&\qquad + *k_3*k_6 + *k_4*k_6\} \\[4pt]
*f_{1a} &= (s + *\lambda)^2 + (*k_4 + *k_5 + *k_6)(s + *\lambda) + *k_4*k_6 \\[4pt]
*f_{1m} &= *k_3(s + *\lambda + *k_6) \\[4pt]
*f_{1d} &= *k_3*k_5
\end{aligned}
\right\} \tag{A.2}
$$

$$
\left.
\begin{aligned}
*f_{2b} &= *k_2(s + *\lambda + *k_5 + *k_6) \\[4pt]
*f_{2a} &= (s + *\lambda + *k_1)(s + *\lambda + *k_5 + *k_6) \\[4pt]
*f_{2m} &= -(s + *\lambda + *k_6)(s + *\lambda + *k_1 + *k_2) \\[4pt]
*f_{2d} &= -*k_5(s + *\lambda + *k_1 + *k_2)
\end{aligned}
\right\} \tag{A.3}
$$

$$
\left.
\begin{aligned}
*f_{3b} &= *k_2*k_4 \\[4pt]
*f_{3a} &= *k_4(s + *\lambda + *k_1) \\[4pt]
*f_{3m} &= (s + *\lambda)^2 + (*k_1 + *k_2 + *k_3)(s + *\lambda) + *k_1*k_3 \\[4pt]
*f_{3d} &= -\{(s + *\lambda)^2 + (*k_1 + *k_2 + *k_3 + *k_4)(s + *\lambda) + *k_1*k_3 \\
&\qquad + *k_1*k_4 + *k_2*k_4\}
\end{aligned}
\right\} \tag{A.4}
$$

Appendix B

EQUATIONS FOR FIVE-RESERVOIR MODEL

For the model depicted by Fig. 3, the coefficients of the Laplace transform corresponding to (3.11) form the array

$$\mathbf{D}(U) = \begin{pmatrix} U + l_1 & 0 & -l_1 & 0 & 0 \\ 0 & U + l_3 & -l_4 & 0 & 0 \\ -l_1 & -l_3 & U + l_2 + l_4 + k_3 & -k_4 & 0 \\ 0 & 0 & -k_3 & U + k_4 + k_5 & -k_6 \\ 0 & 0 & 0 & -k_5 & U + k_6 \end{pmatrix} \quad \text{(B.1)}$$

where for inactive carbon the determinant $D(s)$ is evaluated by setting $U = s$, and radiocarbon $*D(s)$ is evaluated by setting $U = s + *\lambda$ and replacing the k_j and l_j by $*k_j$ and $*l_j$.

Retaining the general symbol U and expanding, we obtain

$$D(U) = U(U^4 + AU^3 + BU^2 + CU + D) \quad \text{(B.2)}$$

where

$$\left.\begin{aligned}
A &= l_1 + l_2 + l_3 + l_4 + k_3 + k_4 + k_5 + k_6 \\
B &= (l_1 + l_3)(k_3 + k_4 + k_5 + k_6) + (l_2 + l_4)(k_4 + k_5 + k_6) \\
&\quad + k_3(k_5 + k_6) + k_4 k_6 \\
C &= (l_1 + l_3)(k_3 k_5 + k_3 k_6 + k_4 k_6) + (l_2 + l_4)k_4 k_6 \\
&\quad + l_1 l_3(k_3 + k_4 + k_5 + k_6) + (l_1 l_4 + l_2 l_3)(k_4 + k_5 + k_6) \\
D &= l_1 l_3(k_3 k_5 + k_3 k_6 + k_4 k_6) + (l_1 l_4 + l_2 l_3)k_4 k_6
\end{aligned}\right\} \quad \text{(B.3)}$$

When this result is compared with (3.12), it can be seen that the five-reservoir determinantal equation, $D(U) = 0$, can be obtained from the four-reservoir determinantal equation (writing U for s) by multiplying the latter by U and then making the substitutions

$$\left.\begin{aligned}
(l_1 + l_3)U + l_1 l_3 &\quad \text{for } k_1 U \\
(l_2 + l_4)U + l_1 l_4 + l_2 l_3 &\quad \text{for } k_2 U
\end{aligned}\right\} \quad \text{(B.4)}$$

Equation (B.2) has five roots λ_k which, as in Section 3, will be numbered in order of increasing absolute magnitude. For inactive carbon the first root, λ_1, is again zero. For radiocarbon, the first root, $*\lambda_1 (= *\lambda)$, does not appear in the final solution because, as before, a factor $(s + *\lambda)$ cancels between each numerator and denominator. Renumbering the roots for radiocarbon $\lambda_6 = {}^*\lambda_2, \ldots, \lambda_9 = {}^*\lambda_5$, we obtain final solutions for n_i and ${}^*n_i$ identical to (3.24) and (3.25) except that the summation limits are raised to 5 and 9, respectively. It remains to tabulate the specific functions $f_i(\lambda_k)$ and ${}^*f_i(\lambda_k)$, where the latter are again composites arising from the virtual radiocarbon sources. Specifically,

$$
{}^*f_i(s) = \left[-l'_7 f_b(s) + l'_8 f_a(s) \right] {}^{*b}f_{1i}(s) + \left[-l'_9 f_e(s) + l'_{10} f_a(s) \right] {}^{*e}f_{1i}(s)
$$
$$
+ \left[k'_7 {}^*f_{2i}(s) + k'_8 {}^*f_{3i}(s) \right] f_m(s) \tag{B.5}
$$

$$
\left.
\begin{aligned}
f_b &= (s + l_3)l_2 U_1 \\
f_e &= (s + l_1)l_4 U_1 \\
f_a &= (s + l_1)(s + l_3)U_1 \\
f_m &= (s + l_1)(s + l_3)(k_3 s + k_3 k_6) \\
f_d &= (s + l_1)(s + l_3)(k_3 k_5)
\end{aligned}
\right\} \tag{B.6}
$$

$$
\left.
\begin{aligned}
{}^{*b}f_{1b} &= -(s + {}^*\lambda + {}^*l_3)\,{}^*U_2 - {}^*l_4\,{}^*U_1 \\
{}^{*b}f_{1e} &= {}^*l_4\,{}^*U_1 \\
{}^{*b}f_{1a} &= (s + {}^*\lambda + {}^*l_3)\,{}^*U_1 \\
{}^{*b}f_{1m} &= (s + {}^*\lambda + {}^*l_3){}^*k_3(s + {}^*\lambda + {}^*k_6) \\
{}^{*b}f_{1d} &= (s + {}^*\lambda + {}^*l_3){}^*k_3{}^*k_5
\end{aligned}
\right\} \tag{B.7}
$$

$$
\left.
\begin{aligned}
{}^{*e}f_{1b} &= {}^*l_2\,{}^*U_1 \\
{}^{*e}f_{1e} &= -(s + {}^*\lambda + {}^*l_1)\,{}^*U_2 - {}^*l_2\,{}^*U_1 \\
{}^{*e}f_{1a} &= (s + {}^*\lambda + {}^*l_1)\,{}^*U_1 \\
{}^{*e}f_{1m} &= (s + {}^*\lambda + {}^*l_1){}^*k_3(s + {}^*\lambda + {}^*k_6) \\
{}^{*e}f_{1d} &= (s + {}^*\lambda + {}^*l_1){}^*k_3{}^*k_5
\end{aligned}
\right\} \tag{B.8}
$$

$$
\left.
\begin{aligned}
{}^*f_{2b} &= (s + {}^*\lambda + {}^*l_3){}^*l_2(s + {}^*\lambda + {}^*k_5 + {}^*k_6) \\
{}^*f_{2e} &= (s + {}^*\lambda + {}^*l_1){}^*l_4(s + {}^*\lambda + {}^*k_5 + {}^*k_6) \\
{}^*f_{2a} &= (s + {}^*\lambda + {}^*l_1)(s + {}^*\lambda + {}^*l_3)(s + {}^*\lambda + {}^*k_5 + {}^*k_6) \\
{}^*f_{2m} &= -(s + {}^*\lambda + {}^*k_6)\,{}^*U_3 \\
{}^*f_{2d} &= -{}^*k_5\,{}^*U_3
\end{aligned}
\right\} \tag{B.9}
$$

$$*f_{3b} = (s + *\lambda + *l_3)*l_2*k_4$$

$$*f_{3e} = (s + *\lambda + *l_1)*l_4*k_4$$

$$*f_{3a} = (s + *\lambda + *l_1)(s + *\lambda + *l_3)*k_4 \qquad \text{(B.10)}$$

$$*f_{3m} = *U_4$$

$$*f_{3d} = -\{*U_4 + *k_4*U_3\}$$

To save some writing, I have introduced the polynomials

$$U_1 = s^2 + (k_4 + k_5 + k_6)s + k_4k_6$$

$$*U_1 = (s + *\lambda)^2 + (*k_4 + *k_5 + *k_6)(s + *\lambda) + *k_4*k_6$$

$$*U_2 = *U_1 + *k_3(s + *\lambda) + *k_3*k_5 + *k_3*k_6$$

$$*U_3 = (s + *\lambda)^2 + (*l_1 + *l_2 + *l_3 + *l_4)(s + *\lambda) + *l_1*l_3 \qquad \text{(B.11)}$$
$$\qquad + *l_1*l_4 + *l_2*l_3$$

$$*U_4 = *U_3(s + *\lambda) + *k_3(s + *\lambda)^2 + (*l_1 + *l_3)*k_3(s + *\lambda)$$
$$\qquad + *l_1*l_3*k_3$$

To obtain specific formulas for the coefficients from Sections 5–7, the following substitutions are to be made:

l_1 for k_1 in (5.20)

l_2 for k_2 in (5.21)

l_3 for k_1 with e replacing b in (5.20)

l_4 for k_2 with e replacing b in (5.21)

l_7' for k_9' in (5.27)

l_8' for k_{10}' in (5.28)

l_9' for k_9' with e replacing b in (5.27)

l_{10}' for k_{10}' with e replacing b in (5.28)

$*l_1$ for $*k_1$ in (5.25)

$*l_2$ for $*k_2$ in (5.26)

$*l_3$ for $*k_1$ with e replacing b in (5.25)

$*l_4$ for $*k_2$ with e replacing b in (5.26)

k_3–k_6 and $*k_3$–$*k_6$ are given, respectively, by (6.54), (6.55), (7.23), (7.24), (6.58), (6.59), (7.25), and (7.26), and k_7' and k_8' by (6.60) and (7.27).

REFERENCES

1. Revelle, R., and Suess, H. E., Carbon dioxide exchange between atmosphere and ocean, and the question of an increase of atmospheric CO_2 during the past decades, *Tellus* **9**, 18–27 (1957).
2. Bolin, B., and Eriksson, E., Changes in the carbon content of the atmosphere and the sea due to fossil fuel combustion, "The Atmosphere and the Sea in Motion, Rossby Memorial Volume," B. Bolin, ed., Rockefeller Institute Press, New York, 1959, pp. 130–143.
3. Broecker, W. S., Yuan-Hui Li, and Tsung-Hung Peng, Carbon Dioxide—Man's Unseen Artifact, Chapter 11 of "Impingement of Man on the Oceans," D. W. Hood, ed., Wiley-Interscience, New York, 1971, pp. 287–324.
4. DeFant, Albert, "Physical Oceanography," Vol. 1, Pergamon Press, London, 1961.
5. Suess, H. E., Radiocarbon concentration in modern wood, *Science* **122**, 414–417 (1955).
6. Craig, H., The natural distribution of radiocarbon and the exchange time of carbon dioxide between atmosphere and sea, *Tellus* **9**, 1–17 (1957).
7. Eriksson, E., and Welander, P., On a mathematical model of the carbon cycle in nature, *Tellus* **8**, 155–175 (1956).
8. Welander, P., On the frequency response of some different models describing the transient exchange of matter between the atmosphere and the sea, *Tellus* **11**, 348–354 (1959).
9. Ekdahl, C. A., and Keeling, C. D., Atmospheric CO_2 in the natural carbon cycle: observations and deductions, Brookhaven National Laboratory Symposium on "Carbon and the Biosphere," May 16–18, 1972, in press.
10. Eriksson, E., Natural reservoirs and their characteristics, *Geofis Internacional* **1**, 27–43 (1961).
11. Keeling, C. D., and Bolin, B., The simultaneous use of chemical tracers in oceanic studies. II. A three-reservoir model of the North and South Pacific Oceans, *Tellus* **20**, 17–54 (1968).
12. Keeling, C. D., Industrial production of carbon dioxide from fossil fuels and limestone, 1973, *Tellus* **25**, No. 2.
13. Irving, J., and Mullineux, N., "Mathematics in Physics and Engineering," Academic Press, New York, 1959.
14. Sokolnikoff, I. S., and Redheffer, R. M., "Mathematics of Physics and Modern Engineering," McGraw-Hill, New York, 1958.
15. Keeling, C. D., Is carbon dioxide from fossil fuel changing man's environment?, *Proc. Am. Philos. Soc.* **114**, 10–17 (1970).
16. Monteith, J. L., Szeicz, G., and Yabuki, K., Crop photosynthesis and the flux of carbon dioxide below the canopy, *J. Appl. Ecol.* **1**, 321–337 (1964).
17. Lindstrom, R. S., The effect of increasing the carbon dioxide concentration on floricultural plants, *Michigan Florist*, Michigan State Florists Association, **398**, 12–13, 19–22 (1964).
18. Pettibone, C. A., Matson, W. E., and Ackley, W. B., Fertilizing greenhouse air, *Agriculture Res.* **12**, 10 (1964).
19. Imazu, T., Yabuki, K., and Oda, Y., Studies on the carbon dioxide environment for plant growth II. Effect of carbon dioxide concentration on the growth, flowering and fruit setting of eggplant (*Solanum melongena* L.), *J. Japan. Soc. Hort. Sci.* **36**, 275–280 (1968).
20. Riley, J. P., and Skirrow, G., "Chemical Oceanography," Vol. 1, Academic Press, London, 1965.
21. Lyman, J., "Buffer Mechanism of Sea Water," Ph.D. Thesis, University of California at Los Angeles, 1957.
22. Edmond, J. M., and Gieskes, J. M. T. M., On the calculation of the degree of saturation of sea water with respect to calcium carbonate under *in situ* conditions, *Geochim Cosmochim Acta* **34**, 1261–1291 (1970).
23. Hansson, Ingemar, "An Analytical Approach to the Carbonate System in Sea Water," Ph.D. Thesis, University of Gothenberg, Sweden, 1972.

24. Owen, B. B., and King, E. J., The effect of sodium chloride upon the ionization of boric acid at various temperatures, *J. Amer. Chem. Soc.* **65**, 1612–1620 (1943).
25. Murray, C. N., and Riley, J. P., The solubility of gases in distilled water and sea water— IV Carbon dioxide, *Deep-Sea Res.* **18**, 533–541 (1971).
26. Keeling, C. D., Carbon dioxide in surface ocean waters. 4. The global distribution, *J. Geophys. Res.* **73**, 4543–4553 (1968).
27. Gordon, L. I., Park, P. K., Hager, S. W., and Parsons, T. R., Carbon dioxide partial pressures in North Pacific surface waters—time variations, *J. Oceanog. Soc. of Japan* **27**, 81–90 (1971).
28. Pales, J. D., and Keeling, C. D., The concentration of atmospheric carbon dioxide in Hawaii, *J. Geophys. Res.* **70**, 6053–6075 (1965).
29. Bray, J. R., An analysis of the possible recent change in atmospheric carbon dioxide concentration, *Tellus* **2**, 220–230 (1959).
30. Olson, J. S., Carbon cycles and temperate woodlands, Chapter 7 in "Ecological Studies," Vol. 1, D. E. Reichle, ed., Springer-Verlag, Berlin, 1970, pp. 73–85.
31. Leith, H., The role of vegetation in the carbon dioxide content of the atmosphere, *J. Geophys. Res.* **68**, 3887–3898 (1963).
32. SCEP, "Man's Impact on the Global Environment: Assessment and Recommendations for Action," Report of the Study of Critical Environmental Problems (SCEP), MIT Press, Cambridge, Mass., pp. 39–55, 1970.
33. Bowen, H. J. M., "Trace Elements in Biochemistry," Academic Press, London, 1966.
34. Whittaker, R. H., and Likens, G. E., Carbon in the biota, Brookhaven National Laboratory Symposium on "Carbon and the Biosphere," May 16–18, 1972, in press.
35. Keeling, C. D., Adams, J. A., Ekdahl, C. A., and Guenther, P. R., Atmospheric carbon dioxide variations at the South Pole, 1973, submitted to *Tellus*.
36. Suess, H. E., Secular variations of the cosmic-ray-produced carbon 14 in the atmosphere and their interpretations, *J. Geophys. Res.* **70**, 5937–5952 (1965).
37. Houtermans, J., Suess, H. E., and Munk, W., "Effect of Industrial Fuel Combustion on the Carbon-14 Level of Atmospheric CO_2," International Atomic Energy Agency, Vienna, Publication No. SM-87/53, 1967, pp. 57–67.
38. Lerman, J. C., Mook, W. G., and Vogel, J. C., C-14 in tree rings from different localities, Nobel Symposium 12, Radiocarbon Variations and Absolute Chronology, I. U. Olsson, ed., Wiley-Interscience, New York, 1970, pp. 275–301.
39. Broecker, W., Radioisotopes and large-scale oceanic mixing, Chapter 4 of "The Sea: Ideas and Observations on Progress in the Study of the Seas," M. N. Hill, ed., Wiley-Interscience, New York, 1963, pp. 88–108.
40. Craig, H., Abyssal carbon 13 in the South Pacific, *J. Geophys. Res.* **75**, 691–695 (1970).
41. Revelle, R., "Atmospheric Carbon Dioxide," Appendix Y4 of Report of the Environmental Pollution Panel, President's Advisory Committee, The White House, November, 1965, 111–113.
42. Craig, H., A critical evaluation of radiocarbon techniques for determining mixing rates in the oceans and the atmosphere, Second United Nations International Conference on the Peaceful Uses of Atomic Energy, 1958.
43. Craig, H., 6. The natural distribution of radiocarbon: mixing rates in the sea and resistence times of carbon and water, "Earth Science and Meteoritics," North-Holland Publishing Co., Amsterdam, 1963, pp. 103–144.
44. Baxter, M. S., and Walton, A., A theoretical approach to the Suess effect, *Proc. Royal Soc.* **A318**, 213–230 (1970).
45. Sverdrup, H. V., Johnson, M. W., and Fleming, R. G., "The Oceans," Prentice-Hall, Englewood Cliffs, New Jersey, 1942.
46. Williams, P. M., The organic chemistry of sea water, Chapter 5 in "Research on the Marine Food Chain, Progress Report, June 1963–December 1964," University of California, La Jolla, California, 1965, pp. 28–45.

47. Riley, J. P., and Chester, R., "Introduction to Marine Chemistry," Academic Press, London, 1971.
48. Buch, K., Harvey, H. W., Wattenberg, H., and Gripenberg, S., Über das Kohlensäuresystem im Meerwasser, *Rapports et Procès-Verbaux des Réunions, Conseil permanent international pour l'exploration de la mer*, **79**, 1–70 (1932).
49. Buch, K., New determination of the second dissociation constant of carbonic acid in sea water, *Acta Acad. Aboensis, Math Phys* **11**, 1–18 (1938).
50. Buch, K., Das Kohlensäure Gleichgewichtssystem im Meerwasser: Kritische Durchsicht und Neuberechnungen der Konstituenten, Havsforskningsinstitutets Skrift, No. 151, pp. 1–18 (1951).
51. Culberson, C., Pytkowicz, R. M., and Hawley, J. E., Seawater alkalinity determination by the pH method, *J. Geophys. Res.* **28**, 15–21 (1970).
52. Harvey, H. W., "The Chemistry and Fertility of Sea Waters," Cambridge University Press, London, 1955.
53. Verniani, Franco, The total mass of the earth's atmosphere, *J. Geophys. Res.* **71**, 385–391 (1966).
54. Eriksson, E., The circulation of some atmospheric constituents in the sea, "The Atmosphere and the Sea in Motion, Rossby Memorial Volume," B. Bolin, ed., Rockefeller Institute Press, New York, 1959, pp. 147–157.
55. Menzel, D. W., and Ryther, J. H., Organic carbon and the oxygen minimum in the South Atlantic Ocean, *Deep-Sea Res.* **15**, 327–337 (1962).
56. Degens, E. T., Biogeochemistry of Stable Carbon Isotopes, "Organic Geochemistry, Methods and Results," G. Eglinton and M. T. J. Murphy, eds., Springer-Verlag, Berlin, 1969, pp. 304–439.
57. Park, R., and Epstein, S., Metabolic fractionation of C^{13} and C^{12} in plants, *Plant Physiol.* **36**, 133–138 (1961).
58. Craig, H., Carbon 13 in plants and the relationships between carbon 13 and carbon 14 variations in nature, *J. Geol.* **62**, 115–149 (1954).
59. Mook, W. G., Stable carbon and oxygen isotopes of natural waters in the Netherlands, "Isotope Hydrology 1970," International Atomic Energy Agency, Vienna, Publication No. SM-129/12, 1970, pp. 163–190.
60. Vogel, J. C., Isotope separation factors of carbon in the equilibrium system CO_2–HCO_3^-–$CO_3^=$, Summer Course on Nuclear Geology Varena 1960, Comitato Nazionale per l'Energia Nucleare, 1960.
61. Olsson, I. U., ed., Radiocarbon Variations and Absolute Chronology, Nobel Symposium 12, Wiley-Interscience, New York, 1970 .

INDEX